CANADA AND ARCTIC
NORTH AMERICA

Other Titles in
ABC-CLIO'S
NATURE AND HUMAN SOCIETIES SERIES

NATURE AND HUMAN SOCIETIES

CANADA AND ARCTIC NORTH AMERICA
An Environmental History

Graeme Wynn

A B C  C L I O

Santa Barbara, California • Denver, Colorado • Oxford, England

Library of Congress Cataloging-in-Publication Data

Wynn, Graeme, 1946-
    Canada and arctic North America : an environmental history / Graeme Wynn.
    p.   cm.
    Includes bibliographical references and index.
    ISBN-10: 1-85109-437-7 (hardcover : alk. paper)
    ISBN-10: 1-85109-442-3 (ebook)
    ISBN-13: 978-1-85109-437-0 (hardcover : alk. paper)
    ISBN-13: 978-1-85109-442-4 (ebook)
    1. Human ecology—North America—History.   2. Human ecology—Arctic
regions—History.   3. Nature—Effect of human beings on—North America—History.
4. Nature—Effect of human beings on—Arctic regions—History.   5. Paleo-Indians—
North America.   6. Paleo-Indians—Arctic regions.   7. Indians of North America—
History.   8. Indians of North America—Arctic regions—History.   9. North America—
Environmental conditions.   10. Arctic regions—Environmental conditions.   I. Title.

GF501.W96   2007
304.2097—dc22
                                                                                        2006029751

11 10 09 08 07   10 9 8 7 6 5 4 3 2 1

This book is also available on the World Wide Web as an eBook. Visit
abc-clio.com for details.

Production Editor: Martha Ripley Gray
Editorial Assistant: Alisha L. Martinez
Production Manager: Don Schmidt
Media Image Coordinator: Ellen Brenna Dougherty
Media Editor: Jason Kniser
Media Manager: Caroline Price
File Management Coordinator: Paula Gerard

ABC-CLIO, Inc.
130 Cremona Drive, P.O. Box 1911
Santa Barbara, California 93116-1911

This book is printed on acid-free paper ∞
Manufactured in the United States of America

For Barbara,
who left New Zealand for Canada

CONTENTS

SERIES FOREWORD

Long ago, only time and the elements shaped the face of the earth, the black abysses of the oceans, and the winds and blue welkin of heaven. As continents floated on the mantle, they collided and threw up mountains or drifted apart and made seas. Volcanoes built mountains out of fiery material from deep within the earth. Mountains and rivers of ice ground and gouged. Winds and waters sculpted and razed. Erosion buffered and salted the seas. The concert of living things created and balanced the gases of the air and moderated the earth's temperature.

The world is very different now. From the moment our ancestors emerged from the southern forests and grasslands to follow the melting glaciers or to cross the seas, all has changed. Today the universal force transforming the earth, the seas, and the air is for the first time a single form of life: we humans. We shape the world, sometimes for our purposes and often by accident. Where forests once towered, fertile fields or barren deserts or crowded cities now lie. Where the sun once warmed the heather, forests now shade the land. One creature we exterminate only to bring another from across the globe to take its place. We pull down mountains and excavate craters and caverns, drain swamps and make lakes, divert, straighten, and stop rivers. From the highest winds to the deepest currents, the world teems with chemical concoctions only we can brew. Even the very climate warms from our activity.

And as we work our will upon the land, as we grasp the things around us to fashion into instruments of our survival, our social relations, and our creativity, we find in turn our lives and even our individual and collective destinies shaped and given direction by natural forces, some controlled, some uncontrolled, and some unleashed. What is more, uniquely among the creatures, we come to know and love the places where we live. For us, the world has always abounded with unseen life and manifest meaning. Invisible beings have hidden in springs, in mountains, in groves, in the quiet sky and in the thunder of the clouds, and in the deep waters. Places of beauty from magnificent mountains to small winding brooks captured our imaginations and our affection. We have perceived a mind like our own, but greater, designing, creating, and guiding the universe around us.

The authors of the books in this series endeavor to tell the remarkable epic of the intertwined fates of humanity and the natural world. It is a story only now coming to be fully known. Although traditional historians told the drama of men and women of the past, for more than three decades now many have added the natural world as a third actor. Environmental history by that name emerged in the 1970s in the United States. Historians quickly took an interest and created a professional society, the American Society for Environmental History, and a professional journal, now called *Environmental History*. American environmental history flourished and attracted foreign scholars. By 1990 the international dimensions were clear; European scholars joined together to create the European Society for Environmental History in 2001, with its journal, *Environment and History*. A Latin American and Caribbean Society for Environmental History should not be far behind. With an abundant and growing literature of world environmental history now available, a true world environmental history can appear.

This series is organized geographically into regions determined as much as possible by environmental and ecological factors, and secondarily by historical and historiographical boundaries. Befitting the vast environmental historical literature on the United States, four volumes tell the stories of the North, the South, the Plains and Mountain West, and the Pacific Coast. Other volumes trace the environmental histories of Canada and Alaska, Latin America and the Caribbean, Northern Europe, the Mediterranean region, sub-Saharan Africa, South Asia, Southeast Asia, East Asia, and Australia and Oceania. Authors from around the globe, experts in the various regions, have written the volumes, almost all of which are the first to convey the complete environmental history of their subjects. Each author has, as much as possible, written the twin stories of the human influence on the land and of the land's manifold influence on its human occupants. Every volume contains a narrative analysis of a region along with a body of reference material. This series constitutes the most complete environmental history of the globe ever assembled, chronicling the astonishing tragedies and triumphs of the human transformation of the earth.

The process of creating the series, recruiting the authors from around the world, and editing their manuscripts has been an immensely rewarding experience for me. I cannot thank the authors enough for all of their effort in realizing these volumes. I owe a great debt too to my editors at ABC-CLIO: Kevin Downing (now with Greenwood Publishing Group), who first approached me about the series; and Steven Danver, who has shepherded the volumes through delays and crises to publication. Their unfaltering support for and belief in the series were essential to its successful completion.

Mark Stoll
Department of History, Texas Tech University
Lubbock, Texas

PREFACE

Environmental histories come in many forms and they all reflect a basic desire to understand the relations between people and nature through time. Scholars have been interested in this broad and important topic for generations and they have approached it from several quite distinct disciplinary perspectives over the years. Yet use of the term *environmental history* to identify a particular field of inquiry is a relatively recent development. By common North American assessments, it dates back only three or four decades to the rise of the environmental movement in the 1960s.

There is much merit in this view. Popular interest in environmental questions reached new peaks in the last decades of the twentieth century, and the outpouring of historical scholarship on the relations between people and nature through time—particularly in the United States—has been immense and impressive. In addition, the parallelism between the coinage "environmental history" and public anxieties about "environmental issues"—that waxed and waned but often seemed to be rooted in past practices (even if these were relatively recent)—served, brilliantly, to draw attention to the field. Quickly, the phrase "environmental history" came to provide a euphonious and readily understood label for a widely diverse and rapidly expanding body of work, even as it served to veil regional, national, and disciplinary differences in emphases, origins, and approaches to the study of human-environment relations.

In efforts both to respond to and to order this diversity, several environmental historians have identified distinct themes in the expanding literature of their field. Like all classifications, these taxonomies are generalizations; they lump broadly similar work together and split certain emphases and approaches apart from others (from which they may not be radically different and to which they may be complementary) in order to find coherence in variety. But they are useful nonetheless. Most opt for a tripartite arrangement of the field.

According to the prominent American environmental historian Donald Worster, environmental history proceeds on three levels. The first documents "the structure and distribution of natural environments of the past." The second "focuses on productive technology as it interacts with the environment."

The third is concerned with the "patterns of human perception, ideology and value" and the ways in which they have worked in "reorganizing and recreating the surface of the planet" (Worster 1990, 1090–1091). Others claim that environmental historians "study how people have lived in the natural systems of the planet . . . how they have perceived nature [and how] they have reshaped it to suit their own idea of good living" (Warren 2003, 1).

Writing for a broader, more scholarly audience, a third American scholar, John R. McNeill, has suggested that environmental histories can be differentiated by the extent to which they are concerned with material, cultural/intellectual, or political matters. McNeill, who paints literature from around the world into his canvas, is at pains to point out that his categories reflect tendencies and emphases rather than absolute and exclusive distinctions, not least because few authors confine themselves to a single variety of environmental history. Yet his groupings have some utility. Studies of McNeill's first type focus upon "changes in biological and physical environments, and how these changes have affected human societies." Cultural/intellectual environmental history is concerned with "representations and images of nature in arts and letters," and political environmental history examines "law and state policy as it relates to the natural world" (McNeill 2003, 6–9).

This book, like most large works in the field, is hard to slot into any one of these groupings. Some might fit it most readily into McNeill's materialist category. It focuses to a considerable extent upon the ways in which human actions have shaped and reshaped northern North American environments. This is an important story, and one that fits well with the concerns of those whose academic lineage (like mine) runs through historical geography back to the venerable tradition of geographical work on the human-environment interface. But this is not the whole story. Neither the naughty world nor our inadequate books about it ever conform neatly and entirely to the categories we invent and adopt to describe this place or various parts of it. And as Donald Worster notes with reference to his three levels of inquiry, one of the great challenges confronting environmental historians is "to make connections among them."

Rather than adopting a rigid materialist or idealist stance, I seek to understand times and places contextually. Thus, the pages that follow are concerned not only to document environmental change but to understand how places came to be. Although material matters—the grinding of glaciers; the clearing of forests; the exploitation of fish, furs, rivers, and minerals; and the spread of human settlement across the land—are to the fore, other concerns creep into this story. Perhaps the most important of these entails recognition of the dynamic and resilient qualities of the natural world. For all the emphasis accorded the role of human actions in changing the face of the earth, the world of nature is not simply a stage on which props and scenery are moved about at human bidding. Natural (particularly geomorphological and biological) processes are often initiated and inflected by human actions and may extend or mitigate the immediate consequences of those actions. Changes in nature also work, in some circumstances, to constrain human ambitions or deflect the course of human-induced change. Humans and nature interact, and the traffic runs both ways. What is more, these flows are conditioned and shaped by technologies of production, movement, and communication.

By the same token, ideas lie behind many human actions, and "soft" social controls (moral regulation) and political considerations (laws, surveillance) limit people's freedom to act as they will. This is to say that even when people act without explicit deliberation, their behaviors are generally conditioned in some way by societal norms, prevailing beliefs, popular discourse, or legal provisions. For the most part, though, the confines of this book dictate that such matters be treated only in passing, when necessary to carry forward and render intelligible the central, if not necessarily larger, story entailed in the saga of the human occupation, settlement, development, and transformation of half a continent.

Recent literature, by geographers and others, complicates the seemingly easy distinction between nature and culture threaded through the preceding paragraphs by insisting that the natural world is apprehended through human perceptions and that nature is thus a social construction. At some level, this is an unassailable assertion. Humans perceive only parts of the world in which they live and they do so in various ways. Some elements of the environment—infrared light, X-rays, the sounds made by certain animals—cannot be detected by human sensory systems. People understand their surroundings differently: Where some see "waste and howling wilderness," others see potential wealth in lumber, and yet others speak of sylvan glades; the homeowner's affection for the modern, neatly ordered suburban housing area in which she lives stands in contrast to the transit planner's despair at the low density of land use, the medical health officer's conviction that automobile-dependent suburban dwellers are responsible for much of the air pollution that ruins lungs and lives, and the environmentalist's denunciation of tract-housing development as a destroyer of biodiversity.

All of this we know and accept without a second thought. Extrapolating from such everyday observations, it seems almost self-evident that people perceive nature in ways that reflect their interests and experiences. Taken a step further, this is to say that people perceive nature through culturally colored lenses. Yet hackles rise when social constructionists claim that "science" is simply another of these cultural lenses. And furies are unleashed when social constructionism is married with philosophical idealism to insist that ideas are all, that everything is a human construct, and that there is no such thing as material nature.

Starting from the view that everything that humans know about the world derives from their interactions with it, and accepting that people interact with the world "from positions marked by the particularities of our circumstances as embodied human creatures" (which is to say in ways colored by culture, tradition, and experience), Katherine Hayles has tried to find a middle ground between objectivism—the conviction that we know reality because we are separate from it and that "the world is as it is, no matter what any person happens to believe about it"—and the solipsism and radical subjectivism of the more extreme forms of social constructionism that become endlessly, regressively reflexive in their insistence that representation is all.

Calling this perspective "constrained constructivism," Hayles argues that the material objects "out there"—the objects that humans constitute as "reality," "the world," or "nature"—are best regarded as an "unmediated flux" existing apart from these conceptual terms. This flux is brought into consciousness, or understood, through various "transformative processes" that include sensory, contextual (including historical and cultural), and cognitive components. These processes Hayles calls "the cusp." They constitute a zone of interaction, of "interplay between representation and constraint," in which human constructions of "the world" or "nature" are worked out, but only with reference to the physical characteristics that limit the range of possible interpretations of the flux, and which allow humans to determine that some of these interpretations are more valid than others by assessing whether they are consistent with human experience of interaction with the flux. "Neither cut free from reality nor existing independent of human perception, the world as constrained constructivism sees it is the result of complex and active engagements between the unmediated flux and human beings" (Hayles 1995, 53–54). In the end, Hayles's fundamental point—that humans know the world through their connections with it—echoes through and underpins much of the discussion that follows.

For the most part, however, this book eschews explicit methodological or philosophical engagement with such issues. It seeks to document human-induced environmental changes and to understand how these came about. Insofar as it derives its pedigree from my disciplinary training, it rests upon a conviction that geography is an integrative endeavor focused upon "the complexities of relationships between environment, economic activity and social organization in particular places." It is thus inherently open and interdisciplinary in approach, seeking to bring together ideas and information from a range of disciplines about the interconnections between human societies and the environments in which they live.

By essentially avoiding explicit epistemological debate about the nature of nature to explore places and the processes involved in shaping them, it does *not* posit that humankind and nature are entirely separate (and separable) entities. Instead, it seeks to work around that claim by following the lead of an earlier generation of geographers who believed their field to be concerned with much more than spatial patterns. In deploying this materialist ecological approach, this book seeks to honor and extend those scholars' efforts to embrace (as the English geographer Jack Langton has rightly and succinctly said), both humans and nature "through the agency of work within the totality of 'habitat, economy and society'" (Langton 1988, 18–19)

Once upon a time, the task of writing history on a national, or even subcontinental, scale seemed relatively straightforward. Politicians and other great men (rarely women) dominated these stories, and their rise and fall provided a more or less orderly parade of events, achievements, and sometimes failures (and failings) to mark the rise or development of the nation. The complexities of society and space were distilled into the lives and times of a few "representative" individuals. Such approaches were challenged and fractured by the rise of social, labor, and women's history in the latter part of the twentieth century, developments that brought questions of class, race, gender, power, and strife to the fore and made it much more difficult for many to accept the "consensus (political or biographical) histories" written and favored by previous generations.

When the environment or nature is added to the mix of topics within the purview of historical scholarship, and natural or bioregional boundaries come to seem more appropriate than political borders as frames for inquiry, the potential loss of coherence in a welter of competing conflicting perspectives is enormous. Typically, those who seek to grapple with the complexities presented by any one of these more recent approaches to the past find their most congenial turf in limited and local studies: They write of particular towns, the experience of widows, the struggle for female suffrage, the particularities of a specific strike, or the changes affecting an acre in time. Synthesis is difficult, because each event, cause, struggle, or place has a different history; each offers a slightly different lesson; and any effort to identify broad patterns evident across space or through time threatens to obliterate or do violence to local stories. Thus—and in spite of the profusion of excellent work in all of these fields—studies of large areas (especially areas of subcontinental proportions) are relatively rare, and surveys that encompass long periods of time are possibly even more uncommon.

There is no entirely effective way around this dilemma. In the pages that follow, discussion ranges over tens of thousands of years of history and millions of square kilometers of "geography," but it does so selectively. To deal with the enormous arc of time that stretches between the last great continental glaciation that buried almost all of northern North America beneath enormous sheets of ice and the present-day concerns about the implications of global warming, the book is divided into five major parts. These follow each other chronologically, but they are neither delineated by the divides (the elections, the wars, the great events of political-territorial establishment) that structure more conventional histories nor bounded precisely by the borders and subdivisions that define more conventional works of geography. Perhaps they are best thought of as epochs that overlap at blurred and indistinct boundaries rather than as periods defined by the turn of the calendar.

The first part, **Deep Time**, deals with the colonization of the northern part of the American continent by flora, fauna, and the peoples who came to be known in time as indigenes (or Native Americans or First Nations) after the retreat of the last great (Wisconsinan) ice sheets. The second, **Contact and Its Consequences**, explores the effects upon native peoples, and upon many elements of the indigenous biota of this northern land and its extensive littoral, of the coming of newcomers, mainly from Europe, in the years after AD 1000. It is focused on the shifting, often indeterminate and frequently protracted "moment" when natives and newcomers encountered and engaged one another. Because this moment began rather later in the north and west than it did in the east, this discussion slides broadly across space as well as time.

Once newcomers began to stay in the territory to which their predecessors had first come as explorers and sojourners, the eastern parts of the continent saw the establishment of **Settlers in a Wooden World**. From the establishment of their initial toeholds on the Bay of Fundy and the Gulf of St. Lawrence early in the seventeenth century, through to the middle decades of the nineteenth century, when they occupied much of the habitable ecumene of eastern Canada, these newcomers sought to make the land their own by converting it into farms, building roads and towns, exploiting its forests, and so on. But they did so as inhabitants of what Lewis Mumford described as the Eotechnic era. Theirs was an age of wood, wind, and water, in which the sources of motive power were limited to human and animal muscles and the kinetic energy of moving air and water.

The fourth part, **Nature Subdued**, deals with the enormous environmental (and economic and other changes) produced by the impact of Mumford's Paleotechnic and Neotechnic revolutions (or the introduction first of a technological complex defined by the use of coal, iron, and steam, and second by the harnessing of electricity and the development of new alloys) and carries the story through the latter half of the nineteenth century to the eve of the World War that began in 1939. **Nature Transformed** treats the environmental history of this northern realm through the last half of the twentieth century, focusing on the ethos of development and the implications of brute-force engineering that reshaped land and life clear across the continent, from Newfoundland to Alaska, particularly in the third quarter of the century, before turning attention to some of the challenges to progress and environmental domination encountered in the quarter century or so before the present. Finally a brief sixth part considers the broader implications of the story recounted in the preceding pages.

Within this broadly chronological framework, the discussion proceeds thematically. Themes differ between parts, but the great and persistent activities that might be very loosely described as the bases of the fundamental staple trades upon which the Canadian economy was built—fishing, fur trading, farming, "foresting," and mining—are treated, as appropriate, in all but the first section of the book. They provide a loose set of structural leitmotivs throughout these pages, although sections on the rise of towns and the environmental consequences of urban growth supersede discussions of the fur trade in later pages, in recognition of urbanization's role in transforming the lives and living spaces of a growing proportion of Canadians. One corollary of this template is that those interested in a particular activity, such as fishing or mining, might trace the environmental histories of this endeavor more or less completely through extended periods of time by reading the relevant chapters in each successive part of the book.

Through all of this, local studies illustrate larger claims, and arguments are made by reference to specific areas or particular regions. Thus, space is treated selectively. It is impossible, in a study of this scope, to afford equal attention to all places at all times. The inevitable choices and selections reflect what is written (most often by other scholars) and what is known (by me) and they have been made to elucidate the processes and patterns of environmental transformation across northern North America.

Some will feel that particular places get short shrift, among them no doubt residents of Canada's smallest province, Prince Edward Island, and those who support the view advanced in a cult work of 1969 that Cape Breton is the thought control center of Canada (Smith 1969). To them and others (whose important and informative work on parks, on landscape representations, on environmental movements, on big game hunting and wildlife protection, and on many other topics that might justly have found their way into an environmental history differently conceived) I can only express regret at my inability to do more (for example, MacEachern 2001; Bordo 1992–1993; Zelko 2004; Loo 2001, 2006; Colpitts 2002).

In the final analysis, I hope that my intentions are clear. I have tried, plainly and simply, to offer a new perspective on the past of northern North

America, or to see Canada (and the adjoining northwestern corner of the continent), as nineteenth-century naturalist Philip Gosse had it, "in a new point of view." As a work in environmental history this book seeks to be something other than a political history of Canada (and Alaska) with trees. In the process it misses or skips quickly by many topics that are central to many other historical studies of this area and its constituent political parts. There is virtually nothing on federal-provincial relations in Canada in these pages, for example, though some have argued that all Canadian history eventually comes down to this. Except as these might be defined to include trade, the external relations of Canada and Alaska are essentially ignored. Important though race, class, and gender relations are to the social histories of these territories, I have not attempted to "read" these concerns off the mirror of nature.

Yet many facts and arguments familiar to readers of political and social and economic histories treating various parts of this northern realm find their way into these pages. There are two reasons for this. First, they are old and familiar markers for those acquainted, even rather vaguely, with more orthodox histories of this area, and thus help as way-finders in an otherwise possibly rather exotic landscape. Second, and more importantly, it seems to me that at some level environmental history cannot and should not be entirely separated from other perspectives on the past. Practitioners of each of history's subcategorical forms shape and emphasize their accounts differently, but the best of them surely do so without eviscerating all other approaches.

If there is one fundamental lesson that environmental history seeks to convey, it is that the world is of a piece and that the natural warrants consideration along with (indeed, alongside) the cultural. By approaching the past of northern North America from an environmental perspective, the pages that follow only begin the process of mapping this intricate and fascinating terrain. I hope that others might be stimulated and provoked by its provisional tracings to find the delicate balance necessary to bring the marvelous complexities of the peoples, places, and times considered here into fuller, more seamless, satisfying, fresh, and enlightening account.

ACKNOWLEDGMENTS

I often think that the best analogy for intellectual endeavor is an ecological one. Consider scholarship as a coral reef. Like the reef, the life of the mind is part of a larger ecosystem, and it is made up of a diverse set of communities. Societies sustain their thinkers, just as symbiotic relations with certain plant-like algae are essential to the survival of corals. Different species of coral build markedly different structures (think of those produced by "brain corals" and "fan corals" for example), just as the works of scientists, engineers, and humanists vary enormously in form and purpose. Coral species tend to cluster into characteristic zones on a reef, set apart by competition and environmental conditions; university scholars also compete for resources and attention, and typically come together in departments and faculties that provide more or less coherent and congenial settings for the kinds of work that they do.

Both reefs and learning grow incrementally, in a great variety of intricate ways. Each addition, to reef or to knowledge, is a tiny part of the whole. Just as each polyp takes its place on and adds to the substrate of earlier material, so every contribution to scholarship augments libraries and generations of accumulated knowledge. If coral reefs are "the rainforests of the oceans," universities are rich breeding grounds of ideas. Furthermore, both coral reefs and productive intellectual spaces are rare, delicate, fragile, and vulnerable. Those who study the oceans tell us that reefs are susceptible to disease, excess shade, increased levels of ultraviolet radiation, sedimentation, pollution, salinity changes, and increased temperatures.

Those who work in universities know that thoughtful scholarship is also highly vulnerable to external influences and changing environmental circumstances. "Bean-counting" productivity measures, bureaucratic research assessments, and other demands that reward the swift completion of highly visible but unambitious projects with predictable results tend to deflect interest (and support of various kinds) from work that requires quiet contemplation and prolonged gestation. Personality conflicts and internecine struggles within units and institutions can be equally damaging. One is tempted to suggest that the natural and human-induced stresses that affect coral reefs have their parallels in systemic and personal stresses within universities. Further, it might be adduced that both reefs and scholars can be remarkably resilient in the face of short-term pressures and catastrophes, but that neither is well adapted to survive certain kinds of long-term alterations to the ambient environments that allowed them to grow.

These reflections remind me of both my dependence and my good fortune. This book rests, as much as any coral, on what has gone before. It depends upon the ideas, insights, and industry of thousands of scholars, only some of whose contributions can be acknowledged directly in the bibliography. Those from whom I have learned and borrowed (I hope accurately and sensitively) will most recognize my indebtedness and I thank them all, first and most sincerely. Innumerable works in the libraries of the universities of British Columbia, Oxford, Cambridge, and Melbourne have been invaluable to this project, and those who have sustained these repositories—and beyond them those who have published the volumes and journals they hold, some now in digital form—have played their parts, often unknowingly and sometimes long ago, in its completion.

After thirty years in the University of British Columbia, I sometimes think that I have become, like the coral, a sessile animal. If so, this surely has much to do with the environment there, and particularly that within the Department of Geography, which has been my academic home through these decades. It is not, and has not been, all warm climates, colorful fishes, and clear waters. But it has been an important nursery area for ideas, a remarkably diverse and stimulating intellectual environment, and a place that has fostered productive symbiotic relationships with colleagues whose interests have broadened my horizons and whose expertise has helped me understand many things. I hesitate to itemize these contributions individually, because the list is long, the obligations are many, and my memory is increasingly fallible.

I must, however, offer particular thanks to Cole Harris, who showed me how, and why, to become a historical geographer more years ago than either of us care to remember; to Matthew Evenden, an environmental historical geographer who pushes me to be better through his own research and who kindly read and commented helpfully on an earlier draft of the manuscript; and to several outstanding graduate students with whom I have had the pleasure and benefit of association. Other colleagues, particularly historians and geographers, and many of them from beyond UBC and Canada, have inspired and assisted and helped in various important ways along the way. I and they know who they are, and I thank them.

I must also acknowledge the important and useful work of several research assistants who helped in various ways (and for various periods) with facets of this oft-times seemingly impossibly sprawling project: Rebecca Smith, Shannon

Stunden Bower, Matthew Schnurr, Dan Michor, John Thistle, Emily Davis, and Ellen Morgan will see traces of their labors in these pages, though I could not use all they produced. For his wonderful cartography, produced as always to the highest standards, under significant time pressure, I am most grateful to Eric Leinberger of the Department of Geography at UBC. I would also thank all those at ABC-CLIO, including editors Steven Danver, Alex Mikaberidze, and Martha Ripley Gray, who assisted in turning my text into this finished product.

Having time (however squeezed it has sometimes been by administrative and other obligations) in which to complete a project of this magnitude is a gift and I am grateful that Canadian universities are still able to provide it more generously than their counterparts in many countries. But the foundations of this generosity are not robust. The business model is being applied to more and more aspects of university governance and (often even more unfortunately in my view) to the day-to-day operations of these institutions. I would urge all those who have persevered this far to be mindful of the importance of sustaining the delicate ecologies upon which scholarship of various types depends and to contribute in whatever ways they can to ensure proper stewardship of the opportunities for a life of learning in this country.

Finally I thank Barbara, for her love and support, her companionship, and her garden; and our children, Louise and Jonathan, grown now into fine young adults, for constantly reminding me that there are other things in life than reefs and books and anxieties about the future of the past.

Graeme Wynn
Professor and Head of Geography
The University of British Columbia

LIST OF FIGURES

PART 1

Deep Time

Let us begin with the ice. Fifteen thousand years ago ice covered almost all of the territory dealt with in this book, and for thousands of years before that, ice had advanced and retreated across the broad outlines of the landmass that now forms the northern part of the North American continent. The sheer magnitude of these episodes almost beggars belief. At various times, ice extended unbroken from the Arctic islands to latitudes south of the present Great Lakes, and from the eastern shores of Newfoundland to the Queen Charlotte Islands. Even near its southern limits—let us choose the location of present-day Vancouver as an example—the ice was often well over a kilometer in thickness. Some estimates suggest that the ice may have been over two kilometers thick in eastern Ontario. Neither vegetation nor humans could thrive in these circumstances.

Ten thousand years later, only vestigial remnants of these extensive sheets of ice remained, and they were north of the Arctic Circle. People, plants, and animals had spread through the territories now known as Canada and Alaska. In this dynamic environment, human and natural histories evolved with remarkable speed and variety.

ICE RETREAT, VEGETATION ADVANCE

We all know that ice melts, and moves. We also know that, measured on a human timescale, its movement is slow—things that take longer than they ought to do progress, we say, at a "glacial" pace. In geological terms, however, the advance and retreat of glaciers, and of those continental-scale masses of frozen water known as "ice sheets," is a fleeting phenomenon. With their movement typically calibrated in centimeters per year, ice sheets and glaciers seem, by comparison with many almost infinitesimally slow geological processes (such as the uplift of mountains or the drift of continents) to sweep across space in no time at all. Geologists tell us that rocks of Precambrian age—formed at least 543 million years ago—comprise large areas of northern and eastern Canada (loosely known today as "the Shield"). Moreover, they affirm, these rocks are remnants of a small continent, Laurentia, the origins of which lie billions of years in the past.

Collisions between Laurentia and other ancient landmasses, in a series of orogenic events that welded together terranes (or distinct rock formations) and uplifted mountains as drifting tectonic plates came into contact, eventually gave shape to a larger North America, the contours of which continued to evolve almost imperceptibly as a result of the geomorphological processes of erosion and deposition. Through these eons of deep time, occasional cataclysmic events, such as earthquakes and volcanic explosions, also contributed to the physical shaping of the northern part of the continent. All of these processes influenced the later environmental history of Canada and Alaska in important ways. They helped determine the courses of rivers, the qualities of soils, the heights of mountains, the locations of minerals, and the extent of plains, and thus shaped the prospects for settlement, the possibilities for agriculture, the patterns of resource exploitation, and the perceptions people developed of this territory. But none was as consequential as the ice—and the changing climate that accounted for its advance and retreat.

Climates are always changing, and they do so at widely different rates. Some changes take millions of years, some tens of thousands of years, and

The continental ice sheet. *In the absence of photographs of the Wisconsinan glaciation, this recent image of the Greenland ice sheet offers some idea of how the landscape of the current western Cordillera might have appeared some 10,000 years ago. (Courtesy of Michael Bovis)*

others rather less. These are almost incomprehensible sweeps of time, and the ecologist E. Chris Pielou has usefully scaled them to understandable dimensions by suggesting that we think of the last billion years as a decade. In these ten years, two glacial ages have run their complete cycles (lasting a month or two) and we live in a third that began, so to speak, about a week ago. Each glacial age is marked by periods of intense cold (known as glaciations) separated by warmer interglacial periods. In the six or seven days of the current glacial age, there have been nineteen or twenty glaciations each lasting about six hours and separated one from another by interglacials lasting about ninety minutes, the duration of a soccer game. We live in one of these, and on this scale we are in the last few minutes of regulation time.

Accounting for these changes is not easy, and glacial ages and glaciations have to be explained differently. Tectonic drift is currently counted as the cause of glacial ages. These occur infrequently, the argument goes, when the continents are arrayed so as to block the movement of warm tropical waters into polar latitudes. Today (but not always in the past) Antarctica covers the southern polar region and the alignment of the continents in the northern hemisphere deflects warm ocean currents away from the Arctic Ocean. Thus polar summers are cool and ice and snow accumulate there. Glaciations, by contrast, alternate with interglacial periods on a more regular 100,000-year cycle. In broad terms, glaciations last between 60,000 and 90,000 years, and warm interludes last 40,000 to 10,000 years. This pattern is attributable to long-term changes in the earth's orbit around the sun, known as the Milankovitch Cycle. This cycle is itself determined by cyclical changes in the shape of the earth's orbit, in the tilt of the earth's axis, and in the precession of the equinoxes.

Together, these cyclical changes conspire to shift the amount of solar radiation falling in the Northern and Southern hemispheres and in high and low latitudes. This in turn alters the contrast between summer and winter temperatures, producing comparatively warm summers and especially cold winters at one extreme and relatively cool summers and mild winters at the other. When the seasonal temperature contrast is relatively modest, as it is through most of the Milankovitch Cycle, ice and snow accumulate and ice sheets grow. When summers are particularly hot, however, snow and ice melt more quickly than they accumulate during the intervening winters. These are the interglacial episodes, the most recent of which began some 20,000 years ago.

Although the current interglacial has run considerably more than half its course—the incidence of solar radiation in the Northern Hemisphere has been declining steadily for the last 10,000 years or so—local climatic conditions during this period were determined by the warming effects of the sun and the cooling influence of the ice. As a consequence the period of peak warmth (known as the hypsithermal) varied over space and time. In the far northwest of the continent, say in the vicinity of the present-day Alaska-Yukon border, it occurred about 10,000 years ago. For the prairie region the warmest and driest years were between 8,000 and 7,000 years ago. The climatic optimum probably occurred less than 7,000 years ago in the southern Great Lakes (southwestern Ontario), 6,000 years ago along the lower St. Lawrence, 5,000 years ago in northern Cape Breton and the Gaspé, and little more than 4,000 years ago in northern parts of Newfoundland (Pielou 1991, 269–271).

Further complicating the picture is the fact that the climate of the last 20,000 years has been marked by perceptible short-term alternations of warm and cold conditions. Occasionally, such changes have been triggered by catastrophic events, such as the sudden release of vast quantities of icy water from the former Lake Agassiz into the North Atlantic. Thus scientists hypothesize that the development of a new outlet from the lake, about 13,000 years BP, may have sent 9,500 cubic kilometers of icy water through the Great Lakes–

St. Lawrence corridor into the North Atlantic, interrupting the ocean's transport of heat from the equator to the Arctic and precipitating the onset of the global cold period known as the Younger Dryas. Shortly after this cool period ended, about 11,500 years ago, another shift in the drainage of Lake Agassiz spilled vast quantities of water through the Mackenzie corridor into the Arctic Ocean. This produced another short-lived period of global cooling known as the Preboreal Oscillation. Then about 8,400 years ago a third shift in drainage patterns spilled almost one-third more water than is now found in all of the world's lakes (that is, over 160,000 cubic kilometers) into Hudson Bay and the North Atlantic, raising sea levels by half a meter and causing a 400-year decline in global temperatures. In the last 10,000 years, however, climatic conditions also seem to have fluctuated periodically, with a 2,500-year interval that may be attributable to cyclical variations in the release of radiation from the sun. Thus the cooling trend that has marked the last several thousand years has not been uniform or continuous. In the last 1,200 years or so a warmer period known as the Little Climatic Optimum gave way to a markedly cooler period, called the Little Ice Age, producing what the French historian Emmanuel LaRoy Ladurie has aptly characterized as "times of feast [and] times of famine" in the Northern Hemisphere. Other shorter cycles associated with "smaller scale solar variations" (in the order of 200 years) and sun spot cycles (11 years) also modified patterns of temporal variation, but in broad terms the Little Ice Age ended with the onset of a warming trend in the mid-nineteenth century (Pielou 1991; Clark et al. 2001).

The last—and for our purposes most significant—major glaciation of northern North America (known as the Tioga phase of the Wisconsinan Glaciation) reached its southernmost extent approximately 20,000 years ago when two major ice sheets coalesced. The Cordilleran Ice Sheet ran the length of the Pacific Coast, from the Aleutian Islands to Puget Sound (Tacoma) and spilled eastward onto the northern Great Plains (see Fig. 1). The much more extensive and more complex Laurentide Ice Sheet (an amalgamation of several discrete ice masses) covered much of the rest of the territory. At its maximum, it fringed the Missouri River valley, pressed across the 40th parallel of latitude south of present-day Lake Michigan and encompassed the southern coast of what is now Massachusetts. Few areas of southern Canada were free of ice: a small area of the lower Fraser River valley, a narrow finger of land east of the Rocky Mountains, and small patches of the present-day Maritime Provinces and Newfoundland. Here and there—as in the Tantalus Range of southern British Columbia and the Queen Charlotte Islands—mountain peaks (known as *nunataks*) remained above the ice and may have served as refugia for certain cold-tolerant plants and animals. The only extensive area of open land was in the far northwest, where most of modern-day Alaska and part of Yukon Territory supported tundra vege-

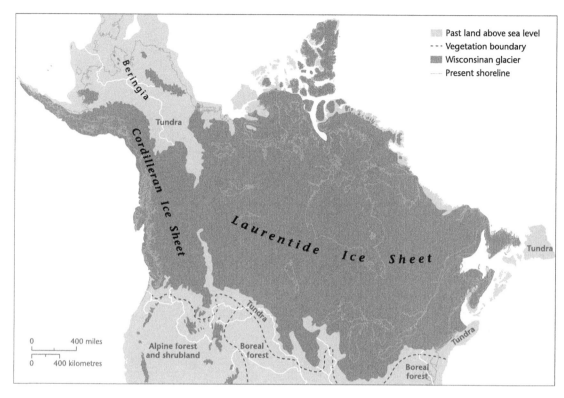

Fig. 1. *Northern North America about 20,000 years ago. (Adapted from* Historical Atlas of Canada *1: pl. 1, and Pielou 1991)*

tation (fungus, herbs, and low shrubs) and populations of woolly mammoth, bison, musk ox, caribou, and horse (*Historical Atlas of Canada*, vol. 1).

As the glaciers and ice sheets expanded and contracted they shaped and sculpted the landscape in numerous ways. In mountainous areas, cirques (amphitheater-like valley heads), U-shaped valleys, arêtes (sharp ridges), and other features were created. Abrasion by the sediment load carried in the sole or bottom of the moving ice, and also the "plucking" associated with freezing in cracks in the bedrock beneath glaciers and ice sheets, were powerful mechanisms of erosion. So, too, were relatively short-lived streams of water flowing at high velocity beneath the ice. Together these processes sculpted the surface of the vast Precambrian nucleus of eastern Canada, selectively eroding softer materials and accentuating structural features in the bedrock. Later, thousands of lakes would form in the depressions thus created. All in all dozens of meters of material may have been scoured from even the hardest of surfaces over which the ice passed.

As the ice began to melt, deposition of this eroded material quickened. Extensive moraines (as for example the Oak-Ridges Moraine in Ontario), long

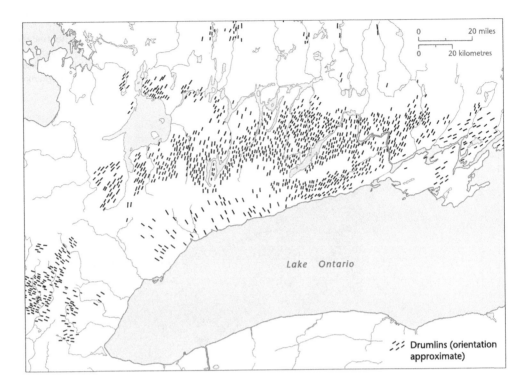

Fig. 2. Drumlins (small, whale-back hills formed by glacial action) of Southern
Ontario. (Adapted from Chapman and Putnam 1984)

sinuous eskers, and extensive drumlin fields (as in the vicinity of Peterborough,
Ontario) marked the landscape of glacial retreat (see Fig. 2). Unconsolidated sediment and clayey "glacial till" were widespread. Inchoate drainage patterns and
the frequent formation of ice dams by the retreating ice created enormous lakes
and waterlogged areas (Chapman and Putnam 1984; Gilbert 1994). The evolution of drainage systems was further complicated by the rebound of the earth's
crust from the depression induced by the weight of ice on the surface—a depression estimated to approximate 200 meters in the St. Lawrence Basin. As iso-
static rebound occurred at different rates, the land surface of the lower Great
Lakes area was elevated more quickly in the northeast than the southwest, inducing further changes in outlet channels and drainage patterns. Vast glacial
lakes—among them Lakes McConnell, Agassiz, Duluth, Barlow-Ojibway, Algonquin, and Iroquois—developed along the edge of the shrinking Laurentide ice
sheet some 11,000 years ago (see Fig. 3). Enormous quantities of sediment were
deposited in these lakes, and if and when they drained almost completely—as
occurred most spectacularly in the case of Lake Agassiz—extensive level areas
of fine, rich sediment remained. The pace and enormity of these changes should
not be understated. In southern Ontario the ice front retreated from south of

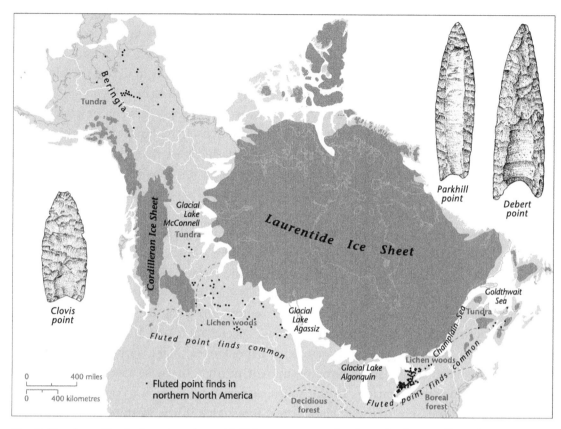

Fig. 3. *Northern North America about 11,000 years ago, with Clovis (widespread), Debert (northeastern seaboard), and Parkhill (Great Lakes) fluted points. (Adapted from* Historical Atlas of Canada *1: pl. 2)*

Lake Ontario to north of the Ottawa Valley in little more than 1,500 years. Similar patterns were repeated in essence through the next 5,000 years over an ever-increasing northern area as the ice sheet shrank and fragmented through 5000 BC. By geological and geomorphological standards, these were enormously dynamic processes, and the environments that they shaped contributed in no small part to subsequent patterns of interaction between humans and nature in this northern realm (Dyke and Prest 1987).

As the climate across northern North America warmed and the ice retreated, plant species, and the vegetation complexes into which they are customarily assembled, spread northward. Tundra typically led the way. When the ice was at its maximum, tundra marked most of its southern boundary. Yet the extent of this tundra zone is still debated. Some suggest that it may have extended 100 kilometers and more from the ice front in some areas. In other locations, spruce and pine seem to have grown in close proximity to both the ice front and such tundra plants as bog blueberry and dwarf willow. Perhaps the most that can be said with certainty (and with Pielou 1991, 87) is that "a strip (but not an uninterrupted strip) of open ground with low vegetation lay immediately south of the great ice sheets at the time of glacial maximum." This vegetation may have looked rather like modern tundra, but it almost certainly differed from it in the details of species composition and so on.

As the ice sheets began to shrink, vegetation was sparse on newly exposed and frozen ground. The main features of much of the landscape were exposed bedrock, coarse-textured glacial till or outwash deposits, and boggy areas. Given the diversity of conditions, the possibility that some newly exposed ground was unfrozen, and the likelihood that sand, soil, and seeds might be blown in from the south, it is impossible to be precise about the invasion of plants that followed retreat of the ice. In some areas an impoverished flora, in which lichens dominated and vascular species accounted for only a quarter or so of the total, likely covered some 15 or 20 percent of the terrain. Beyond this, dwarf shrubs, such as saxifrage, arctic willow, and Cassiope (arctic white heather), appeared with various sedges. At yet greater remove from the ice, mosses and lichens flourished, but the number of vascular species increased and the vegetation became noticeably more woody, with Labrador tea, black spruce, arctic willow, cottonwood, and scrub birch prominent in this otherwise low moss- and herb-dominated vegetation. To the south, tundra gave way to a vegetation complex somewhat akin to the modern boreal forest and, in the west, to an alpine forest and shrubland complex.

As the Cordilleran and Laurentide ice sheets contracted and separated, tundra expanded along the eastern flank of the Rocky Mountains. By 9000 BC it encircled the Cordilleran ice, and extended from the southern tip of Vancouver Island through the western prairies of today to the far northern reaches of Alaska and the Northwest Territories. In the western prairie, it gave way to lichen woodland. Here white spruce was scattered across open landscapes in which light-colored lichens dominated the better-drained sites. Depending upon soil and climate conditions, birch, poplar, jack pine, and tamarack also occurred. In the peninsula of present-day southern Ontario, an open spruce forest invading dwarf-shrub tundra formed another lichen woodland. Spruce, willow, birch, and pine were dominant woody species; wormwood, ragweed, and other weeds also appeared.

Ten thousand years ago (8000 BC), the lichen woodland zone had shifted to occupy an attenuated band of territory north of the Ottawa River, extending from the southeastern fringe of Lake Agassiz, through southern New Brunswick to the Nova Scotia peninsula (see Fig. 4a). In southern Ontario, where lichen woodland had once been dominant, there was a forest of spruce, balsam fir, jack pine, and birch. Although it was broadly akin to today's boreal, this forest—and

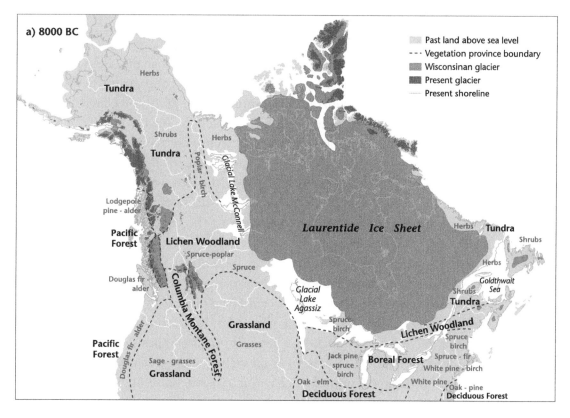

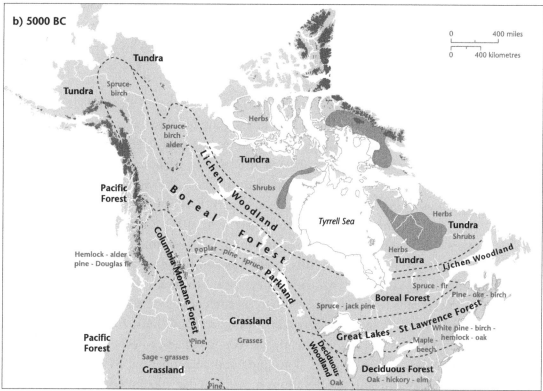

Fig. 4. *Vegetation changes: Northern North America about 10,000 years ago (a), and about 7000 years ago (b). (Adapted from* Historical Atlas of Canada *1: pl. 4)*

the other vegetation complexes that occupied rapidly changing environments at this time—cannot be equated precisely with modern counterparts developed for centuries in relatively constant environmental conditions. Plants colonized new ground at different rates according to their ecological requirements, the ways in which their seeds are transmitted, and so on. As the work of palynologists J. C. Ritchie and G. M. Macdonald has shown, the lightweight, wind-borne seeds of white spruce enabled its advance on the prevailing winds to the east of the shrinking Laurentide ice sheet, at a fairly steady average rate of 300 meters a year, from the latitude of Pennsylvania to the coast of Labrador between 14,000 and 7,000 years ago.

On the western edge of the ice sheet, by contrast, white spruce advanced at the astonishing rate of three kilometers a year, and was carried to the shores of the Beaufort Sea about 9,000 years before the present by strong southeasterly winds blowing clockwise around a persistent, ice-centered anticyclone (Pielou 1991, 92–94). Whatever the precise details of their composition, it is quite clear that these protoboreal forests were far richer in species than the vegetation provinces to the north of them. Sphagnum mosses, horsetail, swamp cranberry, and Labrador tea were common on moist sites, and lichens, asters, goldenrod, and blueberry in drier areas. Red-osier, bunchberry, and grey alder occupied intermediate sites. As revealed by analyses of pollen buried and preserved deep in the sediment of several lakes in this area, spruce and jack pine declined in importance as the climate warmed through the next 2,000 years, to be replaced by white pine (see Fig. 5).

By 5000 BC, the boreal forest zone lay well to the north (see Fig. 4b). On its northern boundary, the lichen woodland zone quickly gave way to tundra surrounding the vestiges of the Laurentide ice sheet in Labrador. To the south of the boreal forest from Cape Breton through Bay Chaleur to Lake Superior, the land was covered by the cool-temperate Great Lakes–St. Lawrence forest (also known as the hemlock–white pine–northern hardwood zone), a mixed coniferous-deciduous forest in which birch, oak, elm, maple, and beech occurred with pines and hemlock. Subject to continental and maritime climatic influences in different parts of its range, this forest was and is highly variable. Moreover, its transitional character (it forms an ecotone between the boreal and broadleaf realms) means that broadleaf hardwoods and coniferous softwoods can be dominant, depending upon local soil and climatic conditions. In the southernmost reaches of the Ontario peninsula, deciduous species—especially oak, beech, maple, elm, and ash—were dominant and accompanied by a rich variety of shrub and herbaceous species (from sumac and dogwood to wintergreen and trillium) as well as forbs and grasses.

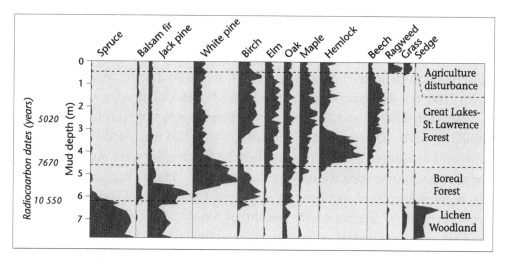

Fig. 5. *Environmental change reflected in the pollen record, Edward Lake, Ontario.* *(Adapted from* Historical Atlas of Canada *1: pl. 4)*

Again, however, it must be stressed that these were dynamic, evolving vegetation complexes (Anderson 1989; Liu 1980, 1981; Scott 1995). The painstaking work of palynologist Margaret Davis, which mapped the changing incidence of dozens of tree species in over sixty sediment cores from eastern North America, reveals how the white pine, eastern hemlock, maple, and chestnut—found together in southern Ontario in the modern period—moved north from different refugia. The hemlock and white pine came from the eastern slopes of the Appalachians, the maple and chestnut from the mouth of the Mississippi. They extended their ranges at different rates, averaging 100 meters a year for the chestnut and 300 to 350 meters a year for the white pine, and according to their different ecological requirements (that allowed white pine, which is able to colonize poor exposed soils, to spread northward a thousand years before the hemlock—which requires richer, moister soils and shade for successful propagation).

By 8000 BC, grasses had replaced lichen woodland in areas west of Lake Agassiz subject—as the climate warmed—to summer moisture stress (or drought) in the shadow of the Rocky Mountains. Here *Stipa* and *Agropyron* grass species were dominant, with fescue and oat grass and sage brush and cacti in the driest areas. Variant forms of lichen woodland, dominated by poplar in the north and spruce in the south, had colonized former tundra north and west of glacial lakes Agassiz and McConnell. On the eastern slopes of the Continental Divide, a pine-spruce forest extended to latitude 50 degrees north. Distinctive Douglas fir–alder (to the south) and lodgepole pine–alder associations dominated the Pacific coast to latitude 60 degrees north. By 5000 BC, the vegetation map of western Canada closely resembled that of recent times. Although the precise mix of species differed across this 6,000-year span, the major vegetation provinces recognized by modern students of the environment conformed quite closely, south to north through the prairies, to the boundaries ascribed to them in AD 1000: grassland, parkland, boreal forest, lichen woodland, and tundra.

Between 5000 BC and 2000 BC, the northern reaches of North America continued to warm, ice sheets and glaciers continued to melt, and the vegetation provinces continued their northward march. The results were most evident in the east. Lichen woodland extended northward to Ungava Bay, and the area of boreal forest increased significantly as its limits shifted across present-day Quebec from 50 to 55 degrees north. Deciduous woodland occupied more of southern Ontario, extending its range along much of the north shore of Lake Ontario. In the west the grassland also extended northward and the parkland belt expanded. All of these trends were reversed through the next 3,000 years as the climate cooled. In Quebec tundra, lichen woodland, and boreal forest zones moved southward. Cool temperate forest reclaimed all but the far southwest of Ontario from the realm of deciduous woodland. Grassland and parkland zones contracted.

2

FIRST PEOPLES/
FAUNAL EXTINCTIONS

At the height of the Tioga phase of the Wisconsinan glaciation, ice sheets spread across Greenland, Scandinavia, and northern Siberia as well as northern North America. Vast quantities of water were locked up as ice. As a consequence, sea levels may have been 130 meters lower than they are today. In the far northwest of the American continent, where the Bering Strait today separates Alaska from Siberia, this was enough to create a land bridge between Asia and America. Between 14,000 and 12,000 years ago, inhabitants of Siberia—hunters and gatherers who pursued roaming herds of woolly mammoths, musk oxen, and other mammals—followed their prey across this bleak and relatively recently exposed piece of land. They had little idea that the path they took would disappear beneath rising seas as the earth warmed generations hence, and even less that they were colonizing what would later be thought of as a new and separate continent. The land they entered was much like that they left behind: flat, hard, and cold.

Analyses of pollen from lake and peat deposits suggest that the environment was colder and drier than that of the present and that until 14,000 years before the present, grasses, sedges, willow, and sage were the prevalent plants. Variously described as arctic tundra or polar desert, this herb-dominated vegetation was probably incapable of sustaining a large or diverse ungulate population. Yet radiocarbon-dated skeletal finds indicate that some large mammals roamed this area well before 12,000 BC, suggesting that grassland or steppelike vegetation may have prevailed in parts of the region. However, the lack of wood for fuel probably inhibited human colonization of this territory through the early part of the Tioga phase. Then dwarf birch, aspen, poplar, and juniper began to appear. They became more prominent through the next 2,000 years. There was likely a sharp increase in alder and spruce about 10,000 years ago.

"Siberian" hunter-gatherers, adapted to life in this harsh environment, made their circumpolar ways eastward into "the Americas"—unself-consciously and, in a sense, inadvertently colonizing the ice-free territory of Beringia. The best archeological evidence of their presence comes from central

Alaska—and particularly from the Nenana and Tanana valleys. There the record indicates that humans entered the area in a warm-climate interlude (or interglacial period) between 12,000 and 11,000 years ago, and that they hunted bison, elk, and sheep, as well perhaps as smaller mammals, birds, and fish. From these locations, generations later, descendants of these early occupants made their way southward, through the similar terrain opened to their passage by the retreating Cordilleran and Laurentide ice sheets. In the centuries that followed, these migratory hunters—known as Paleo-Indians to later American scholars—occupied the central plains and eastern woodlands. By one estimate, 500 years was time enough for their movement from the Yukon Delta to New Mexico.

By 9000 BC they had left evidence of their presence (in what archeologists consider the remarkably similar remains of their encampments) from California to Florida to Nova Scotia. A thousand years after that, artifacts mark their presence in the southern cone of South America. These Paleolithic pioneers were the rootstock from which an increasingly large, diverse, and culturally complex population of indigenous people—American Indians, or in Canada members of the First Nations—descended (see variously Fladmark 1979; Hoffecker et al. 1993; Hoffecker and Elias 2003; Hopkins et al. 1982; Irving 1985; Jackson and Wilson 2004; Meltzer 1989, 1996; Snow 1989; Wilson and Burns 1999).

This, at least, is one story about the first peopling of the Americas. It is not unchallenged. Many native peoples deny that their ancestors were migrants to North America. Their stories, they say, make it clear that their people have always been here, that they have lived in their traditional territories "since time began." On the human scale of things, among peoples embedded in an oral rather than literate culture (and who was literate 10,000 years ago?), the differences between cultural memories and chronological time may be immaterial when it comes to questions such as these. Certainly there is much evidence of affinities between the indigenous peoples of North America and Siberia, and the shared use of a fluted projectile point technology (known as the Clovis Complex) some 11,000 years ago has been taken to indicate a common Paleo-Indian culture among occupants of archeological sites across North America during this period (Haynes 1980). Indeed, quantitative comparisons of lithic (stone tool) assemblages from the Nenana valley and the Clovis site in New Mexico show that they share many diagnostic elements and are highly similar overall.

But none of this is unassailable. Scholars continue to refine and challenge the idea of a single crossing of the Bering land bridge, and every decade for the last half century and more, new arguments and evidence have been put forward to contest what some have called the "Clovis dogma" (Dillehay 1991; Whitley and Dorn 1993). Many now posit three distinct migrations between Siberia and Alaska. Thus, Stanford linguist Joseph Greenberg argues that all the native languages of the Americas are derived from three great linguistic stocks and that these correlate with (and demonstrate) three distinct migrations into the Americas. In Greenberg's view, the vast majority of Native American languages (including those of South America) derive from a single rootstock, known as Amerind, from which divergence began at least 11,000 years ago. This is congruent with a migration across the Bering land bridge 11,000 to 12,000 years ago. Differentiation of a second linguistic rootstock, Na-Dene, the source of Athapaskan-Eyak, Tlingit, and Haida, is estimated to have begun about 9,000 years ago. Thus, Athapaskan speakers are considered descendants of people who crossed into the Americas at about this time. Most of this group settled in the subarctic regions of Alaska and northwest Canada, but some—known today as Navajo (or Diné) and Apache—made their way along the Cordilleran mountain chain to the American Southwest, where they arrived some 600 years ago. Dates for the divergence of the third linguistic rootstock, Aleut-Eskimo, converge at about 3,000 to 2,000 BC.

Those whose languages derive from this stock came to the Arctic fringe of the Americas about 5,000 years ago. Traveling in small boats constructed of wood and animal skins, they depended upon sea mammals for subsistence and spread through this climatically hostile realm to reach Labrador and Greenland some 2,500 years ago. Support for this pattern of successive movements has been found in dental and genetic evidence. Analyses of the morphological characteristics of fossil teeth, which point to north China as the ultimate ancestral homeland of all Native Americans, suggest a significant separation of American from northeast Asian populations about 14,000 years ago and point to three clusters of dental traits that correlate with the linguistic divisions posited by Greenberg (Turner 1986). Similarly, work by R. C. Williams and collaborators on immunoglobulin G (antibody) allotypes is seen, by some, to offer secondary support for the inferences drawn from linguistic and dental data.

Be this as it may, debate continues over when and how the first humans entered the New World. The "Clovis" account rests on the claim that the fluted projectile points, which are diagnostic of this complex, form (as the *Historical Atlas of Canada* 1: plate 2, has it) "the earliest indisputable evidence for the presence of people in North America." It also assumes that the appearance of Clovis artifacts in the archeological record reflects the presence of Clovis people—thus the argument for human migration from Beringia through North and South America in the years after 10,000 BC—rather than the possibility that Clovis traits diffused through preexisting populations. Some would say that archeologists have made too much of similarities in projectile point forms, that they have been insufficiently attentive to possible cultural discontinuities, and (with anthropologist Thomas Dillehay 1991, 13) that there has been a "typological fixation and oversimplification of data and interpretation" in research on these questions. As the geographer Karl Butzer noted in 1998, periods of great climatic, biotic, and landscape change, such as the late stages of the Wisconsinan glaciation, are generally far from ideal for the development of undisturbed, stratified archeological sites. Rising sea levels, rapid erosion, major changes in drainage systems, and other geomorphological processes likely all played havoc with the preservation of archeological remains. Too much can be made of the lack of evidence of a human presence in North America before ca. 10,000 BC.

Debate over these questions has quickened in recent decades because a number of archeological sites have yielded materials that challenge the assumed Bering Strait–Clovis chronology. The most significant of these sites are in South America, at Monte Verde in Chile and Pedra Furada in Brazil, but Bluefish Caves in the Canadian Yukon and Meadowcroft Rock Shelter in Pennsylvania have also produced archeological finds dated to more than 12,000 years ago. Many supporters of the Clovis hypothesis reject these claims, and others based on finds from some two dozen North American locations, as the products of disturbed sites or invalid radiocarbon dates. But doubts about the rapid advance of fluted projectile point makers through the Americas after 11,500 BP continue to accumulate. Using accelerator mass spectrometry and other techniques to date organic remains found under the wind-formed rock varnish covering petroglyphs on talus boulders and surface artifacts on desert pavements, the geographer Ronald Dorn has identified a number of sites in Arizona and California at which indisputably human-manufactured specimens are clearly of pre-Clovis age. Indeed four independent assays of a petroglyph in the Petrified Forest National Park in Arizona place its creation some 16,000 to 18,000 years ago.

Because the Clovis hypothesis rests, at its core, upon the southward migration of newcomers from Beringia, critics have also begun to reexamine the human-ecological assumptions upon which models of this migration process were built. Their arguments are destabilizing. None of the most widely accepted models of a Clovis-first Paleo-Indian migration are robust. One of the more prominent, associated with the palynologist and geochronologist Paul Martin, posits an advancing wave of population pressing through the Western Hemisphere in approximately a thousand years and driving several species of megafauna to extinction along the way. It assumes a population growth rate akin to that for the farming-fishing population of Europeans and Polynesians on nineteenth-century Pitcairn Island. This is far in excess of fertility rates documented for hunter-gatherer societies. So, too, this model's hypothesized migration rate of sixteen kilometers per year seems excessive, drawn as it is from the forced displacement of iron-age pastoralists in nineteenth-century southern Africa. Moreover, high fertility and high mobility rates rarely if ever coexist in human populations. An earlier (1966) formulation of the Clovis migration by C. V. Haynes Jr. adopted more realistic parameters of reproduction and movement, placing the population growth rate at about one-third and the migration rate at about 40 percent of those in the aforementioned calculations. According to its author, it satisfactorily explained the North American spread of Clovis sites through half a millennium. But it does not come close to accounting for the accepted dating of several South American sites, underestimating their occupation dates by a thousand years and more (Grayson 1989; Meltzer 1989, 1996; Melzer and Mead 1985).

A third, less widely noticed formulation of the migration problem by F. A. Hassan in 1981 attempted to take into account the ecological carrying capacities of North American environments, as well as the most credible demographic data and migration rates associated with early hunter-gatherer societies. Preferred rates of both reproduction and movement were considerably below those utilized in other models, with the latter reflecting in part the necessity of adaptation to significantly different environments in moving across the expanse of North America. The results were striking. Under these conditions, the colonization of North America would have required between 8,000 and 10,000 years. Working backward from the established dates of various archeological sites, this implies an initial migration into the Americas "between 25,000 and 20,000 years ago or somewhat earlier" (202).

None of this is conclusive. Models are simplifications based on assumptions. Researchers continue to attempt refinements—but they often focus their efforts on only one part of the complex problem and take other aspects of it for granted. Thus, Steele, Adams, and Sluckin (1998, abstract) developed a model of "Paleo-Indian dispersals" in which migration rates are "modified by local habitat values," in order to move beyond what they see as the widespread treatment of North American space as "a single homogeneous surface." After mapping vegetation cover at a broad scale between 13,000 and 10,000 radiocarbon years before the present, they succeed in simulating observed variations in the occurrence of diagnostic artifacts, but only by "assuming that intrinsic population growth rates were fairly high."

In 1997 two scholars from Brazil adopted a different approach to the question. They argued, on the basis of new mitochondrial DNA sequence data, for a single early migration of people from Siberia into the Americas (Bonatto and Salazano 1997a, 1997b; see also Torroni et al. 1992). The crux of their findings is that DNA sequences of peoples from the three great language groups of North America—the Amerind, Na-Dene, and Eskimo-Aleut—are much closer one to another than they are to any Asian group but the Chukchi of Siberia, and that the Na-Dene, Eskimo-Aleut, and Chukchi have closer affinities one with another than they do with the Amerind population. This suggests that all of these people descended from a single population. Further, this analysis places the date of human arrival in North America at some 30,000 to 43,000 years ago. This is a good deal closer to "time immemorial," and opens the prospect of a substantially more measured southward migration than that implied by the timing of a single pulse out of Beringia to the far reaches of South America (a distance of some 7,500 miles) over the course of several centuries.

It also re-envisages Beringia as a hearth rather than a bridge. By this account, migrants from Asia moved into the Arctic steppes of eastern Beringia between the Tenaya and Tioga phases of the Wisconsinan glaciation and remained there for some time. Some members of this population then moved southward through an ice-free corridor in the northern prairies before the coalescence of the Cordilleran and Laurentide ice sheets rendered this route impassable about 20,000 years ago. Thus, the southward-moving Amerind population was genetically isolated from other inhabitants of Beringia for several thousand years. Able to expand geographically in generally favorable climatic conditions, this group maintained most of the original DNA sequence diversity, while the Beringian population, spatially confined in a deteriorating climate, experienced a reduction in sequence diversity. Subsequently, the Chukchi were isolated in Siberia by the rise of the Bering Sea, and the Na-Dene and Eskimo-Aleut populations diverged, perhaps through the specialized adaptation of different parts of the population to inland and coastal existence.

These arguments are interesting, not least for what they contribute to the debate over the antiquity of the "Taber child" unearthed in the Oldman River Valley of Alberta in 1961. These remains of a four-month-old child were excavated by a leading geologist of the Canadian Quaternary, Archibald MacSween Stalker. Accurate radiocarbon dating was impossible with the techniques available at the time, but the bones were buried beneath 10 meters of alluvium and a layer of glacial till—suggesting that they predated the last glaciation by some time. On the basis of his knowledge of ice movements in the region, Stalker estimated that the remains were "older than 22,000 years, according to stratigraphy at the bluff" from which they were unearthed, and that correlation with nearby sites suggested that they were more than 37,000 years old (Stalker 1969, 428).

Attempts to date the Taber bones using improved radiocarbon techniques in the 1980s suggested that they were only 4,000 years old (but these findings might have been contaminated by preservatives applied to the bones in the 1960s). Some geologists hypothesize that the bones could have been deposited in the ancient sediments in which they were found by more recent mudflows (see Sundick 1980; Wilson, Harvey, and Forbis 1983). The issues remain undecided. Yet they hold out—with other similarly inconclusive evidence from DNA sequences and newly discovered sites (such as that in the Valsequillo Basin [Puebla, Mexico], with human footprints reputedly dated to 40,000 years BP)—the tantalizing prospect that human beings were present south of the ice sheets well before the opening of the ice-free corridor between the Laurentide and Cordilleran ice masses (Pielou 1991, 112–114; Stalker 1969, 1977, 1983; Brown et al. 1983).

Increasing attention is also being focused on the possibility that the first Americans were "ancient mariners" who moved south along the Pacific coast in small craft, fishing, gathering, hunting, and taking refuge in coastal caves from Alaska to California and beyond (Hetherington et al. 2004). Evidence is difficult to find. Paleolithic coastlines have long been submerged by the rising sea levels that accompanied the melting of the Wisconsinan ice sheets, and even the most sophisticated sonar equipment, robotic dive craft, and computer mapping and visualization techniques do little to make the search for submarine artifacts easy. Yet tantalizing fragments of new evidence are being found, in coastal caves, if not in any quantity on the ocean floor (Koppel 2003; Pynn 2004; see also Dillehay 2003). A site on Santa Rosa Island, California, has yielded indications of human occupancy about 13,000 years ago. Remains from Prince of Wales Island in southeast Alaska are dated to 11,600 years ago. Early in 2004 it was announced that two spear points from a limestone cave on the Queen Charlotte Islands have been dated to between 12,100 and 11,800 years ago.

Again, there is room for difference over the significance of these finds. Some would subsume them under the Clovis-first hypothesis by pushing the timing of Paleo-Indian migration across Beringia back a couple of thousand years (to 13,000 or 14,000 years ago) and suggesting that the stream of newcomers bifurcated, with some following rivers east and south, and others taking a coastal route. Yet this move is confounded by claims based on statistical analyses of the cranial measurements of some of the oldest skeletons found in the Americas, that the first arrivals in the New World ("Paleoamericans") were from south Asia and the southern Pacific rim.

These conflicting accounts of the timing and routes of human colonization in the Americas will not be resolved quickly. Much is at stake here, from the reputations of individual researchers to questions about the relative utility of Western scientific and indigenous understandings of the past. The process of bringing data found in new sites and revealed by new research technologies into focus alongside artifacts and understandings derived from long-known locales will likely produce decades of debate and reinterpretation. But the crucial (and relatively incontrovertible) points of significance here are that a substantial number of faunal species went extinct as the ice sheets retreated and that in the

Mastodon. *One of the large mammals found in North America during the Pleistocene (glacial) epoch, the mastodon (*Mammut americanum*) resembled the woolly mammoth but belonged to a distinct, now extinct, species. Approximately nine feet in height, mastodons were better adapted to chewing leaves than to grazing; they are known to have eaten conifer twigs, leaves, and swamp plants as well as grasses, and were probably more common in eastern than western North America, where they found suitable habitat in spruce woodlands. Mastodons were killed by indigenous hunters as well as natural predators (including scimitar cats,* Homotherium serum*). They died out about 10,000 years ago. (North Wind Picture Archive)*

wake of both these events much, though certainly not all, of present-day Canada was peopled from the south.

The extinctions were astounding. The consensus is that at least thirty-five genera of mammals disappeared in a relatively short span of time. Among them were mammoths and mastodons, giants of the tundra and grasslands. So, too, were giant ground sloths; camels; horses; varieties of moose and antelope; giant (300 lb.) beaver; and wolves, bears, and felines all larger and more powerful than their present-day counterparts. Lemmings, salamanders, giant relatives of the guinea pig, several birds, and various other creatures also became extinct at this time. For Paul Martin, Paleo-Indians and Paleo-Indians alone were responsible for these extinctions. Spreading far and fast—Martin's estimates of Paleo-Indian population expansion and migration behavior were intended to provide a mechanism for this claim—Paleo-Indians were superpredators who mounted a fierce onslaught on animals that knew no fear of humans.

Skilled and deadly hunters with their three- to six-inch-long, flaked-stone spear points (the diagnostic artifacts of the Clovis complex), Martin argues, Paleo-Indians wrought havoc with the late Pleistocene ecology of the Americas. By his account, it was all quite simple: In one scenario, a hundred Paleo-Indians on the Alberta prairie, killing a dozen animals per person each year and moving south at a rate of several miles per annum became a population of 100,000 within 300 years. In that time they spread 2,000 miles south and killed over 90 million 1,000-pound animals. Little wonder that two-thirds of all New World species that weighed over 100 pounds at maturity (the so-called megafauna) disappeared during this period.

But much is wrong with this picture. Martin's assumed reproduction and migration rates are untenable. His Paleo-Indians are far more fixated on big game kills than seems likely to have been the case. There are surprisingly few archeological sites in which signs of humans are associated with extinct species. This account provides no adequate explanation for the survival of some animals (bison) and the demise of others (mastodons) or for the loss of approximately the same proportions of avifauna and megafauna. Nor does it allow for the ecological effects of rapid climate change. We still know too little about the details of climate change and its consequences during these years. Given the extent and rapidity of vegetation change (driven by climate warming that may have amounted to 7 or 8 degrees Celsius overall in many parts of northern North America, with greater seasonal variability marked by hotter summers and colder winters in many areas, however, it is entirely probable that the effects may have been calamitous for animals unable to adapt to swift changes in habitat and food availability. And the loss of one or two species may well have had a cascading effect upon ecosystems in unusual flux at the time. None of this is to deny the influence of Paleo-Indian hunting on mammalian populations, but it is to caution against the assumption that the late Paleolithic extinctions were the work of human influences alone (see Teacherserve n.d.).

3

NORTHERN NORTH AMERICA
A THOUSAND YEARS AGO

With the climate, the vegetation, and the availability of game changing rapidly—albeit barely discernibly within the span of individual lifetimes—human occupants of the northern parts of North America adapted to shifting circumstances. As populations grew and the areas utilized by them increased through the ten thousand years and more of the prehistoric period (between the retreat of the ice and European contact with indigenous inhabitants of this region, people settled into place and cultures diversified. Diffusion and local innovation replaced migration as the major impetus to cultural change, and technologies were both elaborated and increasingly precisely adapted to local environments. By AD 1000, most of present-day Canada and Alaska had been occupied for thousands of years by many different peoples. Our knowledge of them remains incomplete. Largely derived from the analysis of physical remains—stone implements, rock art, burial places, and the location of archeological sites—and such persistent cultural features as language patterns, it rests heavily on inference and is undoubtedly both undiscerning and imprecise, particularly with respect to cultural matters (see *Historical Atlas of Canada*, vol. 1, for an overview).

Yet it can be said that most of those who occupied these northern regions lived by hunting, gathering, and fishing. Agriculture came late to this realm and was never widespread. Overall, population densities were low. Most people lived in small mobile groups and all depended upon the close and detailed knowledge of their environments acquired over the centuries by ancestors who had occupied their territories for generations before them. Typically, each band developed a seasonal round of activity within an approximately defined territory. Within this area, particular species of animals would be hunted at certain times and places, perhaps determined by their migration routes, plants would be gathered at appropriate times of the year, and fish and other resources would likewise be exploited according to the seasons. Trade linked groups of people and different geological and ecological regions. Stone suitable for flaking into tools, such as obsidian, silica, or chert, was probably the most common trade good. Minerals (copper, galena) and shells moved long distances (even thousands of kilometers), probably passed in succession from band to neighboring band.

The numbers of these early inhabitants cannot be known with certainty. Until a few decades ago, widely accepted estimates suggested a total of 80,000 people for all of present-day Canada on the eve of European contact. Today, some would revise that figure upwards of two million. Even relatively conservative calculations, based on the careful assessment of various types of evidence, produce a total of 400,000, which might be taken as a scholarly consensus. A dozen linguistic families and many more languages and dialects—derived and elaborated from the Amerind, NaDene, and Aleut-Inuit rootstocks associated with the first peopling of the continent—have been identified and their ranges mapped approximately, providing yet more evidence of the elaboration and diversification of indigenous cultures through this period.

In most accounts the peopling of northern North America beyond Beringia and the ice-free corridor east of the Cordillera is taken to begin with the occupation of the tundra and lichen woodland zones south of the Laurentide ice sheet in the years between 9500 and 8200 BC. Fluted projectile points resembling the Clovis points common on the plains of the western United States, but smaller and more similar in fluting technique to those of Alaska and Yukon, have been found in southern parts of the Canadian prairie and dated to the period when this was a lichen woodland zone. In the east, substantial numbers of finds more closely comparable to Clovis models have been associated with the hunting of migratory caribou in the lichen woodland–boreal boundary zone of southern Ontario about 8600 BC. Broadly similar—although recognizably distinctive—points, probably also made by caribou hunters, have also been found at Debert and a few other sites in the present-day Maritime Provinces.

Through the next several thousand years, increasingly complex patterns of stone-tool technology and associated cultural groupings emerged from the widespread movement of peoples and cultural traits through the expanding land area of northern North America. Canadian archeologists identify inhabitants of this northern realm between 8500 and 6000 BC as Plano people. Descended from fluted point makers and primarily caribou and bison hunters, these people lived in tents and manufactured several tools found only in the north.

Beginning about 8000 BC, Plano Culture was influenced by the Early Archaic Culture, a complex developed from fluted-point antecedents in the woodlands of eastern North America. By 6000 BC it was so changed that it is considered to have been superseded. Distinguished by the development of spear-throwing technology and, on the coast, by the use of toggling harpoons, members of the various Archaic culture groups that emerged through 4000 BC exploited a wider range of food sources than their Plano predecessors. More flexible in their subsistence requirements and diversifying culturally as they adapted to particular environments, Archaic groups were likely able to sustain higher population densities and larger overall populations than their predecessors. By one American estimate, Archaic subsistence techniques could support ten times more people in a given territory than those of big-game dependent Paleo-Indians. Although there is scant direct evidence of this, it seems likely that members of the Shield Archaic group that emerged during this period developed birchbark canoes and snowshoes to allow their survival in the subarctic river and lake-spotted country that they occupied.

During this period, manufacturers of distinctive long, narrow, sharp-edged flakes known as microblades—and thus known to later scholars as bearers of Microblade Culture—expanded from central Alaska and spread east and south to occupy sites in the northern interior and the Pacific Coast. Archeological analyses generally construct these areas (with some hesitation) as separate Northern Interior Microblade and Coastal Microblade realms, on the grounds that bifacially flaked stone tools and burins (boring tools) are found in the interior but not on the coast. Until the 1980s, it was also generally argued that makers of microblades (which clearly have an Asiatic origin) had lived in Alaska and Yukon territory since at least 13,000 BC. However, more recent work in central Alaska places the appearance of microblades some 10,700 years ago. This date is more congruent with the argument advanced by Joseph Greenberg and other linguists for the movement into precisely these areas of a second wave of migrants across the Bering isthmus—speakers of Athapaskan, Haida, and Tlingit, whose languages derive from Na-Dene. In any event, by 4000 BC the broad outlines of persistent regional patterns of indigenous culture were becoming clear.

Through the next 3,000 years the pace of environmental change slowed, although gradual climatic warming continued in parts of the northeast until approximately 2000 BC, before temperatures declined (and vegetation patterns adjusted) to approximate those of today. Sea levels also stabilized at something close to present benchmarks. Partly as a result of this relative stability, the archeological record is richer than that for earlier periods. It reveals increasingly differentiated and elaborate technological and cultural complexes. In many areas, population densities were higher, and there are indications that some groups developed more sedentary ways of life, or at least occupied favored sites on a seasonal basis over many years. In the east, Maritime Archaic hunters expanded their range into Labrador and up the St. Lawrence, leaving behind, variously, remnants of longhouses, burial mounds, ornaments, woodworking tools, and a diverse array of ground and chipped stone tools, weapons, and needles made of bone. In the shield country new stone and copper tools and ornaments appeared, as did tools manufactured from nodular chert. To the south, members of the Laurentian Archaic Culture hunted deer and small animals, fished, and gathered plant foods using a wide range of tools, many of which derived from contact with groups to the east and south.

On the plains and in the forest/parklands that flanked them to the north, bison hunters were developing a rich and distinctive culture that depended heavily, for subsistence and other purposes, on stampeding bison over cliffs or into ravines and human-constructed pounds where they could be slaughtered. All of this implies a relative high level of social organization, and the inference is strengthened by the existence of long-utilized medicine wheels that point to a strong ceremonial/belief system. On the Pacific Coast enormous shell middens and other archeological remains mark the gradual elaboration during this period of a rich and relatively sedentary culture utilizing abundant salmon, eulachon (oolichan), and shellfish resources, engaging in considerable trade, and developing early forms of the highly wrought decorative art later associated with inhabitants of this region. The major migration of this period involved the spread through the high Arctic of Paleo-Eskimos, who occupied the coasts and islands from Alaska to Labrador and southern Greenland and pressed inland particularly in the territory between the Mackenzie River and Hudson Bay. Small, highly mobile groups advanced rapidly west to east, subsisting on caribou, musk oxen, and fish in some areas, and seals and other sea mammals elsewhere.

These developments provided the foundations upon which yet more complex cultural patterns were elaborated in the next two millennia, mainly through local innovation and the diffusion of ideas and artifacts rather than by the migrations of peoples. Chief among these developments were the introduction of pottery to eastern Canada from the south and to the Yukon coast from the west about 1000 BC; the widespread adoption of the bow and arrow; and the introduction of agriculture—centered first on the cultivation of corn, with later additions of squash, sunflowers, beans, and tobacco probably occurring several centuries later—into southern Ontario after about AD 500.

Each of these was consequential. The introduction of pottery not only changed the archeological taxonomy applied to eastern Canadian native peoples ("transforming" the Laurentian Archaic into Point Peninsula, Saugeen, and Meadowhood cultures, for example, although there was a good deal of cultural continuity in subsistence patterns, belief systems, and other facets of technology between the people subsumed under these classifications) but also facilitated the storage, preparation, and transport of food. The bow and arrow improved the efficiency of the hunt, especially for small game, and likely intensified indigenous use of land mammals. With agriculture came the development of semipermanent, palisaded villages associated with the development of slash-and-burn cultivation. Each of these developments likely contributed in one way or another to increases in population.

The major human migration of this period occurred only toward its very end, when a Neo-Eskimo culture that had evolved a sophisticated maritime technology in the Bering Strait area began to push eastward. For many years, scholars following the arguments in Robert McGhee's doctoral thesis saw this as a migration in pursuit of bowhead whales. But McGhee now sees things differently, believing that his earlier portrayal of residents of the western Arctic as an "adaptive species" responding to new environmental opportunities was both too simplistic and at odds with more recent evidence. Put simply, the timing of both the migrations and the onset of climatic deterioration, as well as the existence of hundreds of miles of sea ice between the western and eastern Arctic, renders the original argument less than robust. It is more likely, McGhee now concludes, that knowledge of the availability of iron and other metal (derived from meteorite debris in the vicinity of Cape York, Greenland) drew people from the western Arctic on "journeys of exploration across the barren channels of the Central Arctic." In any event, these early eastward moving migrants, known as the Thule people, were efficient and mobile hunters and they displaced their Dorset predecessors (descended from Paleo-Eskimos) from most of Arctic Canada within a few generations (McGhee 2005, 121–125).

By AD 1000 most of northern North America was occupied by people whose predecessors had lived for dozens, and even hundreds, of generations in much the same places and in much the same ways as they did (see Fig. 6). This is not to suggest for a moment that indigenous cultures were static and unchanging. Far from it. They were the products of skillful adaptation, over millennia, to changing environmental circumstances and the opportunities presented by cultural contact and the diffusion of ideas and technologies. Simple though indigenous technological complexes may appear to modern eyes, the stone and bone tools that were at their core were remarkable accomplishments in context. Some have characterized them as one of the great forward strides of humanity.

All parts of the native "tool kits" in use on the eve of the European encounter with the Americas were products of astute observation of nature. Indigenous toolmakers had to know their materials, to understand how they could be worked, and to recognize where they could be obtained. They also needed a sophisticated appreciation of the natural world to fit their material technology to the efficient exploitation of its possibilities and the satisfaction of their needs. From the fluting of stone points to create effective spears, through the development of the notched spear thrower (atlatl) to increase the velocity and thus the efficiency of the projectile, to the use of floats attached to harpoons to hunt whales on the Pacific and Arctic coasts, indigenous hunting

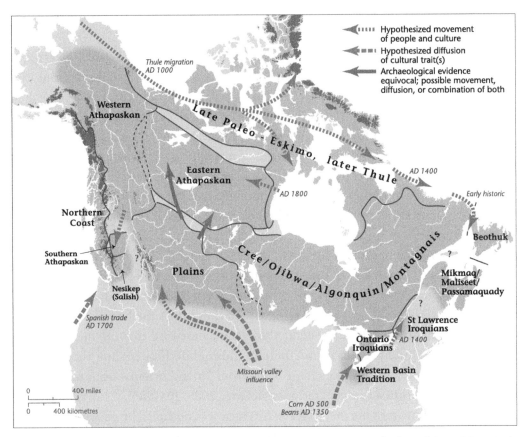

Fig. 6. *Indigenous peoples of northern North America about 1000 years ago. (Adapted from* Historical Atlas of Canada *1: pl. 9 and* Handbook of North American Indians, *vol. 5 (ed. David Dama) and vol. 6 (ed. June Helm)*

technologies were significant cultural achievements that reflected high levels of environmental understanding and a good deal of ingenuity. So, too, with the snowshoe and the birchbark canoe, responding and adapted to the particular conditions of the eastern sub-Arctic, and the dog-drawn sled or travois developed by natives of the plains long before the arrival of the horse among them.

Fire was another important tool, and one that had a more significant environmental impact than most indigenous technologies. Indigenous peoples set fires on the margins of the eastern woodlands to drive deer to the kill. Fanned by the wind, these sometimes escaped and consumed or charred the vegetation of considerable areas. The grasslands of the west also burned frequently, both from lightning strikes and from fires set by the native inhabitants. These blazes curbed tree growth and, by improving the grazing for bison and antelope, may have limited the spread of aspen into the grassland. Aboriginal burning may also have maintained many of the open meadows within the parkland and boreal forests and along watercourses.

Native people of New France. *Four indigenous people of New France (Canada): two men at top in war attire, with shield, bow, and arrows; mother holding paddle and nursing infant; and man clothed for winter, wearing snowshoes. From Samuel de Champlain. 1619.* Voyages et descouvertures faites en la Nouvelle France. *Paris: C. Collet. (Library of Congress)*

By and large, however (and setting aside the possible special circumstances in which human intervention might have caused a cascading collapse of megafaunal populations reduced to a state of criticality by environmental change at the end of the Pleistocene [see Buchanan 2001]), the capacity of early native peoples to alter their environments in significant, lasting ways was quite limited. Ranging across relatively extensive territories, following the movements of game, fishing lakes or streams at certain times of the year, and responding to the seasonality of plant growth, they made comparatively little mark upon the earth.

PART 2

Contact and Its Consequences

Contact—the most commonly used term to describe the encounter between natives and newcomers in the Americas—is a forceful and elusive concept. On the one hand, it conveys a sense of transformation. Because the word is sometimes used to pinpoint the instant that a blast is detonated, characterizing the meeting between people from beyond the continent with those indigenous to it as "contact" rightly signifies an explosive, volatile, unbalancing moment. Yet space, time, and social and ecological processes all confound this simple understanding.

"Contact" between New and Old worlds (so-called) cannot be assigned a date, or attributed to a particular time and place. Across the vast expanse of northern North America, European incursions into native spaces took place on several fronts. Whether on the northeastern foreland of the continent, in Hudson Bay, along the St. Lawrence, via the Hudson River, from north or south on the Pacific Coast, or in dozens of localities in between, none of these encounters was an exact replica of any other. They involved different groups of people—Norse, English, Scottish, French, Dutch, Spanish, and Russian from the Old World, settlers and their descendants who had come to call North America home, and innumerable distinct bands and peoples indigenous to the territory. They occurred at different times, from the eleventh century to the twentieth, with all that that implies in the way of technological development, accumulated knowledge, shifting attitudes, and changing understandings of self, society, nature, and humanity on both sides of the equation of encounter.

People engaged one another for different reasons and with different intentions—out of curiosity; with a desire for trade, territory, and/or resources; in search of alliances; bent on conflict; or hopeful of saving souls. Whatever their purposes, these engagements frequently altered the relationships between peoples and environments, relationships upon which traditional societies were often crucially dependent for survival. Adjustments imposed upon peoples by these circumstances often precipitated a further cascade of changes to established ways of life and patterns of interaction with the natural world.

Moreover, societies—human worlds—are not inert substances rendered suddenly and irresistibly unstable when brought together. They are complex, evolving, resilient, vulnerable constructs, shaped both by the sentient actions of those who comprise them and by processes and circumstances beyond the control and creation of their members. Indeed, it is well to remember, in seeking to grasp the many and varied consequences of contact, that its influences often spread fast and far (in ways sometimes indiscernible to contemporaries) from the locations in which people first met face-to-face.

In an effort to bring some order to the complicated and intricate set of processes involved in the almost interminable range of encounters between cultures, the Swiss historian Urs Bitterli has defined three forms of engagement between Europeans and non-Europeans. He calls these "contacts," "collisions," and "relationships." **Contacts** were short-lived—they were truly "first meetings" between European and indigenous peoples, and were typically fleeting and superficial. But they were not all of a piece. Some were pristine (when participants had no prior knowledge of each other); others were shaped by some degree of awareness of one group by the other (through hearsay or physical evidence). There was often a strong ceremonial quality about these engagements, but ceremonial forms differed among peoples, and the rituals of encounter were almost always, and inevitably, acted out across a wide communications gap; gestures, sign language, and pantomime were far more characteristic than dialogue, and the meanings imputed to the actions of others remain difficult to pin down.

Although "first meetings" were, in Bitterli's estimation, usually (albeit not invariably) peaceful, contacts often turned into **collisions** "in which the weaker partner, in military and political terms, was threatened with the loss of cultural identity, while even its physical existence was jeopardized and sometimes annihilated altogether" (Bitterli 1989, 29). With all of the early modern world in mind, Bitterli is quick to acknowledge the diverse forms of collision, and the multiplicity of outcomes—shaped by geographical circumstances and the disparities in power between the parties—to which they led. Basically, however, the move from contact to collision in this scheme marks a shift from peace to war. European chroniclers of this change frequently characterized it as a reflection of indigenous treachery; from the other side of the encounter it was often a response to the threat of loss, of property or a way of life.

Indeed, it may not be a significant exaggeration to say that the seeds of collision were implicit in contact, because newcomers, who sought peace when they were weak and few, almost invariably moved to increase their power and authority as their strength and numbers grew. In Bitterli's assessment, the most important "source of conflict was the seizure of land by Europeans" (ibid., 31). Other factors that precipitated this change were European interference in disputes between indigenous groups (either by forming alliances or supplying weapons); the sharpening of rivalries or the exhaustion of resources consequent upon trade; and the escalation of incidental violent skirmishes.

The devastating effects of disease might also be considered a form of collision. On a world scale, diseases took their toll on European as well as non-European peoples. Malaria, yellow fever, and other afflictions killed thousands of European newcomers to the tropics. But the introduction and transmission of foreign diseases—smallpox, measles, influenza, tuberculosis, and so on, against which indigenous peoples had built up no resistance—to New World territories was far more invidious and disastrous. By all accounts, disease epidemics killed innumerably more indigenous Americans than did military campaigns. Whatever precipitated them and whatever form they took, collisions were clearly devastating, dehumanizing events that contributed to the decimation and subjugation of indigenous peoples and societies.

Relationships rested on commerce and religion, and the instruments of their development on the European side were traders and missionaries. By Bitterli's reckoning relationships involved "a prolonged series of reciprocal contacts on the basis of political equilibrium or stalemate" (ibid., 40). They depended on mutual trust and on supply and demand for trade goods, political prestige, technological know-how, and so on—although in Bitterli's droll account, the eternal life that missionaries undertook to supply was "seldom really in demand" (ibid.). By and large, relationships were peaceful, so long as the exchanges at their core remained of interest to both parties. Yet by placing particular interactions upon a reciprocal and to some extent privileged basis, they often fostered collisions in adjoining territories and with neighboring groups. Through time, relationships also led to acculturation, as cultural beliefs, practices, and forms were affected by the exchange of goods and ideas.

This was a two-way process, although it clearly had more persistent and far-reaching effects on indigenous than European cultures. Few, if any, indigenous groups adopted European culture in its entirety, to be sure, and there were great differences in the ways in which individuals within particular societies responded to the challenges and opportunities presented by European ways of life, but these patterns of partial and differential acculturation were in themselves profoundly destabilizing, producing complicated displacements and overlaps, and patterns of doubt and syncretism that divided societies and left individuals disaffected.

These groupings help, but only to a degree. They are hardly pure, exclusive, or all-encompassing classifications. First encounters were not always nonviolent. Trade spread disease. Colonial maps erased native claims to space. There is no place in this formulation to consider the administration of native/colonial territories by newcomers. Neither strictly a collision—because it was intended in part at least to address problems of cultural conflict—nor a relationship constructed out of supply and demand—although it was clearly a relationship framed for the long term and seeking some form of political equilibrium—the exercise of political authority over indigenous bodies and territories, manifest in regulations defining citizenship, governing access to resources, and the creation of reserves, and so on, is clearly crucial to understanding contact and its consequences. Still, with these limitations in mind, Bitterli's schematic account of the contact process provides a moderately useful grid against which to consider the near millennium-long unfolding of the indigenous encounter with peoples of Old World descent across northern North America. By enumerating the main elements of this process in somewhat abstract form, it offers a rough chart—not a precise template—with which to explore the most significant consequences of contact for native peoples and their environments.

4

BETWEEN LAND, SEA, AND ICE
Inuit, Beothuks, Aleuts, and Newcomers

Since 1960, when archeological remains in Newfoundland confirmed long-familiar but uncertain accounts of contact with parts of the continent known as Helluland, Markland, and Vinland by adventurers from Greenland, it has been clear that the first encounter between Europeans and indigenous people in northern North America occurred about AD 1000—although the authenticity of the so-called Vinland map, discovered in 1965, which purports to offer cartographic evidence of the Norse presence in North America, was long debated (Fitzhugh 1985; Quinn 1988; McGhee 1987).

According to the Icelandic sagas written at least two centuries after the events to which they refer, there were a series of voyages of exploration and colonization westward from Greenland. By one of these accounts, the *Greenlander's Saga,* Leif and Thorvald Eiriksson, descendants of people from Norway and the Baltic who had settled in Iceland in the ninth century and Greenland in the tenth, established a short-lived settlement at *Leifsbudir* very early in the eleventh century. Their voyages to this site, now tentatively associated with the dwellings and workshops excavated near L'Anse aux Meadows, in northern Newfoundland, almost certainly followed the coast of Labrador, and possibly visited Baffin Island. In these areas, the sagas recount, the Greenlanders encountered unfamiliar people of modest stature, who used "whale teeth for arrowheads and sharp stones for knives" (Dickason 1992, 87, 91). These "small ill-favored men," who were apparently eager to exchange food for European weapons, they likened to the trolls of northern European legend, and named *skraelings* (to mark their short, withered appearance). Beyond this, the sagas reveal little of these indigenous inhabitants.

By most reckonings, the so-called skraelings were probably Dorset people who had lived in this area for at least a thousand years (but see McGhee 1984). Utilizing the harpoon and microblade technology of the Arctic small-tool tradition developed by their Early Paleo-Eskimo predecessors, they lived in partly subterranean winter houses and used ground slate tools. Several items of European provenance have been found in Dorset-age archeological sites, but the extent, and effects, of trade between the Norse and these indigenous people remains unclear. The European presence on the Labrador coast was sporadic and short-lived. The sagas record voyages to this area in 1010 and 1011 but say little of North America thereafter. There is no archeological evidence of direct contact between Norse and native peoples in North America, and most of the items of European origin recovered from native sites likely reached these areas through trade. Indigenous peoples acquired small quantities of metal, but there is no indication in the archeological record that it significantly influenced their ways of life. Nor are there any signs of cultural influence or disease impacts stemming from the brief European presence.

Still, the cultural geography of this area of the continent was marked by a good deal of instability in the five hundred years or so surrounding Norse contact. By AD 1000 Dorset people had begun to retreat from the southernmost extent of their range (which once encompassed Newfoundland). At this date they were concentrated in the vicinity of Hudson Strait–Baffin Island, with some presence in areas west and north of there. Radiocarbon dates from the vicinity of L'Anse aux Meadows show occupation of that site by Middle Dorset people until about AD 750 when they were replaced by a very different indigenous group. These were most likely prehistoric ancestors of the Beothuk. Hunters who made seasonal use of coastal resources, these people were possibly representatives of a Maritime Archaic Culture once spread more extensively along the coasts of Labrador but displaced by the southward advance of earlier Dorset peoples. On the west, members of the Dorset Culture were also under pressure from Thule peoples, whose use of boats, floats attached to harpoons, the bow and arrow, and dogsleds made them more efficient hunters than the Dorset. As Thule hunters moved eastward with remarkable speed, Dorset peoples were killed, displaced, or assimilated. By 1100 Thule people had established settlements in Victoria Island and northern Greenland; by 1200 they were in Baffin Island; by 1350 in southern Greenland; and by 1500 on the Labrador coast.

At the same time, the Norse settlements began to run into difficulties. Remote and ill-adapted to the pastoralism that sustained the Scandinavian outposts in the New World, L'Anse aux Meadows was soon abandoned. By 1100—and probably before—Little Passage/Beothuk people occupied the site once more. The relatively warm period between AD 700 and AD 1000 gave way to the Little Ice Age. Failing to adapt to changing conditions, facing hostility from Thule migrants themselves being forced to adjust to the worsening climate, and increasingly cut off from Europe, the three hundred or so Norse settlements in Greenland in the twelfth century declined (McGovern 1980–1981; McGovern et al.

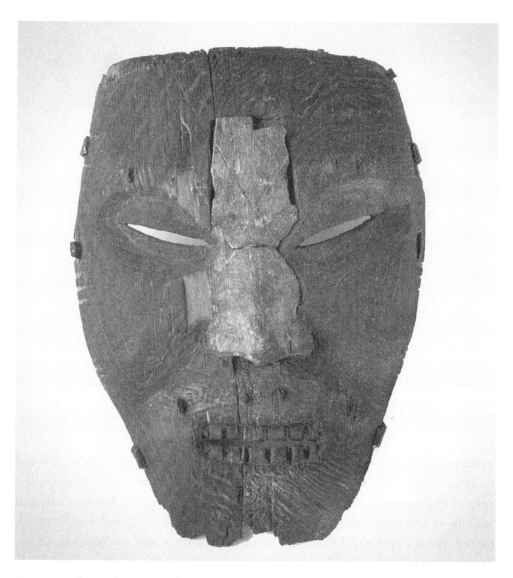

Dorset mask. *Masks were widely used for ceremonial purposes by the indigenous peoples of northern North America. Generally they represented the spirit world or mythical beings. This wooden mask, carved from driftwood a thousand years ago by Dorset people of the far north probably had magical and religious importance. (Canadian Museum of Civilization)*

1988). By 1400 the western (and more northerly) settlements had been abandoned. Within a century the eastern settlements met the same fate.

By the sixteenth century Thule Culture was also evidently in transition across much of the Arctic. Unable to maintain permanent settlements in large villages with houses built of stone and bone, inhabitants of the region used the

Thule dart point. *This dart or harpoon point, made of bone, was manufactured by Thule people who moved eastward across the high Arctic approximately a thousand years ago. Their success, in harsh conditions, depended upon the skillful use of locally available materials and intimate ecological knowledge. (Canadian Museum of Civilization)*

summer months to hunt whales; they moved inland in the autumn to fish and hunt caribou; in the winter they returned to the coast, where they lived (in the central and eastern Arctic at least) in snow houses (igloos) and hunted seals and walrus; in the spring they returned to the rivers to fish. Using wood-framed, skin-covered kayaks to hunt seals amid the ice floes and caribou swimming across lakes or rivers, and larger umiaks to pursue larger marine mammals, as well as sleds for winter travel overland, they were a mobile and resilient people. Acquiring smelted metals from former European settlements in Greenland and (later) guns from European whalers in the Hudson Strait, they developed a new, Inuit, way of life in the fifteenth and sixteenth centuries by abandoning parts of their earlier cultural lexicon, adapting traditional technologies to a worsening climate, and borrowing goods and materials from contact with Europeans.

In the sixteenth century Inuit people from Labrador were moving south and east to trade with growing numbers of European fishermen drawn to the waters off Newfoundland. Baleen (the stiff plates of keratin in the mouths of certain species of whales, the frayed inner edges of which mat to form a strainer for filter feeding) was one of the main items that they offered for trade, and the whales from which this came were soon a major focus of European interest in the resources of the north. By 1578 English authorities reported twenty to thirty Basque whalers operating in the waters around "the new found land." This was likely a minimum. By one recent account, Red Bay, Labrador, on the Strait of Belle Isle, was the whaling capital of the world between 1550 and 1600, and fifteen or more ships congregated there every year to exploit right whales in late summer and bowhead whales, moving south ahead of the advancing pack ice, in the fall (Tuck and Grenier 1981, 1989).

There were at least a dozen other whaling ports between Middle Bay, Québec, and Cape Charles, Labrador, during this period. Over six or seven decades, thousands of whales were killed, and their blubber rendered down to oil, along this coast (Barkham 1980, 1982, 1984; Proulx 1986, 1993). Some have suggested that the catch from the Labrador area was in decline late in the sixteenth century as a consequence—although this was more likely attributable to changes in climate and water temperature than to overexploitation. Nonetheless, it forced whalers further afield, up the St. Lawrence and ever northward into the waters around and beyond Hudson Strait.

For the better part of three centuries, the hunt for whales coupled with the quest for a Northwest Passage from the Atlantic to the Orient sustained contact between Inuit and Europeans and carried it into the central Arctic. Some of these encounters were perhaps more properly described as collisions than as peaceful contacts. In three successive voyages into the Baffin Island area in 1576, 1577, and 1578, Martin Frobisher had several skirmishes with indigenous bands. During one of these, he was struck by an arrow tipped with iron (which must have been of European origin). On his second voyage he took an Inuit man, woman, and child back to England with him. Those left behind in the Arctic appear to have preserved "sundry odds and ends" from this encounter, including pieces of brick and brass rings—as well perhaps as memories of those removed and never returned (Dickason 1992, 91).

Subsequent meetings almost invariably left traces behind, although many of the details are undocumented, and the consequences for indigenous cultures can only be inferred. Certainly exchanges took place. Europeans sought information, assistance, and supplies from Arctic dwellers. Inuit appreciated the utility of iron and other goods that Europeans possessed and they did not. Sometimes the transfer of commodities was inadvert. In the forbidding northern environment, European endeavors often came to grief. Ships were abandoned in the ice, equipment and supplies were left on shore. But however it occurred, the transfer of European goods to indigenous societies had consequences—although these have often been difficult for outsiders to perceive and understand. Thus, when HMS *Investigator*—one of the vessels sent into the Arctic to find Sir John Franklin after his disappearance in search of a Northwest Passage in 1845—was abandoned to the ice of Mercy Bay in the northern part of Banks Island in 1853, the ship and the many tons of materials and supplies that its crew had cached ashore proved a rich source of introduced goods to the local people known as the Copper Inuit. Indeed, local informants told the Arctic adventurer Vilhjalmur Stefansson in 1910–1911 that their forebears had salvaged goods from these sites into the 1890s. For all that, trained ethnologists working among the Copper Inuit people in the second decade of the twentieth century characterized them as a completely traditional hunting-and-gathering society, and they were referred to many times, in the decades that followed, as almost completely unaffected by European contact.

Yet research in the 1970s and 1980s revealed a very different story. Impressed by the quantity of *Investigator* materials that entered the local economic system—by meticulous assessment, more than a ton of iron, copper, and brass; two thousand tin cans; and large amounts of wood ranging from ash and pine to oak, maple, and mahogany—anthropologist Clifford Hickey began to wonder how the infusion of so much "wealth" might have affected a society of eight or nine hundred people spread over an enormous territory. In a subtle, if necessarily somewhat speculative, analysis, he argued that it was far more consequential than earlier commentators, including Stefansson, had allowed in thinking that its effects were limited to the depletion of musk oxen on Banks Island. In Hickey's view, many of the unique characteristics of Copper Inuit society identified by early ethnographers were direct responses to the infusion of *Investigator* materials into the economy after 1850, as local people altered several aspects of their society to maintain others.

Crucial to this interpretation are the facts that the introduced goods were quickly distributed through Copper Inuit society, that less than a fifth of the population had direct access to the abandoned ship and its supplies, and that people who had been widely dispersed across Copper Inuit territory during the warmer parts of the year came together in large sealing camp settlements during the winter. Thus the exchange of introduced goods was a one-way flow of valuable items: "year after year certain individuals were showing up in the winter camps with renewed quantities of exotic [and very valuable] materials" (Hickey 1984, 25). Because gift giving was traditionally based on reciprocity, and repayment was expected more or less immediately (except from kin), this must have created a significant dilemma for the Copper Inuit people.

In theory, several nonexclusive responses to such structural tensions are possible. Society might differentiate, with gift givers gaining status and power over gift receivers. Kin relationships might be redefined, or gift-giving relationships might be formalized, to allow debt repayments to be deferred over longer periods. Or the economy might be intensified, by putting new goods or additional quantities of traditional trade items into circulation. By Hickey's account, the Copper Inuit attempted to maintain egalitarian relationships by pursuing variants of the latter two options; "by formally delaying 'foreclosure,'" they maintained their ideology by altering their economy (ibid.).

In effect, socioterritorial groupings became more formalized and significant, with local groups becoming "production corporations" for the acquisition and exchange of particular materials. At the same time, the exploitation of such resources as "native copper, soapstone, driftwood, and wood from the tree-line," as well as caribou, increased, because these were items required by the "caribou-poor" (but *Investigator*-rich) northern Bankslanders (ibid., 26). To facilitate this, the exploitation of maritime resources by more southerly socioterritorial groupings decreased, because all Copper Inuit had access to polar bears and seals. Indeed, Stefansson recorded both the enormous numbers of seals on the ice near Inuit settlements and the decline of seal hunting within the lifetimes of some of his informants. In Hickey's view, returns from seal hunting could contribute little to "the maintenance of reciprocal socioeconomic relationships" between south and north, and so people began to leave the sea ice earlier in the spring to seek more valuable trade items only available toward the southern limits of their territories (ibid., 2). And thus the marked strength of the "sea-land" dichotomy, as well as many of the other distinguishing features attributed to "traditional" Copper Inuit society by twentieth-century ethnographers.

Whatever the precise details, this account neatly reveals the ramifying social, economic, and often difficult-to-discern ecological consequences of even the most limited contact between Europeans and indigenous peoples. Here the European presence was indeed fleeting. Crew members from the *Investigator* gave presents to an Inuit group near Berkeley Point on Victoria Island in exchange for geographical information in 1851, but this seems to have been the limit of personal interaction between the groups. Yet inert materials, abandoned in one of the most remote locations on earth, worked their transformative effects into and through a far-flung society to reshape indigenous practices in profound ways.

If nothing else, the case of the Copper Inuit surely serves as a reminder that far from being static, passive, and vulnerable, indigenous societies were adaptable, resilient, supple, innovative, and versatile in their capacity to respond to many of the opportunities, challenges, and changes with which they were faced as a consequence of European contact. Given space enough and time to adjust they sought to do precisely that. In the case of the Copper Inuit, indeed, the adjustment was fast and fluent enough to convince serious scholars, examining the society less than three quarters of a century after the processes of change began, that they were confronting a culture that had remained unaltered for centuries. And this, too, is a reminder of the difficulties inherent in trying to understand the environmental and other histories of societies in rapid transition when the records of change are partial, incomplete, and often unreliable.

For all that, it is clear that both Inuit and whales were profoundly influenced by the largest and most enduring whale fishery in Arctic North America, conducted in the Davis Strait–Baffin Bay area from the seventeenth century into the twentieth. Focused on a discrete stock of bowhead whales that wintered off northeastern Labrador, moved northward in spring, crossed Baffin Bay to Lancaster Sound in June, and then moved south again in August to winter on the edge of the open Atlantic, this fishery engaged Danish, Dutch, French, German, British, and American whalers, who preyed upon the whales with growing intensity through the eighteenth century. In the nineteenth century, exploitation increased even more markedly. According to Gillies Ross, the foremost student of Arctic whaling during this period, there were more than 3,000 voyages to the Davis Strait whaling grounds between 1800 and 1914. Their effects were far-reaching. Inuit people were drawn into the commercial operations of a whaling fleet increasingly dominated by British vessels. Both material and cultural dimensions of Inuit life were modified considerably. Individual Inuit were taken to the United Kingdom, voluntarily and otherwise. And the stock of Davis Strait bowhead whales was reduced dramatically.

Nineteenth-century accounts of the Inuit of Baffin Island—several of them written by members of whaling expeditions—reveal some of the ways in which local material cultures and subsistence practices were modified by contact with international whalers. Inuit were hungry for European goods and were willing to work for them. Skilled hunters, boatmen, and dog drivers, they were quickly incorporated into the industry and became vital to its survival once vessels began to overwinter in the Arctic ice in the middle of the century. Inuit men were hired to hunt and provision the ships in the winter. In the spring they and their dog teams were employed moving blubber and whalebone from the ice-edged margin of open waters to the frozen-in ships.

On Aberdeen skipper William Penny's 1853–1854 voyage, when eighteen whales were caught on the spring floe edge, twenty-one miles from his ships, the Scottish captain calculated that twenty-two Inuit dogsleds engaged in carrying the blubber back to the ships traveled a total of 14,000 miles in their repeated voyages back and forth. In the summer many Inuit men served in whaleboat crews or used their own boats and gear, earlier acquired as wages in kind, and hunted whales under contract with whaling masters. The availability of year-round work led many Inuit families to relocate to one or another of the winter harbors selected by the whaling captains as bases for their operations.

Because this disrupted subsistence hunting, whaling masters often found it necessary to provide food for families whom they employed. "My department," wrote Margaret Penny, who accompanied her husband, William, on a whaling voyage in 1857–1858, "is to feed the natives & pretty hard work it is when you know that there are three hundred of them" (quoted in Ross 1997, 173).

Ship captains also paid their workers with used whaling equipment, guns, ammunition, tools, clothes, and so on. By the end of the century many Inuit hunters had "rifles, telescopes, sheath-knives, jack-knives, hatchets, saws, drills, awls, steels and files" (Ross 1985, 240; Ross 1997, 30, 88, 108–110). Women cooked in metal pots and copper kettles, used metal scissors and steel needles, and sewed with cotton thread. In summer Inuit wore the clothes of Europe: cloth trousers, shirts, jackets, and caps for the men; cotton skirts and woolen shawls for the women. Some men sported bowler hats, waistcoats, pocket watches, and chains; women wore brooches, ribbons, fancy hats, and even foundation garments. Sometimes local materials were used to make facsimiles of European items—animal-skin dresses with pearl buttons, bonnets of deer hide, and the like.

According to Matthaus Warmow, a Moravian missionary who traveled to Baffin Island with William and Margaret Penny in 1857–1858, the local Inuit were "imitating the Europeans in all respects," and had begun to "copy our mode of life." In his view, "the frequent intercourse of the Esquimaux with Europeans had, to a considerable extent, weaned them from their original habits and created artificial wants." This did not please the missionary, who thought indigenous people were "better off in their original state" and concluded that the process of acculturation that he saw about him not only made the Inuit harder to convert "to the kingdom of God," but also left them vulnerable: "they are neither properly European nor Esquimaux, and will speedily die out, in consequence of the change." Guided by such convictions, Warmow was quick to regret the "wretched," "miserable," and poverty-stricken condition of Inuit people in Cumberland Sound.

His description of an Inuit family in an igloo at Kekerten, written for his European brethren, was full of pathos. Old and young sat, heads drawn down between the shoulders in response to the cold, upon a few well-worn reindeer skins in the light of a single lamp, "very dim and scarcely burning." One woman kept the flame alight by constant attention, another sat "chewing at skin, destined to be made into boots or clothing." They had "nothing to eat, except occasionally a handful of seaweed." All in all, Warmow reported (with scant regard for indigenous volition), he found "the evil results" of efforts "to cast the natives in an European mould" all too evident—although he was quite prepared to introduce Christianity among them (quoted in Ross 1997, 109–111).

Yet it would be untoward to dismiss this assessment of the condition of the Inuit out of hand. Warmow was in Cumberland Sound in 1857–1858 on a "visit of observation" on behalf of his church, largely because William Penny had entreated the Moravians to send a missionary into the area to offset the harmful influence of whaling, which had begun there less than twenty years before. And Penny himself noted that the Inuit who had congregated near the overwintering whalers on Kekerten Island were "very ill off for food" when climatic and oceanographic conditions isolated them from mainland food supplies (Ross 1997, 122, 50).

Estimates of the size of the Davis Strait bowhead whale stock are purely conjectural for the early years and somewhat unreliable for later ones, but fairly recent and robust accounts place the number of bowheads making the broadly counterclockwise annual migration into Lancaster Sound early in the nineteenth century at about 11,000. These numbers were probably attributable, in large part, to the concentration of early whaling on the eastern side of Davis Strait. This occurred for practical reasons—the Greenland coast was closer to, and thus more easily reached from, Europe. But the concentration of whaling effort on the Greenland coast allowed safe passage northward for some whales and thus helped to sustain the stock. By one estimate, over 3,500 voyages to the "East Side" in the eighteenth century captured fewer than 10,000 whales. After 1820, when whaling masters shifted operations to the Baffin Island or western side of the strait, and the number of vessels engaged in the industry increased, pressures on the stock became intense. In 1833, the most productive year ever recorded in this region, over seventy British vessels killed more than 1,600 whales in Davis Strait. Relatively few vessels matched the 1833 average catch of 22 whales in subsequent years. By 1840 falling yields had pushed whaling crews into Cumberland Sound and encouraged them to employ Inuit labor. Early in the 1850s ships and crews began to winter in the Arctic to exploit the ice-edge stocks in May and June. In 1857 twenty-nine whaling vessels sailed for Davis Strait, and six overwintered. By this time, catches from one-season voyages averaged only two or three whales per vessel; Captain Penny's overwintering venture yielded seven whales per ship per year.

Whaling in Davis Strait continued through the next half century. But the ships became fewer and returns per unit of effort declined despite the establishment of permanent shore stations staffed by resident managers and dependent upon Inuit labor. In 1911, five steam whalers from the Scottish port of Dundee killed only one whale. The commercial whaling industry in Davis Strait, in Hudson Bay, and in the Beaufort Sea was no more. Early in the twentieth century, Inuit hunters who had become substantially dependent on European weapons and other goods were hired to bring "scraps"—walrus, caribou, seal and bear hides, narwhal and walrus tusks, and oil—to the few small and entirely marginal shore stations scattered across the far north.

·   ·   ·

John Cabot's voyage to Newfoundland in 1497 is widely regarded as bringing the riches of the Grand Banks to English notice, although the English historian David Quinn made a case (in 1961) for an earlier discovery by sailors from Bristol, and this argument has been pushed somewhat controversially by Kirsten Seaver in recent years. However, it was Cabot's voyage that brought Terra Nova—the expansive territory between the Strait of Belle Isle and New England—to wider European attention (Hayes 2002; Allen 1992; see also Pope 1997). Within a decade French and Portuguese fishing vessels were working the waters of the region; in another five years they were joined by fishermen from the Basque country on the border of France and Spain; a quarter century later whalers from the same corner of the Bay of Biscay were in the Strait of Belle Isle (Janzen 1999). By the 1570s, said one English merchant, these countries sent 400 vessels and more than 12,000 people to participate in the cod and whale fisheries each year.

Typically, fishing vessels and their crews arrived in the spring, fished inshore waters from small boats, dried their catch on land, and departed in the fall. Fishing intensified through the decades that followed, although the nationalities of those participating in it changed in relative importance, and some vessels fished the Banks and carried their catch back to Europe without drying it ashore. By the end of the sixteenth century, there were fewer Basque and Portuguese venturers in Newfoundland. A century later English fishers dominated the East Coast and the French worked parts of the Petit Nord and the South Shore (Pope 2003, 2004). To maintain the wharves and flakes and sheds that crews built in Newfoundland coves to facilitate their summer fishing and to dry and store their catch, caretakers and families began to remain on the island through the winter.

In 1680 there were fewer than 1,000 French and probably only about 1,500 English overwinterers (with an appreciably larger, although still small, proportion of women among the English). By the 1760s there were over 20,000 English people involved in the Newfoundland fishery every year, with rather more than half of them resident on the island during the winter. In 1790 the total exceeded 30,000 with fewer than two in every five migrating to work on the island each summer. Almost all were male. There were barely 3,000 women resident in the colony late in the eighteenth century. The French fishery was smaller and more

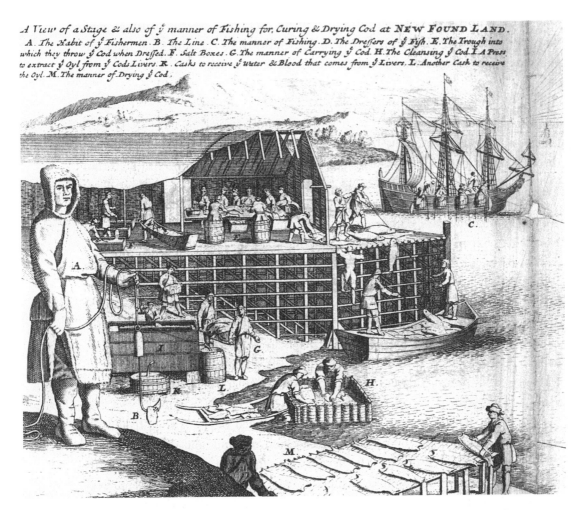

A View of a Stage & also of ÿ manner of Fishing for, Curing & Drying Cod at NEW FOUND LAND.
A. The Habit of ÿ Fishermen. B. The Line. C. The manner of Fishing. D. The Dreſſers of ÿ Fiſh. E. The Trough into which they throw ÿ Cod when Dreſſed. F. Salt Boxes. G. The manner of Carrying ÿ Cod. H. The Cleanſing ÿ Cod. I. A Preſs to extract ÿ Oyl from ÿ Cods Livers. K. Casks to receive ÿ water & Blood that comes from ÿ Livers. L. Another Cask to receive the Oyl. M. The manner of Drying ÿ Cod.

The dry fishery, Newfoundland. *No one ever saw a scene quite like this. The seventeenth-century artist who made this drawing has taken liberties to produce a composite intended to encapsulate as many aspects of the early fishery as possible. Thus we have a "fisherman" (A); an overlarge baited hook (B); an oceangoing vessel from which men are fishing (C); an "unlettered" dory at a stage, onto which men are hauling cod; a group gutting and cleaning fish in the shed on the stage (E, F); and the fish being prepared and salted (G, H, etc.) before they are laid out on a flake (M) to dry. Despite its close focus on the techniques of the fishery, this illustration also suggests, in the background, that most of the forest or other vegetation has been removed from the land immediately beyond the shore. (National Archives of Canada)*

heavily migratory than the English. In most years it engaged approximately 300 vessels, slightly more than half the English numbers, which climbed, briefly, over 600 late in the 1760s and reached a second, slightly lower, peak twenty years later (Humphreys 1970; Janzen 2002; Mancke 2005). Shifting markets, European competition, and war drove the Newfoundland fishery into sharp decline in the last decade of the eighteenth century. By 1800 the British migratory fishery had all but ceased.

Changing circumstances regenerated markets for codfish in the nineteenth century, but resident fishers supplied much of this demand. In the first forty years of the century, Newfoundland's population quadrupled to exceed 80,000 with far-reaching ramifications for indigenous peoples, the definition of property rights, and the environment (Norrie and Szostak 2005). Settlements occupied countless coves from Cape Ray in the southwest to Cape St. John in Notre Dame Bay. To the legally designated French Shore, facing the Strait of Belle Isle and including the eastern side of the Petit Nord, some 10,000 fishers from France came each summer (*Historical Atlas of Canada* 1: plates 20–28; and Myers 2001 for an ecological argument about catch-rates and settlement).

Through much of this period, British politicians conceived of Newfoundland, metaphorically, as "a great ship moored near the banks in the fishing season, for the convenience of English fishermen" (Prowse 1895, xix). It was also Beothuk territory (see Fig. 7). A hunting and gathering people whose numbers prior to European contact with Newfoundland can only be guessed at—but which were probably in the order of 2,000—the Beothuk had adapted to the challenges of existence in what Ralph Pastore, their foremost historian, called "a rather impoverished piece of boreal forest in the mouth of the St. Lawrence River" (Pastore 1994, 31). They ranged widely and seasonally across the island exploiting the resources of the interior (caribou, beaver, and arctic hare), the shore (shellfish, seals, and walruses), the waters (cod, salmon, capelin) and the air (great auks, murres, ducks, and geese). This generalized subsistence pattern was necessary, and fragile, not least because storms, winds, currents, and insect eruptions might reduce the availability of caribou and harp seal, its main components, and there were few "fall-back" food supplies available (Marshall, 1996).

Early explorers noted the native presence in "New Founde Land," and contact undoubtedly took place. A whaling lance of Basque provenance was recovered from a Beothuk burial site in Notre Dame Bay; native captives taken to Rouen in 1509 may have been Beothuks—although this is uncertain; and in 1534 the French explorer Jacques Cartier used a Beothuk word to describe the great auk. Early in the seventeenth century the leaders of an abortive English colonizing venture discovered a copper kettle of European manufacture in an abandoned Beothuk dwelling and reported an encounter in Trinity Bay in which they understood the natives to display "signes of joy, & gladnes" (quoted in Pastore 1994, 31).

By and large, however, face-to-face engagements were few, and they grew scarcer during the seventeenth century. By 1700 they were a rarity. A hundred

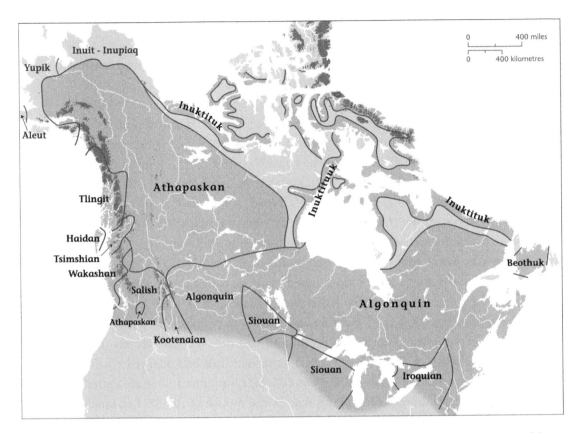

Fig. 7. *Indigenous linguistic groups in northern North America, about 500 years ago. (Adapted from* Historical Atlas of Canada *1: pl. 18 and* Handbook of North American Indians, *vol. 5 (ed. David Dama) and vol. 6 (ed. June Helm)*

years later, in 1793, a British parliamentary committee heard that barter between Beothuks and the English had occurred in former times, "by our people leaving goods at a certain place, and the Indians taking what they wanted and leaving furs in return" (ibid.). This unusual pattern of withdrawal by people living in an environment as limited and difficult as that in Newfoundland has puzzled historians who note the eagerness with which native groups sought to trade for European goods elsewhere.

Yet the Beothuk had little need to clamor for trade. The archeological record shows that they fashioned nails and other pieces of European metal into projectile points and tools of various types through the historical period. In sixteenth- and seventeenth-century Newfoundland they were able to acquire these goods by scavenging more easily than by exchanging country produce for them. Migratory fishermen left their shore works untended when they departed in the fall. Nails, awls, kettles, barrel hoops, and staves, fish-hooks, pieces of rope, and various other items were there for the taking. When overwintering began, pilfering became more difficult. Initially few in number, overwintering Europeans almost certainly feared the indigenous presence, and they were inevitably resistant to the loss of the property over which they had care; hostility between natives and newcomers increased. But until the end of the seventeenth century there were not many overwinterers, and most shore stations were unoccupied between the fall and the spring.

Excavations at Boyd's Cove and Inspector Island in Notre Dame Bay suggest that the Beothuk may have prospered between 1650 and 1730. Still able to exploit traditional food sources by hunting and fishing—even if at fewer sites than a century or two before—and with metal tools to make their lives easier, in Notre Dame Bay at least they entered a period of cultural florescence. Consuming a varied diet they were, concluded the archeologist who excavated the Boyd's Cove site, "a people at ease in their environment and obviously exercising a fair degree of control over use of its resources" (Stephen Cumbaa quoted in Pastore 1989, 67).

A century later the Beothuk world had contracted dramatically, and Beothuk society had folded in upon itself. Important trading (and perhaps genetic) contacts with native people in Labrador were disrupted by the developing European economy and consequent population shifts in the Strait of Belle Isle. Piece by piece the coast of Newfoundland was occupied by year-round settlements: The English were in Trinity Bay by 1675 and fifty years later more than 250 people overwintered there; Bonavista Bay was occupied by newcomers before 1700; seventy-five years after that the English were in Notre Dame Bay year-round. In precisely parallel fashion, the French settlement at Placentia excluded indigenous Newfoundlanders from the rich resources of Placentia Bay and the southern Avalon in the third quarter of the seventeenth century. Then the plight of the Beothuk was further compounded by the movement of Algonkian-speaking Miq'mak people from Nova Scotia onto the south shore of Newfoundland. By the mid-eighteenth century their presence had "denied yet another area of the island—and its caribou and seals—to its aboriginal inhabitants" (Pastore 1989, 69).

Cut off from vital access to coastal resources, the Beothuk attempted to survive on the meager possibilities of the interior. By the turn of the nineteenth century they lived in the Exploits River Valley, a struggling, shadowy and, in the eyes of many of the newcomers, sinister population. The violent, and for the Beothuk sometimes fatal, clashes with fishermen and settlers that had begun decades earlier continued occasionally, until philanthropically inclined churchmen and officials began to show a new interest in the native peoples of eastern Canada in the 1820s. Considering the Beothuk a mysterious group, they tried to communicate with and civilize them. Some were captured to help the process. None of this was effective. Captive and free, the Beothuk succumbed. Taken into white society in 1824, Shawnadithit died of tuberculosis, the last of her people, in 1829.

In the years since, sensational accounts have claimed that the Beothuk were hunted to extinction (Horwood 1959; see also Upton, 1977). But their tragic fate had more systemic causes (Tuck and Pastore 1985; but see also Renouf 1999 and Holly 2000). Archeological remains reveal that by the end of the eighteenth century, the Beothuk had been reduced to almost absolute dependence upon the caribou. They were a "beleaguered and dwindling" people, living on "the ragged edge of desperation" (Pastore 1989, 68). As newcomers took over sites where native peoples had traditionally garnered important resources, generalized subsistence strategies closely adapted to ecological conditions in Newfoundland were disrupted. Immiseration and debilitation followed.

There is perhaps no more poignant marker of this decline than the early nineteenth-century child's shroud and moccasins now held in the Newfoundland Museum. The covering, made of several much used, heavily patched and repaired pieces of caribou hide, was a recycled piece of adult clothing. The moccasins were likewise shaped from old materials. As Pastore observed somberly, they "did not even have enough animal skins to clothe a child properly" (ibid.). The Beothuk world had collapsed well before Shawnadithit, her mother, and her sister were kidnapped. Its very foundations had been undermined by the growing number of newcomers who failed to understand, and remained largely indifferent to, the fate of those whose delicately balanced traditional way of life was dislocated and disrupted by European occupation of this island territory.

• • •

Commercial interests in the marine resources of the North Pacific Coast also transformed indigenous lives and economies. Siberian adventurers, alerted to the enormous numbers of sea mammals in the Aleutian Island chain by the survivors of Vitus Bering's final voyage into these waters, began exploiting sea otters in the 1740s. Through the remainder of the century, forty-two Russian, companies made over a hundred voyages to the region and acquired almost 187,000 pelts. By the 1780s Russian settlements—villages and forts—were being established (on Kodiak at Three Saints Harbour in 1784 and St. Paul's Harbour in 1791, on the mainland at Slavorossiya in 1795, and elsewhere) to advance the trade.

Russian numbers were small—perhaps 500 in the 1780s and half that total late in the 1790s after consolidation of several separate interests. But their impact was large. Without firearms, the Aleuts were quickly forced into servitude.

By historical geographer James Gibson's account, half of male Aleuts aged between eighteen and fifty were forced to hunt under Russian supervision while large numbers of their female relatives were held hostage. As the trade spread eastward, many Aleuts, prized for their skill in hunting with harpoons from *baidarkas* (kayaks), were removed from their homes and relocated to Russian posts. By the end of the eighteenth century the Aleut population had declined by half. Sea otters also grew scarce. By 1780 they had disappeared from the shores of the Kurile Islands, by 1789 they were "rarely seen around the Aleutians," and in the 1790s returns from the northern Gulf of Alaska began to decline (Gibson 1976, 10).

Hunting pressures only intensified when British, Spanish, and American traders moved onto the Northwest Coast after James Cook's third voyage. The Tlingit, who lived in the present-day Alaska panhandle region, were antagonistic toward the Russians (and their Aleut and Kodiak hunters) for plundering their sea otters and infringing on their territory. As the German-born naturalist Georg Von Langsdorff, hired as a scientist with the first Russian voyage to circumnavigate the world, noted when on this coast in 1805:

> That they should have an unalterable enmity against the Russians is the more natural, since they have not only been driven from their hereditary possessions by them, but they have also been deprived of their great means of wealth, nay, even of subsistence derived from the sea-otters, the fish, and other marine productions, with which the coast they inhabit so richly abound[s]. (Quoted in Gibson 1992, 17)

Much to the chagrin of Russian interests, English and American ships collected 10,000 to 12,000 pelts a year in trade with the Tlingit during the 1790s. Moreover, they were able to sell them at high prices in the Canton market, to which Russian traders were denied direct access. The key motifs of eighteenth-century Alaskan history, concluded one of is chroniclers, were "courage, greed, violence and hardship" (Fortuine 1989, 102–103). For native peoples of the Aleutian Islands and the Alaska Peninsula who had borne the brunt of contact in this region, these themes translated into tragedy. By 1798 the Aleuts "were by and large a subject race" (ibid.). Despite their undoubted courage in attempting to resist the avaricious newcomers, they and their indigenous neighbors had seen their populations decimated by violence, starvation, and disease.

In the nineteenth century Russian interests expanded southward after the Russian-America Company was granted a monopoly of Russian trade and settlement in America in 1799. For a brief period they included establishments on the Sandwich (Hawaiian) Islands and for almost three decades a presence in present-day California (centered on Fort Ross, a few miles north of the Russian

River). But the maritime fur trade focused on the Gulf of Alaska remained the core of the company's business, and New Archangel (Sitka)—a settlement of approximately 400 Russians and 300 Creoles during the 1830s and over 2,000 in the 1850s—was its administrative hub and commercial entrepôt. Exploitation was intense and yields fell precipitously. The number of pelts returned from the Kodiak district fell by almost 90 percent (to about 100) in the four decades before 1830. In the 1850s the Russian-America Company's annual take of sea otter pelts ranged between zero and 2,000.

Fur seals, less valuable but more numerous and easier to hunt than sea otters, were also exploited rapaciously, and accounted for over two-thirds of all the pelts exported by the Russian-America Company before 1861. Sea otters, by contrast, accounted for less than 15 percent. Between 1786 and 1832 two small islands of the Pribilof group yielded well over three million fur seals, but the annual take declined from an average of 94,000 pelts in the late eighteenth century to about 39,000 in the 1820s. So large was the kill in the early years that skins were inadequately dried, and thousands accumulated in warehouses or left on the islands just rotted away.

From 1821, to be sure, the company made efforts to regulate the harvest by declaring a moratorium (*zapuski*) on hunting in particular locales every five years, by setting a quota (so sanguine that it could not be met), and by proscribing the killing of bulls and pups except for food or oil. But this was too little, too late. Annual yields fell into the 1850s. Only in 1847–1848, with a prohibition on the killing of fur seal cows, was the decline of the stock abated. By 1867, when Alaska was sold to the United States, fur seal numbers probably approached late eighteenth-century levels. Still this was only part of the story. Sea lions, from which Aleut obtained food and hides to build their *baidarkas*, gut from which to manufacture waterproof capes, and materials for making footwear and hats, were also heavily exploited. By the mid-1840s there were too few hides available to line the containers in which the company shipped their furs to Russia—and presumably to meet local native requirements.

Indigenous hunters remained integral to the Russia trade during these years. Aleut and their *baidarkas* were relocated to all parts of the Bering Strait and Gulf of Alaska, and to California to hunt sea otters, fur seals, and sea lions. Forced to venture out in heavy seas, many perished; on one often-cited occasion, over one hundred perished from paralytic shellfish poisoning in southeastern Alaska when they paused in Peril Strait while paddling home from Sitka Sound to Kodiak. Reports indicate that violence and abuse of native peoples by company servants was punished in one way or another—but also that it took place. Women and children were not taken hostage in the nineteenth century, but they were left in native settlements without their men for months and even years on end. Unable to hunt or fish for themselves, they became dependent upon European food, supplies of which were often scarce and unreliable. Some of the clothing provided native peoples in return for their labor was poor in quality and the company statutes asserted its "right or obligation to prevent the growth of luxury among the Aleuts, such as the use of bread, tea and similar items" (Fortuine 1989, 304).

Russian demands for labor, Russian attitudes toward the people of the Alaskan peninsula and its neighboring islands—in short, the Russian presence in the Gulf of Alaska—picked apart cultures, families, and traditional modes of survival. If the economic circumstances of the Aleut can be characterized, in the words of one of their most knowledgeable students, as "part serf, part proletariat and part Aleut," their cultural condition surely demands a less optimistic judgment: They were a broken and demoralized people (Jones quoted in Busch 1985, 101).

5

NATIVES, NATURE, AND TRADE IN THE INTERIOR

On the mainland, contacts between native peoples and newcomers quickly evolved into trading relationships. Early in the sixteenth century, the explorers Giovanni da Verrazzano (1524) and Jacques Cartier (1534–1541) exchanged gifts with indigenous peoples. True to imperial form, Cartier erected a cross on the Gaspé, thirty feet high, embellished with a shield bearing three fleurs-de-lys and a board inscribed (in translation) LONG LIVE THE KING OF FRANCE. By Cartier's own account, this action provoked local resistance. The chief, he wrote, "made us a long harangue, making the sign of the cross with two of his fingers." Then he "pointed to the land all around about, as if he wished to say that all of this region belonged to him, and that we ought not to have set up this cross without his permission" (Pastore 1994, 28; on the taking of possession more generally see Seed 1992). Cartier then carried two of the chief's sons off to France. They were returned to their village on a second voyage in 1535, which took Cartier to Hochelaga, the site of present-day Montréal. Again indigenous leaders were kidnapped on Cartier's departure. Six years later he returned once more, ostensibly to take control of these territories "by friendly means or by force of arms" (ibid.) and to establish a settlement.

In this he failed. The locals were hostile—according to the report of Spanish Basque fishermen who traded with a group of indigenes from Canada at Grand Bay, "they killed more than thirty-five of Jacques' men"—and in 1542 Cartier and the survivors of his expedition withdrew (ibid.). A second French attempt at colonization later that year lasted less than twelve months. Not for sixty years, until the establishment of a fur-trading post at Tadoussac, the voyages of Samuel Champlain, and the colonizing ventures of Pierre de Monts early in the seventeenth century, did the French officially reassert their interest in the territory that Cartier had loosely christened "Canada" (Heidenreich 2001).

Yet other, less formal connections proliferated in the last half of the sixteenth century. Stimulated by an emerging market for beaver pelts, from which European hat makers manufactured newly fashionable felt headwear, whalers and fishermen began to trade for furs on the coasts of North America. Details

Claiming Canada for France. *In 1908, as Canadians celebrated the tercentenary of Champlain's settlement at Québec, a Montréal newspaper offered this rendition of Jacques Cartier's earlier (1534) enactment of the French claim to possession of this part of the New World. In ways broadly similar to those employed by other imperial agents in other settings at other times, the French sought to establish their authority with ceremony. By Cartier's own account, however, the occasion was less blissful than this depiction suggests: The indigenous leader made "a long harangue" that even Cartier understood to indicate resistance to his actions. (National Archives of Canada)*

about the beginnings of this exchange remain obscure: It is difficult to know how many fishing vessels worked the coasts of Nova Scotia and the St. Lawrence estuary at this period, how involved they became in trading for furs, or when the trade might justly be described as commercial rather than incidental. New World furs certainly reached Spain in 1525, Basques traded for deer hides and sea lion skins in the 1540s, and a Portuguese report suggested that thousands of animal skins reached France from the New World each year. In the following decade French notaries began to record fur trading as a commercial activity.

Recent work in the archives of several French departements by the Québec historian Laurier Turgeon has illuminated something of what this trade meant to native peoples. As early as 1565 vessels leaving La Rochelle for the cod fishery carried an assortment of trade goods, including beads and "mirrors, harness bells, pendant earrings, scissors, German knives and axes, billhooks, haberdashery and Flemish embroidery material." More significantly, Micheau de Ho-yarsabel, a Basque sailing from Ciboure and Bordeaux, carried over 500 red cop-per kettles, weighing about 12 pounds a piece, across the Atlantic to trade with "the savages of Canada," between 1584 and 1587. His 1584 cargo also included almost 2,000 knives, 50 axes, and several swords, as well as glass goods and hab-erdashery of various sorts (Turgeon 1998, 600). Much of the trade for which these items were destined occurred near the mouth of the Saguenay River, at Tadoussac, and on Ile aux Basques, where the archeological record reveals con-temporaneous Basque and Amerindian occupation.

Highly valued European trade goods spread quickly through native societies. Passed along existing trade routes, and perhaps forging new linkages among na-tive peoples, they moved far in advance of Europeans themselves into the inte-rior of the continent. Indeed, some have suggested that European copper reached the Ohio River valley within a decade of its being traded on the St. Lawrence. Fragments of copper, brass, and iron were among the earliest European items to reach the area north of Lake Ontario in the sixteenth century and were afforded an importance "out of all proportion to their economic or technological signifi-cance" in burial rituals. As the ethnohistorian Bruce Trigger has observed, "tra-ditional native values were transferred to European commodities," which led to an efflorescence of mortuary ceremonialism (Trigger 1985, 162). With familiarity and abundance, however, introduced goods were scored on more commonplace registers and became more fully integrated into the daily lives of native peoples. Basque-type kettles and reworked fragments of the distinctive copper of which they were made have been found in several Miq'mak sites in northern Nova Sco-tia and New Brunswick, as well as in southern Ontario; they frequently appear in association with beads that can be dated precisely to the period between 1580 and 1600. Durable and widespread though they were, the distinctive metal goods and ornaments derived from the Basque trade began to give way to similar items of Norman provenance when Champlain and others renewed French interest in Canadian territory early in the seventeenth century.

In the end, however, the regional origins of these goods in Europe had little influence upon the utility or significance that native peoples found in them. As the trade in kettles and axes and knives and so on expanded, and as they be-came more abundant in native societies, these goods triggered a complex and cascading series of changes in indigenous economies and ways of life. Where new and rare European artifacts were once incorporated into native societies in non-European roles, they soon also, and increasingly, became objects of compe-tition, conflict, and transformation.

According to the seventeenth-century merchant Nicolas Denys, introduced trade goods quickly became "an indispensable necessity" to the indigenous peoples of Acadia (present-day Nova Scotia, New Brunswick, and Prince Edward Island). By 1670, he claimed, they had "abandoned all their own utensils," either "because of the trouble they had . . . to make . . . [and] to use them" or because of the ease with which highly convenient, and thus invaluable, items could be obtained "in exchange for skins which cost them almost nothing." By this account, the European copper kettle ranked, among the Miq'mak, as "the most valuable article they can obtain from us" (Denys 1672, 440–441). Nor was this unusual, at least according to Marie de L'Incarnation, an Ursuline nun resident in Québec City, who reported in 1665 that French soldiers had taken "fully four hundred kettles, and the rest of their riches" from a group of Mohawk people (quoted in Turgeon 1997, 10). Contemporary accounts of indigenous societies, by traders and missionaries such as Denys, Marc Lescarbot, Chrestien Le Clerq, Pierre Biard, Jean de Brebeuf, Gabriel Sagard, and others, early recognized that trade goods often had multiple meanings and implications in native worlds.

Copper kettles certainly did. Indeed, ethnohistorian Calvin Martin has suggested that they had four incarnations or "lives" among the Miq'mak—even beyond their everyday utilitarian use in food preparation. First, they were highly prized commodities of trade; Miq'mak exchanged them and other European goods with their Algonkian neighbors to the south for corn, tobacco, beans, and pumpkins that they did not produce. Judging by their frequent occurrence in mortuary sites, where three or more might be carefully integrated into the burial as protection for grave goods and the bones of the deceased, copper kettles evidently (and second) held an important place in Miq'mak ceremonial life. They were also (third) ascribed spiritual identity: Confronted with a badly oxidized kettle exhumed from a grave and finding that it no longer rang loud and true when tapped, a local man reported to Denys that "it no longer says a word, because its spirit has abandoned it to go to be of use in the other world to the dead man . . ." (quoted in Denys 1672, 439–440). But it is the fourth role that Martin identifies for the Miq'mak copper kettle—its "geographical-social function"—that is most pertinent here, because in his view it was nothing less than a solvent of traditional settlement and land-use patterns.

By this argument, precontact Miq'mak society followed a seasonal round that divided the year between summers on the seacoast and winters in the interior. Typically, individual households camped, at each season, "beside a body of water . . . in the vicinity of the band river" (Martin 1975, 127). At each of these locations they had a wooden cooking trough (or kettle) painstakingly hollowed from the trunk of a fallen tree with fire and stone or bone hand tools and heated by the immersion of fire-heated stones in water to bring it to a boil. As Denys sought to understand the implications of these practices for indigenous society, he reported that natives "were obliged to go to camp near their grotesque kettles . . . [which] were the chief regulators of their lives as they were able to live only in places where these were" (Denys 1672, 443). There is too much technological determinism in this phrasing. Making wooden kettles was undoubtedly hard and time-consuming work, but they were fashioned where families chose. Still, Denys was right to recognize the sense of stability implied in the yearly migration to and from particular beach or hunting areas. He was also quick to realize the implications of European "kettles easy to carry." By 1672 the Miq'mak had "entirely abandoned" use of "immovable kettles" and were "free to go to camp where they wish" (Denys 1672, 443, 406).

All of this took place as a consequence of contact, and the development of trade. In Martin's contentious interpretation, the value and convenience of trade goods and the growing market for furs created among the Miq'mak "an overpowering urge to range indiscriminately, [and] to hunt beyond the usual bounds" of traditional family territories until—as Denys observed—"furbearers and browsers were hounded to near extinction" (Martin 1975, 127; see also Martin 1974 and Denys 1672, 187, 199, 209, 220, 402–403, 426, 429–433). The resulting pressures on subsistence and survival led, in Martin's view, to the development among the Miq'mak, late in the seventeenth century, of a new system of land tenure in which particular hunting territories were allotted to individuals. In little more than a century, according to this reckoning, the diffusion of convenient cooking pots through Miq'mak society intensified human mobility, facilitated "the reckless slaughter of wildlife," brought a society earlier marked by "permanence and stability" to the edge of chaos, and led to significant changes in both its social organization and the apportionment of its resources (Martin 1975, 127; Martin 1978).

Whatever the precise details of this case—and the data are less informative than one would like them to be at several crucial points—the sorts of reverberations charted here have been the focus of much research—and debate—among scholars interested in the impact of the fur trade (and other forms of contact) upon native society. The argument goes back a long way, to the beginning of the last century, at least, when American ethnologist Frank G. Speck argued—in a direct challenge to the Marxist claim that precapitalistic societies were communistic—that indigenous peoples between Lake Winnipeg and Nova Scotia had traditionally adhered to a form of land tenure that he called the family hunting territory system. Under these arrangements, Speck insisted, members of nuclear families possessed certain absolute rights over resources (animals) in defined and assigned parts of their band territory.

Canadian anthropologist Diamond Jenness and others (including the American Julian Steward) disagreed, arguing that the fur trade had "catalyzed the partitioning of band territory into family hunting grounds" (Martin 1975, 118). Their arguments were confirmed in the 1950s by Eleanor Leacock's work in Labrador, which offered convincing evidence that the shift, among the Montagnais, from communal hunting to hunting on land on which individuals claimed rights stemmed from growing dependence on introduced (non-native) foodstuffs and the diffusion of arguments for individual tenure emanating from white society. Refinements, qualifications, and nuances have been added to our understanding of these matters in the half-century since.

Hunting groups, if not hunting territories, existed prior to contact, at least in some areas. Internal cultural or sociodemographic changes may have been as important as external pressures in moving people to adopt exclusive hunting territories. Ecological changes—produced by fires or fluctuations in animal populations—could shift the focus of hunting and subsistence efforts and lead to changes in the ways people regarded both lands and resources. The picture, in short, is complex and varied. But on its main dimensions there can be no equivocation: The trade that followed contact had wide and deep ramifications for the structure and organization of native societies, for their subsistence and ceremonial activities, and for their relationships with nature.

Without tracing these outcomes in detail, it is clear that the destabilizing effects of European trade goods reached far and fast across the continent. The fur trade that began in rather desultory fashion along the eastern seaboard in the sixteenth century was a major enterprise in the Great Lakes region in the seventeenth, a dominant force through much of the western interior in the eighteenth, and the focus of both fierce competition and monopoly practices from the Atlantic to the Pacific and into the Arctic in the nineteenth. Under French control it expanded from Tadoussac and Montréal to encompass the Great Lakes, to tap the headwaters of the Mississippi and the Missouri, and to press north and westward of Lake Winnipeg. After the conquest of New France by the British in the 1760s, traders from the St. Lawrence continued to press inland, beyond the areas utilized by their French predecessors, to reach Great Slave Lake by 1790. Under the auspices of the British company licensed in 1670 to trade in the territory draining into Hudson Bay, the fur trade spread its tentacles inland along the Moose, Hayes, Nelson, and Churchill river systems into areas entered from the southeast by Montréal traders.

As European interests pressed further and further into the country, and the demand for furs increased under the spur of competition between the Northwest Company (out of Montréal) and the Hudson's Bay Company, more and more indigenous peoples became involved in the trade. Ever-greater quantities of European goods—kettles, iron tools, blankets, tobacco, alcohol, and guns—passed into native hands. And as always, the most valued items moved from

Grand Portage, Lake Superior. *This illustration of Grand Portage on Lake Superior depicts a pivotal node of the early eighteenth-century fur trade out of Montréal, where western brigades met supply crews from the St. Lawrence. Later this function was transferred to Fort William. At another scale, though, this illustration captures many of the essential elements of fur trading posts across the interior: the location on a water route; the tight cluster of buildings (for the storage of trade goods, the accommodation of traders, the maintenance of canoes and equipment, and so on) within the surrounding defensive wall; the entrance portal; and the lookout tower. (National Archives of Canada)*

indigenous group to indigenous group well ahead of the entry of Europeans into native space. Items of French manufacture reached Lake Winnipeg and the shores of James Bay via the St. Lawrence in the first half of the seventeenth century. By the beginning of the eighteenth, English goods introduced through recently established trading portals on Hudson Bay could be found alongside items from France on much of the northern plains. Inhabitants of the north's major drainage system used trade goods from both countries almost half a century before Alexander Mackenzie navigated and gave his name to their river in 1789.

Although all depended upon hunting (to which some added fishing) for subsistence, the native peoples of this northern realm were divided by language—with Athapaskan speakers north and west and Algonquian speakers to the south and east of the Churchill River, and members of the Siouan language family on the plains south and west of Lake Winnipeg—and occupied markedly varied environments. Athapaskans (Chipewyans) were people of the tundra and lichen woodland. Algonquian-speaking Cree occupied the boreal forest and aspen parkland. Assiniboines (Siouan speakers) migrated between the prairie grasslands and the parkland. For all, traditional life ways were closely tied to the seasonal movements of the large animals upon which they preyed: the

Buffaloes at rest. *This lithograph, made early in the twentieth century by the American Louis Kurz, long after the herds had been decimated, suggests the earlier abundance of plains buffalo (bison) across the landscape of the western interior. (Library of Congress)*

Chipewyan following the caribou on their migration from tundra to lichen woodland; the Cree (who hunted small game—such as wolf and beaver—in, and fished the streams of, the boreal forest through the summer) moving into the parkland to hunt the plains bison that sheltered there in the winter; the Assiniboine following the bison from parkland in the winter to the western plains in the summer.

For the most part, these people made their tools, weapons, and clothing from local materials: Bows and arrows, stone-tipped lances, deadfall traps, snares, impoundments, and cliff-drives (buffalo jumps) enabled them to catch game; pelts and hides provided clothing and shelter; scrapers and blades were shaped from bone and stone; birchbark covered canoes and lodges; and wooden snowshoes, dogsleds, and toboggans facilitated winter movement. There is perhaps no better encapsulation of this full and intricate dependence on local resources than that offered by fur trader Alexander Henry, describing the value of the buffalo to the natives of the plains: "The wild ox alone supplies them with every thing which they are accustomed to want. The hide of this animal, when dressed, furnishes soft clothing for the women; and dressed with the hair on, it clothes the men. The flesh feeds them; the sinews afford them bow-strings; and even the paunch . . . provides them with that important utensil the kettle" (quoted in Ray 1987, 34).

By later estimates, the buffalo provided the Blackfoot with over one hundred specific items of material culture. But such varied use does not necessarily imply a lack of wastefulness. At Head-Smashed-In Buffalo Jump, near Fort Macleod, Alberta, the height of the cliff over which bison were stampeded (10–12 meters, or 35–40 feet) is almost matched by the accumulation of bones and soil formed by the remains of the animals killed there. Analyses of similar sites elsewhere suggest that as many as one in four animals killed at such jumps were hardly utilized. Other evidence from the historic period indicates that buffalo were sometimes slaughtered for their humps or the fetuses the cows carried in late winter.

Despite recent claims to the contrary, waste was not unknown to native peoples. But in considering their place in nature as a whole, it can be said that the people of the northwestern interior exploited a broad range of resources, depended upon a close knowledge of the environments through which they moved, and traded with neighboring groups for items that they were not able to produce themselves. They constituted mobile, small-scale, essentially egalitarian societies not much given to the accumulation of possessions.

Among these peoples, as among those in the east, the quickening of European trade had far-reaching effects. As Arthur Ray, a leading student of native-European relationships in the fur trade, has observed, the exchange of pelts for imports meant the replacement of a "traditional technology with an exotic one" (Ray 1987, 73). Natives were far from pawns in this process; indeed, Ray describes them as "discriminating consumers" (ibid.). They were adept at exploiting the competition for furs, explicit in their preference for light and robust equipment, and disinclined both to accept inferior trade goods and to supply more furs than were necessary to satisfy their own demands for trade items.

Yet almost all facets of native life were affected by the exchanges. Cooking food in kettles that could stand the heat of fire was far less laborious than boiling it by the immersion of heated stones; metal tools (knives, scissors, and needles) made the preparation of traditional clothing much easier; woolen blankets were portable and durable, and if less warm than garments of fur, they were far

Buffalo driven over cliffs. *Native people of the plains had long depended upon the bison (or buffalo) for many of their needs and the practice of stampeding herds over cliffs (or buffalo jumps) was a traditional one. As this illustration suggests, the acquisition of the horse made the drive more effective and more dramatic. Thousands of buffalo were killed at favored sites, suggesting that the carnage was greater than need required. But once turned to flight, there was probably little that could be done to stop large herds from their "head-smashing" fate. (National Archives of Canada)*

less laboriously acquired. European axes and chisels made it much easier for native peoples to break open frozen beaver lodges. Late in the eighteenth century, this traditional form of beaver hunting was superseded by the introduction of steel-spring traps—baited and easily set out in number (along traplines), they increased the efficiency of the hunt as they transformed it.

Guns had equally revolutionary effects, and the numbers introduced were far from trivial; between 1700 and 1730, well over 500 firearms were traded on average each year by the Hudson's Bay Company through its establishments at York Factory, Albany Fort, and Churchill Fort. Even the inaccurate smoothbore, flintlock muskets of the eighteenth century were much more effective than bows and arrows in slaying game; they could be used at greater distances and usually produced a quick kill. It is hardly surprising, then, that many elements of the "technological invasion" brought about by trade quickly assumed far more than utilitarian significance in native societies. As John Ewers, a student of the Blackfoot of the western interior, once noted,

> the gun became more than just another, or even a better, weapon to the Indian. As a trophy it symbolized the highest warlike accomplishment. It found a place in Indian etiquette, in medicinal practice, and even in Indian religion. As Indians came to regard the gun as a necessity, it provided a closer link between the red and the white trader. (Ewers 1972, 14)

The gun was a particularly potent increment to native societies when combined with a second European introduction—the horse. Horses came to the Canadian prairies from the south. Brought to the Americas by the Spanish in the sixteenth century, they had indigenized in the continent of their long-since extinct and much smaller ancestors and were acquired by the Blackfoot and other people of the northern plains by capture or through trade with more southerly native groups in the first half of the eighteenth century—probably in exchange for European goods introduced through the fur trade (Binnema 2001). They conferred dramatic new mobility. They allowed men armed with lances and bows and arrows to gallop with the buffalo and dispatch them more efficiently than they might on foot. They facilitated the driving of buffalo into pounds and over cliffs. When repeating rifles reached the West in the latter part of the nineteenth century, they combined with the horse to increase the deadliness of the hunt even further. A more efficient hunt meant more time for other things, and many have traced an efflorescence of cultural display and ceremonial activity among the Blackfoot, Blood, and Piegan bands to this conjuncture. Certainly horses became important symbols of individual, and collective, wealth among the people of the plains.

Expanding markets, increasingly intense competition between Montréal and Hudson's Bay Company traders until the consolidation of the Canadian fur trade as a Hudson's Bay Company monopoly in 1821, and new capacities afforded native societies by trade goods together had far-reaching effects upon both indigenous peoples and fur-bearing animal populations. Soon after trade began, several native groups adjusted traditional seasonal rounds of subsistence activity, finely tuned to local ecologies, in order to capture more beaver, marten, and other animals with valuable pelts. As less-used traditional skills atrophied, indigenous people grew more dependent upon trade goods. And as competing interests sought new ways to draw trade away from their rivals, competition was heightened. Because indigenous demand for most durable utilitarian trade goods (kettles, knives, blankets) remained relatively inelastic, beyond the satisfaction of immediate needs, traders sought new enticements to exchange.

Trading posts proliferated throughout the West in the late eighteenth and early nineteenth centuries (see Fig. 8). In 1774 there were seventeen in the northwest (beyond 78 degrees west); in 1821 there were 125; and in the interim some 600 new posts were built (most of them obviously short-lived) in a confusing scramble for position and patronage (*Historical Atlas of Canada* 1: plates 38–40, 57, 60–66). With widely scattered new establishments making direct access to trading posts considerably easier, old established patterns of trade among native groups that had seen furs pass from producers to European traders through middlemen broke down. The imperative recognized by the chief factor at York Factory in the 1720s—"now is the time to oblige the natives, before the French draws them to their settlement"—was reciprocal and, as the century wore on, ever more intense (quoted in Ray 1987, 73).

Like ice cream vendors on a hot beach in the famous demonstration of the principles underlying the location of economic activity by the American statistician and economist Harold Hotelling, Hudson's Bay Company and Montréal traders typically located their trading posts in close proximity along rivers and lakeshores—"new entrants" and "established participants" metaphorically backing into each other to maximize their potential access to the market (Darnell 1990). By 1800 few native inhabitants of the northwest lived much more than fifteen miles from a trading post. In these circumstances, tobacco and alcohol became vital lubricants of exchange in the half-century before 1821, not least because they could be consumed in short order and were addictive—both characteristics that carried them through the demand ceiling limiting European-native trade in "consumer durables" and made them effective incentives to trade. But alcohol, in particular, had other, severe consequences. More than 21,000 gallons of it were said to have reached the interior in 1803 alone. Traders began to note that "these Natives are given very much to Quarrellg. when in Liquor" (Hudson's Bay Company trader James Isham, quoted in Ray 1987, 76). Seeds sown for the encouragement of trade grew to choke respect for traditional authority, to demoralize individuals, and to darken the futures of entire peoples.

Half a century of intensifying competition between trading interests also ravaged the animal resources of the northwest. The growing battalion of European fur traders occupying the area—estimates suggest there may have been as many as 1,500 in any year—created a demand for food. Local supplies of fish and animals (deer, moose, caribou) were exploited. So, too, were the bison of the plains, whose flesh was dried to produce pemmican. But it was the beaver, whose pelt was the prize item of the fur trade, that suffered most. Relieved of the need to travel long distances to trading posts (or the alternative of trading through intermediaries), natives had more time to hunt and trap. Because relatively short market journeys could be repeated easily, the proliferation of trad-

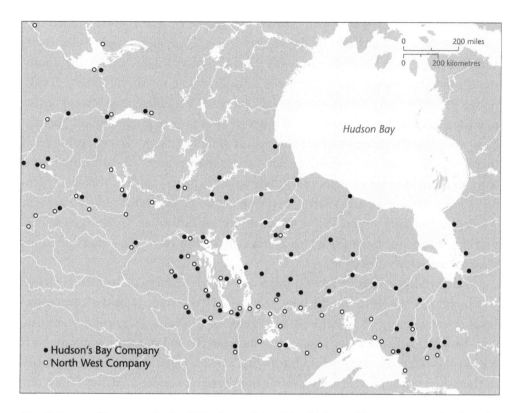

Fig. 8. *Fur trading posts in the "Northwest" in 1821. (Adapted from* Historical Atlas of Canada *1: pl. 62)*

ing posts lessened constraints (imposed by the capacity of individual canoes) on the number of furs brought to market. New technologies—especially the steel trap and the gun—made hunting more effective.

And trade was its own snare. Moving away from the traditional subsistence economy, growing increasingly dependent on imported goods (not least ammunition), trading for nondurables such as alcohol, and finding local food supplies more difficult to acquire for one reason or another, many native people were left with little alternative but to increase their take of beaver to survive on increasing quantities of store-bought or traded goods. All of this fit well with the views of the fur traders, who considered native peoples better employed hunting furs than food. When these cumulative pressures were exacerbated by deliberate efforts to exploit fur-bearing animals before they fell to competing interests—as happened south of James Bay and later in the Columbia district of the west—the results were grave.

In many southern parts of the Petit Nord (between Lake Superior and James Bay) many food and most fur-bearing animals had been hunted to the verge of extinction by 1840. Twenty years earlier the people of Rainy River needed buffalo hides delivered to them by the Hudson's Bay Company to make moccasins and clothing. Such immiseration was virtually unavoidable under the conditions generated by the fur-trading economy in these northern forests. Traditional attitudes to resources recognized ecological variability and shared access to animals across large territories; though band members might exploit particular areas, others were permitted to use these tracts to catch caribou or beaver if these animals were (temporarily) scarce in their own areas. With limited pressure on a broad range of resources, these arrangements—predicated on the importance of mutuality and sharing between bands as required by local scarcities —were ecologically sound and sustainable. But with intense pressure on particular stocks across large areas—as generated by the fur trade—traditional arrangements were vulnerable.

These native societies faced a form of what the American human ecologist and bioethicist Garrett Hardin famously called "the tragedy of the commons" because they lacked the institutional structures to deal with general scarcities. Any band recognizing a particular resource scarcity and reducing pressure on it within its traditional area would likely find others, faced with similar scarcities in their own areas, moving in to exploit their resource. Indeed, Arthur J. Ray, one of Canada's leading students of the fur trade, suggests that these circumstances induced by the fur trade undercut traditional patterns of resource sharing between bands, reduced band mobility, and led to more defensive, protective identification of particular bands with individual territories. At the same time and in turn, these developments led to a growing dependence upon small game populations—subject to extreme and relatively short-term ecological fluctuations.

Whatever the precise mechanisms of resource depletion and adjustment, the disastrous competition between Montréal and Hudson's Bay Company interests that spawned it impoverished significant numbers of native peoples and heightened their dependence on Europeans and their trade goods. Recognizing this, the impossibility of profit under existing arrangements, and the long-term threat to fur-trading operations implicit in the exhaustion of beaver stocks, George Simpson, the governor of the Hudson's Bay and North West companies merged under the name of the former in 1821, moved decisively to change things. Trading posts were closed (there were only forty-five in 1825), and the numbers employed in the fur trade were slashed. The Great Lakes route to Montréal was abandoned. Where the Hudson's Bay Company's monopoly was strong, the alcohol trade was abolished, steel traps were banned, and strict limits were put on the numbers of furs that could be purchased at any post. Simpson also tried to deflect pressure on beaver stocks by limiting summer hunting and by encouraging trade in the pelts of other fur-bearing animals.

Arthur Ray has described these initiatives, accurately, as early conservation schemes. Their principles were sound, but their implementation often faltered. Competition—from American traders and others—undermined efforts to reduce pressure on beaver stocks, and company employees were not always scrupulous in following Simpson's lead. Moreover, natives, finding it difficult to survive in the changed circumstances of the nineteenth century, continued to hunt for food and pelts; neither their immediate subsistence needs nor their views of the natural world (in which both animals and humans were animated with symbolic and religious significance) fit well with the long-term goals and ecological underpinnings of Simpson's policies.

On the western plains, the onslaught on the buffalo (bison) herds that were integral to the subsistence of Blackfoot and other native inhabitants of the grasslands quickened after 1820. Reductions in the number of fur trade posts encouraged some of those formerly engaged in the trade to move into the increasingly commercial buffalo hunt. Initiated to provision the fur trade with supplies of pemmican and dominated by Metis from the Red River area early in the nineteenth century, the commercial buffalo hunt grew ever more rapacious as the demand for bison robes expanded spectacularly in the middle decades of the century. After local herds were diminished or driven away from the Red River, Metis hunted further and further westward in the 1820s and 1830s.

Their annual pursuit of the bison involved growing numbers of people, who ranged onto the plains in a spectacular cavalcade of horses, oxen, and Red River carts that became even larger once the Canadian hunt was linked, through St. Paul, Minnesota, to the American market for buffalo robes in the 1840s. Enormous numbers of buffalo were killed. Conducting a scientific exploration of the western plains for the British government in the late 1850s, Captain John Palliser learned that native inhabitants of the region knew a small tributary stream of the Qu'Appelle River as *Oskana-Ka-asateki*, translated as "The Creek Before Where the Bones Lie" or "Pile o' Bones" and rendered in English as Wascana (Riddell 1981). But buffalo were already scarce in these parts, and their numbers declined across the grasslands in the years that followed as a growing army of market hunters challenged native and Metis access to the resource.

The construction of the Union Pacific Railroad in the United States in the 1860s divided the great North American buffalo herd in two, brought more hunters to the west, and carried the robes they produced to market. Between 1872 and 1874 over three million buffalo were killed. In the midst of the slaughter, Red Cloud, a Sioux leader, is reputed to have observed: "Where the Indian killed one buffalo, the hide and tongue hunters killed fifty" (quoted in Krech 1999, 141–142). Late in the 1870s the southern (U.S.) herd stood on the verge of extinction. Then the Northern Pacific Railroad reached North Dakota and hunting pressure focused on the northern herd. By 1890 the wood buffalo of the far north were the only significant wild bison herd in Canada. In little more than half a century the enormous buffalo herds so central to the indigenous economies of the grasslands and northern parklands had been reduced to a few fragile remnants (Isenberg 1994; Hornaday 1899).

All of this raises important questions about the influence of European contact upon indigenous attitudes to nature. For those who subscribe to the view, widely popular since the 1960s, that native Americans lived in harmony with nature, that they were the continent's first ecologists, that they respected other living creatures and trod lightly upon the earth, any deviation from such environmentally sensitive behavior is taken as a reflection of European disturbance of indigenous and natural harmonies. Native peoples respected environmental limits and protected the delicate organization of natural systems until avaricious, capitalist Europeans arrived to shoulder aside such views and command spendthrift exploitation of the natural world.

But there is much awry with this picture. It essentializes indigeneity, ignores the historical record, is willfully inattentive to circumstances, and applies modern ideas, anachronistically, to the conditions of the past. There is no question that native peoples of the precontact period lived close to the land (or environment) in a basic and important material sense. They knew its relief, depended upon detailed knowledge of the habits and movements of its creatures, endured its seasonal fluctuations, felled its trees, killed its animals, and felt its presence in ways unfamiliar to modern urban dwellers. They understood that actions had consequences—that burning the grassland encouraged new growth that attracted buffalo; that yields could be improved by tending clam beds, or that "gardening" would help favored food plants to flourish. They also thought about the natural world in ways quite different from those typically adopted by modern students educated in the Western scientific tradition. Indigenous origin stories characteristically blur the boundaries between humans and animals. Many native groups saw nature as animate and believed in reincarnation; rituals were utilized to propitiate the spirits of plants and animals before they were used, and so on.

But none of this should be elided with, or understood as, equivalent to modern-day conservationist thought and ecological principles. Nor should contemporary sensibilities foster the characterization of native peoples as primitive innocents in these matters. Indigenous peoples interacted with the natural world in a great variety of ways. They made use of it for their purposes in ways that conformed to their own understandings of the world and their places within it, and they were adaptable and resilient in facing the challenges of existence in generally unpropitious environments. To ignore this and to suggest that native peoples trod so lightly on the earth as to leave no marks upon it through the centuries before European contact simply and inappropriately (as historian Richard White has written in the American context) "demeans Indians. It makes them seem . . . like an animal species and thus deprives them of culture" (quoted in Krech 1999, 26; see also Hirschkind 1983).

In recent years, ethnohistorians have done a great deal to show that the stable cultural patterns, believed to result from the close adaptation of peoples to environments and posited for pre-European North America by anthropologists working in the tradition of Franz Boas, were far less constant than these anthropologists understood them to be. Scholars have also challenged the persistent view that relations among indigenous peoples were harmonious until their lives were disrupted by contact. The movement of goods, the migration of peoples, struggles over territory, and even genocidal warfare all occurred in the prehistoric period. But trade between natives and newcomers unquestionably increased the volatility of relations between and among native groups. Examples are numerous, but all point to the ways in which new, European-orientated trading relationships destabilized existing patterns of power and resource use among indigenous communities.

Thus guns obtained from the fur trade shifted the balance of power among peoples of the western plains from the Shoshone (whose territory lay south of the current Canada-U.S. boundary, and who acquired the horse before their northern neighbors) to the Blackfoot Confederacy (Blackfoot, Blood, and Piegan) in the eighteenth century (Binnema 2001). As fur trade posts proliferated through the northern forest, some Cree bands settled in their shadows and became substantially dependent upon the casual labor available in and around the posts, collecting firewood, maintaining structures, and providing provisions. These "Home Guard Indians" developed close, and tender, ties with the fur traders. Many of their women were taken as traders' wives, according to "the custom of the country," and from these liaisons the Metis emerged, in time, as a distinct group of people (van Kirk 1980; Brown 1980). Other Cree and some Assiniboine bands became specialized traders, middlemen who sustained a position as intermediaries between those bands that produced furs and the Europeans with whom furs were exchanged for imported goods.

In early accounts of this pattern by the Canadian political economist H. A. Innis and others, "middlemen" were essentially portrayed as professional traders, doing little trapping themselves and utilizing their position to maximize returns by buying cheap and selling dear. In Bruce Trigger's judgment they were rendered as "economic stereotypes only minimally disguised in feathers" (Trigger 1985, 184). Not surprisingly, such formalist interpretations (which regard all peoples as subject to economists' theories of market behavior) have been challenged. Followers of Karl Polanyi, known as substantivists, insist that precapitalist economic behavior is more thoroughly shaped by political and cultural influences than by market forces. In this view, advanced most strongly in the present context by Abraham Rotstein (and also by Richard White, 1991), the fur trade was a form of treaty or embassy trade, based on relatively fixed rates of exchange (rather than prices fluctuating according to the laws of supply and demand), marked by formal rituals of gift giving, and carrying with it at least an implicit sense of formal (political-military) alliance between trading partners.

According to Arthur Ray and Donald Freeman, however, the substantivist model does not fit the later Hudson's Bay Company trade. By their account, based on statistical analyses of trading records, most nonmarket features of this eighteenth-century trade were hollow relicts of the past, rather than influential determinants of trading practices. There were trading rituals, but they were quite informal. The Hudson's Bay Company did not provide military aid to those who traded at their posts. Fixed exchange rates were honored in the breach, with both parties endeavoring to work the margins in their favor. And native peoples calculated the costs of travel into the prices they were prepared to give and receive at particular posts. This trade was less traditional than transitional (see also Ray 1974, 1980, and 1985).

Still, the emergence of indigenous middlemen in the fur trade was both a stimulus to the expansion of the European trading network and a source of instability among native peoples. The movement of French and English traders far into the interior of the continent was driven not only by competition between them, but also by the hope that they could obtain furs more cheaply by "going to the source" and "cutting out the middleman." Almost invariably, however, new groups of native peoples took advantage of the strategic advantage afforded them by the establishment of a trading post in their territory by adopting the middleman role. In this they were quite prepared to play off rival European trading factions to obtain better prices, to mark up the prices of goods traded to other indigenous groups, and to defend their intermediary position vigorously, even, if need be, violently. By Arthur Ray's account, guns acquired in the fur trade allowed some Assiniboine and Cree to prevent the access of other indigenous groups to Hudson Bay and James Bay (and thus European traders) in the late seventeenth and early eighteenth centuries. The newly acquired weapons also encouraged these groups to extend their influence west and northwest.

Considerable conflict and bloodshed resulted, as territorial claims were extended and populations displaced. Shortly before European explorers began to travel through the northern and western interior, Chipewyans were driven northward, Beaver and Sekani people were displaced to the west, and the Gros Ventre were forced southward. Consequences cascaded from these movements.

The intimate knowledge of particular territories developed by native peoples was devalued. So, too, were intricate connections between people and place severed. Sometimes traditional knowledge could be adapted to new settings—but this was less likely if displacement involved movement into significantly different ecological zones. Thus, long-established modes of living on the earth were disrupted. Traditional rituals and attitudes toward nature were undermined—and in some cases, at least, the onslaught on beaver and other game animals may have quickened as a consequence.

6

TRADE AND THE HURON

In the first half of the seventeenth century the fur trade of the western St. Lawrence was substantially controlled by the Huron (also *8endat,* pronounced Ouendat or Wendat) confederation of four Iroquoian-speaking tribes, which together acted as intermediaries in a complex, far-reaching pattern of exchange built upon earlier trading networks. The Huron emerged from the migration and melding of formerly distinct groups during a period of major change and instability in the sixteenth century, probably produced in part by the movement of European trade goods into the area by way of Lake Nipissing to the north. They lived on a peninsula in what is now northern Simcoe County in Ontario (from whence they may have derived their name, because 8endat is translated as "islanders" or "peninsula dwellers") and were the northernmost practitioners of corn cultivation among the Iroquoian-speaking peoples of the Great Lakes basin.

Agriculture both imposed and allowed the development of a relatively high-density, semi-sedentary population. By most accounts, among which those of Conrad Heidenreich and Bruce Trigger are the most useful, there were probably about 21,000 people living in the 500 or so square kilometers of Huron territory early in the sixteenth century. They occupied a number of palisaded villages, each of which included several longhouses in which an average of six families, or thirty-five to forty people, lived and stored their food and possessions. Villages varied in size from the very small—with three or four longhouses—to the substantial—with populations of more than 1,500 people. In spatial terms they ranged from two to fifteen acres in extent, with most probably encompassing about five acres, including twenty to thirty longhouses, and housing between eight hundred and a thousand people.

Clearings, created by slashing and burning the forest, surrounded the villages. In these fields, ranging from a few to several hundred acres in size, corn was planted (with beans and squash) in hills amid stumps and rocks (Heidenreich 1971). Great effort went into making these openings for agriculture. "Clearing [which was done by men] is very troublesome" for them, wrote Recollect missionary Gabriel Sagard, because "they have no proper tools" (quoted in Ray 1987, 23). Stumps were burned and roots might eventually be removed, but with the fertility of the soil declining year by year, most fields were probably abandoned within twelve years of being created. During this time they were carefully tended by Huron women, who cleaned and prepared the ground, planted and harvested, and kept the rest of the patch "cleansed of noxious weeds, so that it seems as if it were all paths, so careful are they to keep it quite clean" (Ray 1987, 24). These women also gathered a range of wild roots, nuts, and berries; made clay pots; wove mats of reeds and corn leaves; cooked; sewed; and tended to children.

Farming may have provided almost three-quarters of the Hurons' food requirements, but fishing and hunting remained important. Both of these were male activities. Fishing, a major source of protein, took place almost year-round, but focused on the whitefish and trout stocks of Georgian Bay late in the fall and the netting of walleye, pike, and sturgeon at such places as the narrows between lakes Simcoe and Couchiching in the spring. Hunting occurred in the fall, and took men south or east in pursuit of game. Deer, the main prey and especially valued for the hides they yielded, were taken in snares, in traps, with bows and arrows, and, most often, in large well-organized communal drives— yet early missionaries sometimes remarked on the lack of meat in the Huron diet. Huron men also built village palisades and longhouses, constructed birch-bark canoes, and manufactured tools and weapons from bone, stone, wood, and twine.

The most thorough ecological analysis of Huron agriculture, by Conrad Heidenreich, which pays close attention to soil types and yields, suggests that a Huron population of about 21,000 probably cultivated at least 6,500 acres a year. Much more of the vegetation around each village bore the marks of human disturbance for decades as abandoned fields were colonized by weeds, grasses, bracken ferns, sumac, berries, poplar, birch, and eventually white pine. Because villages were also abandoned periodically, when the distance to potential new field sites made relocation more sensible than development, the cumulative impact of Huron people on the landscape of Huronia was considerable. In all, Heidenreich estimates, some 50,000 acres, about a third of all the potentially cultivable land available to the Huron, was utilized on a cyclical basis. Given inescapable limitations in their knowledge of soil quality and management and the unpredictability of perturbations produced by weather and pests, this implies that the Huron were probably utilizing the land at close to capacity, relative to their ability to produce food from it.

Late in the sixteenth century, the Huron traded with neighboring groups to the north and southwest of their homeland. Corn—which provided winter sustenance—was crucial in exchanges with Algonquian-speaking hunter-gatherers to the north. According to the Jesuit missionary Jean Brebeuf (who probably exaggerated on this), it made Huronia "the granary of most of the Algonquians" (quoted in Thwaites 1896–1901, 8: 115). Members of the Nipissing, Petite Nation, and other bands exchanged cornmeal (and fishing nets) for furs and fish they had caught as well as buffalo robes and native copper that they had acquired in trade with groups to the north and west. Small quantities of tobacco and wampum—which the Huron acquired from their closest Iroquoian-speaking neighbors, the Tionnonate (or Petun) and Neutral peoples, also passed from Huron hands into those of their Algonquian trading partners. Northern furs moved south in turn, in exchange for these and other items such as raccoon-skin robes from the Ohio country and conch and other shells from the Gulf of Mexico.

When the Huron and the French first met, in 1609, it was not to trade but to cement an alliance. An Algonquian group brokered the meeting, near the site of present-day Québec City. It involved several days of speeches and gift giving and brought the three groups together for a raid on the Iroquoian-speaking Mohawk, whose territory lay south of the St. Lawrence. For six years, until Champlain traveled into Huron territory, trade played a minor part in relations between these groups, which were dominated by posturing, raiding, and skirmishing among the Algonquian, Huron, and Iroquois. Here, the substantivist case for trading relationships shaped by cultural and political rather than strict market forces is stronger than it is for the later western interior. Indeed, Conrad Heidenreich, one of the most careful students of early seventeenth-century Huronia, finds no evidence that the Huron participated directly in trading furs with the French before 1615. After the French-supported raid on the Onondaga, and Champlain's encounters with Petun, Neutral, Nipissing, and Ottawa peoples during the winter of 1615–1616, however, the Huron quickly came to assume a central place in the French fur trade.

Exploiting both local beaver stocks and, more importantly, their position as suppliers of corn to northern hunter-gatherers (which allowed them to serve as middlemen in trade with native peoples who might otherwise have sought direct access to European trade goods), the Huron effectively carved out a pivotal place for themselves in the developing fur trade. For a dozen years until Québec was taken, temporarily, by the English, flotillas of Huron canoes traveled along the Lake Nipissing–Ottawa River route to that settlement, trading with Algonquian groups on the way. The trade fluctuated, and estimates of its magnitude are tentative, but all in all it is likely that about sixty canoes and 200 to 250 men made this journey, bringing 10,000 or more furs (approximately 60 percent of all the furs traded on the St. Lawrence at the time) into French hands each year. Each summer Huron and French traders participated in an elaborate ritual that included speeches, feasts, and the exchange of gifts. Native traders clearly disliked (and avoided) haggling over the value (price) of individual furs; for them, trade was clearly an integral part of their political and social alliances with the French and was conceived and conducted, so far as possible, in traditional ways.

After suffering a series of disruptions into the 1630s, this trade returned to earlier levels by the end of the decade. By Bruce Trigger's estimate, it probably yielded 12,000 to 16,000 pelts a year—about half of all the furs exported from Québec at this period. Almost all of these pelts came from the Hurons' Algonquian neighbors, as even in 1630 there were relatively few beaver left in Huronia. As both indigenous groups suffered significant population decline as a result of introduced epidemic diseases in the 1630s, trading at these levels must have required each of them to invest a considerably greater proportion of available labor in support of the fur trade than they had in the 1620s. For the Algonquians this probably meant more time spent killing beaver and less on hunting and fishing. To compensate for this shift, they likely depended more heavily on corn and beans from the Huron. This would have required more time and effort from Huron men in clearing more land and from women both in the fields and grinding meal. Thus, the temptation to substitute imported goods for traditional tools and other items that were time-consuming to manufacture was likely enhanced. But satisfying this temptation meant trading yet more furs. All of which spelled escalating dependence on the fur trade.

At much the same time, late in the 1630s, Iroquois groups from south of the Great Lakes—the Seneca, Cayuga, Oneida, and Onondaga—began to raid and harass Huron villages on the edges of Huronia year-round, while eastern members of the Iroquois Confederacy—the Mohawk and some Oneida—ambushed Huron trading canoes bound for Québec from the banks of the Ottawa and St. Lawrence rivers. As early as 1640, Jérôme Lalemant, the superior of the Jesuit mission to Huronia, noted with alarm that Iroquois raiders were killing or carrying off women and children from village fields and stealing furs and trade goods from village longhouses at night (Trigger 1985, 263). Driven by the depletion of furs in their own territories, empowered by guns acquired from English traders in the Connecticut Valley late in the 1630s, and anxious to obtain the pelts they needed to acquire more powder, guns, and other trade goods, the southern Iroquois were formidable opponents. They were made even more powerful when Dutch traders, fearing the loss of their trade in furs with the Mohawk to the English, broke the embargo on trading firearms that both the French and Dutch had implemented from the beginning of the trade. By 1643 the Mohawk had 300 guns. Recognizing that successful raids on their northern neighbors by the newly powerful Mohawk diverted furs bound for Québec to their own posts, Dutch officials condoned the formerly illicit trade in firearms.

In response, the French began to arm some of their native trading partners. But fewer French weapons reached native hands, and they were inferior to the Dutch arms held by the Iroquois. By 1648 the Huron had about 120 firearms, the Iroquois more than 500 (Trigger 1976, 627–632; 1985, 262).

The 1640s were years of mounting population loss, disruption, and starvation for the Huron. In 1642 the Iroquois destroyed a small village on the eastern flank of Huronia. The tactic was repeated regularly thereafter. Surviving Huron were captured and taken south with the other spoils of battle. Fear spread, planting was disrupted, and in 1647 the annual Huron trading trip to Québec was canceled in expectation of an Iroquois attack. The following year a large group of Iroquois marauders destroyed two sizable villages. In March 1649 another Iroquois attack essentially smashed the Huron Confederacy. Villages and hamlets were destroyed and seven hundred Huron were killed. Many of those who survived were undernourished and the prospects of a crop that year seemed slim. Panic prevailed.

Within weeks the Huron had abandoned and burned their remaining settlements. In family units or small groups they dispersed. Many merged with their Algonquian, Tionnonate, and Neutral trading partners; others joined the Iroquois. Some joined Jesuit missionaries in short-lived retreat on Gadhoendoe (Christian Island) in Georgian Bay, but even this site was abandoned by 1651. Having dispersed the Huron, the Iroquois turned their weapons on the Tionnonate. After one of their major settlements was destroyed late in 1649, they abandoned their traditional lands for the western Great Lakes. Then the Neutrals came under attack, and (as Bruce Trigger has written) quickly disappeared "from the pages of history" (1985, 272). Algonquian groups to the north also suffered Iroquois attacks and relocated. Early in the 1650s, the Iroquois reigned supreme across the broad territory of what is now southern Ontario.

In the strife precipitated by competition for furs and the struggle for power and influence among indigenous peoples and European interests, native societies were disrupted and destroyed. At the same time, the trading network centered on New France—a trading network that had, in parallel with similar ventures by the Dutch and English to the south, significantly altered indigenous lives and severely reduced beaver populations across large parts of the Great Lakes basin in less than a century—was shattered and the economy of New France brought to the verge of collapse. These were dramatic and extreme—but not entirely atypical—outcomes. The substantivist qualities of the Huron-French trade, which knotted a commercial connection into a pre-existing web of intergroup antagonisms, may have exacerbated its ultimately destructive effects.

Yet trade relations between European and indigenous peoples across the northern half of the continent generally involved more than an extension of the orbit of contact and the introduction of indigenous peoples to new goods, new opportunities, and new demands that altered their modes of life to some degree. They almost invariably induced conflict (or collisions) among native groups that reverberated across space and through time to produce a kaleidoscope of population movements and demographic shifts that recast societies and remade the connections between people, place, and environment. And trade rarely worked alone.

7

PLAGUES, PREACHERS, AND THE TRANSFORMATION OF INDIGENOUS SOCIETIES

Plagues and preachers were at least as important as trade in the transformation of indigenous societies and their relations with nature. Indeed, environmental historian Alfred Crosby has gone so far as to argue, in his landmark study of the processes of ecological imperialism, that European successes overseas rested on the "ills" that they introduced to New World settings. "It was their germs, not . . . [the] imperialists themselves," he wrote, "that were chiefly responsible for sweeping aside the indigenes and opening the [territories called] Neo-Europes to demographic takeover" (Crosby 1986, 196, 201). By this account, introduced contagious and infectious diseases were "the shock troops" of the imperial conquest. Smallpox—a terrifying "disease with seven-league boots"—measles, whooping cough, and even influenza, spread far and fast through native societies with devastating effect and often well in advance of any actual European presence among native peoples (ibid.). Respiratory, enteric, and venereal infections followed in the wake of newcomers and exacted their deadly tolls.

Because the indigenous peoples of North America were relatively homogeneous, genetically, "viral infections were preadapted to successive hosts rather than encountering a wide variety of new immune system responses" (Thornton 1997, 311). Because the diseases were new, native peoples of the Americas had no immunity to them, and they spread through populations quickly, in what have come to be known as "virgin-soil epidemics." Because introduced diseases were little understood, indigenous responses to the deaths caused by them (mourning practices; flight) often exacerbated their consequences by facilitating their transmission from person to person or their diffusion through space. Because these introduced maladies were so different from the afflictions to which they were accustomed, it is now widely accepted that Old World diseases had devastating effects upon Native American peoples.

Charting the incidence and impacts of introduced diseases upon native peoples is no easy task, especially in the early years of contact, for which there are few detailed contemporary European accounts and little good information about indigenous population numbers. Yet it is clear that the consequences were extremely severe. Writing from Acadia early in the seventeenth century, the Jesuit Father Pierre Biard recorded that local Miq'mak were "astonished and often complain that . . . they are dying fast, and the population is thinning out." Moreover, continued Biard, they insisted that before their "association and intercourse" with the French, "all their countries were very populous, and they tell how one by one the different coasts, according as they have begun to traffic with us, have been more reduced by disease" (quoted in Martin 1978, 53).

Neither the missionary nor his native informants entirely understood what was taking place. Some Miq'mak seemed to believe that the French were poisoning them. Others suggested that the items for which they traded were "counterfeited and adulterated" and that spoiled foodstuffs—peas, beans, prunes, bread—sold them by the French corrupted their bodies. Biard himself sometimes attributed the decline in native numbers to their spendthrift habits and their liking for "various kinds of food not suitable to the inactivity of their lives" (quoted in Thwaites 1896–1901, I: 177; Martin 1978, 53–55). He also ascribed similar views to the Miq'mak leader Membertou, who allegedly claimed that since the French arrived his people did "nothing all summer but eat." In the same passage, however, Biard records his informant reflecting that they paid "for their indulgence during the autumn and winter by pleurisy, quinsy [acute tonsillitis] and dysentery, which kills them off" (quoted in Thwaites 1896–1901, 1: 177).

Eighty years later, Father Chrestien Le Clerq heard from the Miq'mak of the Miramichi of "a time," the date of which cannot be ascertained, "when their country was afflicted with a very dangerous and deadly malady which had reduced them to extreme destitution in every respect and had already sent many of them to their graves" (Le Clerq 1691/1910, 146–152). By Biard's account, pieced together from information imparted to him by native informants, the "Souriquois" (also referred to as "Gaspesians" and today known as Miq'mak) had been reduced to about three thousand by 1616. No one knows what their population might have been before Europeans first came to these shores. The ethnohistorian Virginia Miller (1976) once posited a late prehistoric population of 50,000 for this area. But this is extreme and no longer given a great deal of credence. The New England archeologist Dean Snow offers a more measured estimate, based in part on an assessment of the carrying capacity of the environment, of 12,000. This is now more generally accepted. Neither Miller nor Snow provides a firm mechanism for the undoubted fall in Miq'mak numbers, although both eschew the likelihood that smallpox was the major cause of population decline. Miller doubts that that potent killer crossed the Atlantic in the sixteenth century, preferring instead to follow Biard in the argument that dietary changes "weakened native resistance to endemic diseases and to localized outbreaks of European ones" (Thwaites 1896–1901, 3: 105; Martin 1978, 53–55). Snow on the other hand sees the massive epidemics of 1616 through 1619 as the first such occurrence in coastal New England. Others, including the demographer Henry Dobyns (1983), have also recognized these outbreaks and their deadly effects and suggested that they were more likely an eruption of bubonic plague than smallpox.

Arrayed against, and forming a backdrop to, these debates are claims (for which there is only very limited and uncertain archeological evidence) that there were at least five major disease outbreaks in the northeast during the sixteenth century, including a smallpox pandemic in the 1520s, a measles outbreak in 1596, and three serious eruptions of unidentified diseases in the years between. Some suggest that these large claims, by Henry Dobyns and others, owe more to "faith than epidemiological principals [sic]" (Hackett 2002, 23; Henige 1986). All of this leaves open—to hesitant answer at best—the question whether an early (sixteenth-century) outbreak of chronic (such as tuberculosis) or vectored (such as bubonic plague or typhus) disease, or even the common cold or influenza, might have occurred in Acadia and been contained there. On balance, though, there is little evidence for significant outbreaks of introduced disease in the northeast in the sixteenth century.

There is no question about the eruption of smallpox in Massachusetts early in the 1630s, however. This first *recorded* instance of the disease in northeastern America was said to have destroyed entire native settlements—"not so much as one soul escaping Destruction" (quoted in Crosby 1986, 202)—and settlers of nearby Plymouth Plantation described its human toll in the most poignant terms. The indigenous people afflicted by it, wrote William Bradford, "fell down so generally . . . as they were in the end not able to help one another, no not to make a fire nor fetch a little water to drink, nor any to bury the dead" (ibid.). Alfred Crosby sees this outbreak "raging" through New England, moving west into the St. Lawrence–Great Lakes area and perhaps beyond, and "whipsaw[ing] back and forth through New York and surrounding areas in the 1630s and 1640s" to produce a 50 percent reduction in the populations of both the Huron and Iroquois confederacies (ibid.). Certainly the Huron and their neighbors were sorely afflicted by disease in the 1630s. But a more detailed consideration of the evidence regarding the patterns of disease in the northeast and the onset of afflictions among the Huron suggests a more complicated, and less conclusive, picture than that sketched by Crosby.

Because the European population of northeastern America remained relatively small and geographically dispersed into the eighteenth century, crowd diseases such as smallpox and measles did not become endemic there in the seventeenth century. With reductions in the sailing time required to cross the Atlantic after 1600, however, ships brought diseases into Boston, New York, and Québec on a regular basis. Thus, successive epidemics were triggered from external sources. Yet once they were ashore, introduced diseases often diffused rapidly from ports to frontier towns to indigenous populations, and were then carried further afield by the movements of native peoples. Sometimes diseases crossed from colony to colony—from Dutch to English to French territories—by such means. In the fall of 1634 an epidemic disease spread through all of the Huron villages. Rather than spreading from New England via the Mohawk as Crosby suggests, however, this infection probably reached the St. Lawrence from France in June. It killed significant numbers of Montagnais and Algonquian people at Trois Rivieres in July and then infected Huron traders who carried it back to Huronia. There, so many were laid low that fishing was curtailed and the harvest abandoned. Yet relatively few died. This, and the fact that Jesuit missionaries who contracted the diseases recovered within a few days, suggests that it was more likely measles or influenza than the mild form of smallpox that some have hypothesized.

Two years later influenza ran through several Huron settlements (as well as areas under Dutch and English influence) during the winter, and took a significant toll of lives in the spring when food supplies were low and people were at their weakest. Estimates, which are really no more than modestly informed guesses, suggest that between five hundred and a thousand people may have died. In the following summer an unidentified and more virulent illness that may have been scarlet fever spread among the Huron from the south, diffused into the Ottawa Valley, and moved on to Trois Rivières, killing more than previous epidemics had done. Then, in the summer of 1639, smallpox reached Huronia. It appears to have arrived from the St. Lawrence, whence it had been brought either from New England by Algonquians who had been among the Abenakis, or from France by ship. Its consequences were dramatic. By the spring of 1640, when the disease had run its course, Jesuits living among the Huron estimated that their population had declined by two-thirds, from 30,000 to 10,000. Modern estimates based on the careful consideration of several sources, suggest that both of these figures are high, and that the former should be reduced to about 21,000, with the latter at about 9,000.

Yet even this recalibration points to a population decline of close to 60 percent over seven years, with disproportionate numbers of older adults and children among the dead. Debilitating and dispiriting as this undoubtedly was, it was by no means unusual. Iroquois, Petun, and Neutral peoples suffered in like degree, even though epidemic cycles and the impacts of particular diseases differed among them. "It . . . seems reasonable to conclude," writes Bruce Trigger in his major study of the Huron People during this period, that they and the Iroquois, who had roughly similar numbers before the epidemics began in 1634, both "lost approximately one half of their population by 1640" (Trigger 1976, 602).

The contagion was not limited to Iroquoian speakers. Because Algonquian-speaking Ottawa and Nippissing were deeply engaged in trading with the Huron, and because they traded in turn with indigenous groups of the Lake Superior and James Bay areas, the influenza outbreak of 1636 and the smallpox of 1639 also ran inland. When the Huron were dispersed by the Iroquois a decade later, early sixteenth-century patterns of exchange among native peoples north of the Great Lakes were shattered, but this did not isolate those who lived in the area south of James Bay and north of Lake Superior (the "Petit Nord") from European diseases. Although few Algonquian groups traveled the Ottawa River after 1650 for fear of Iroquois raids, trading networks were quickly rebuilt along northern routes that connected the St. Lawrence with the interior via James Bay. Thus the trade in furs was sustained—and new conduits were opened for the passage of disease. When the establishment of English traders in Hudson Bay in 1670 led the French into more active engagement with the interior, the potential for disease transmission from the St. Lawrence was only heightened.

It was soon realized. In 1670 smallpox reached Sault Ste. Marie from Québec, from whence it had spread upriver and down in the fall of 1669. Scattered sources suggest that local outbreaks of unidentified diseases (perhaps colds or influenza) affected native peoples of the Lake Huron–James Bay area and others to the northwest in the years that followed, but little can be said of these events. It is clear, nonetheless, that half a century after the outbreak at Sault Ste. Marie, smallpox entered the Petit Nord proper from another quarter, the Hudson's Bay Company's York Factory. This was unusual, given the length of the voyage into the Bay, which was rarely less than fifty days and in this case was eighty-nine, but smallpox had been abroad when the *Hannah* left London, and soon spread to the Homeguard Cree around the post after her arrival in early September. Many died in the following winter, but because there were few contacts with more distant peoples at this season, the disease does not appear to have spread. Distance (or more precisely the time required for, and the timing of, disease introductions) spared inhabitants of this remote region from recurrent virulent epidemics well into the eighteenth century. Through this period, the Petit Nord remained on the periphery of the European disease pool.

Geographer Paul Hackett has provided a masterful account of the way in which this region shifted "from epidemic isolation to progressively greater inclusion within the disease frontier of formerly distant pools" in his book

"A Very Remarkable Sickness": Epidemics in the Petit Nord, 1670 to 1846 (2002, 237). This account is impressive, not only for its detail, but also for its insistence on the variety of local experience and on the importance of changes in settlement patterns, transport technology, and human movement within and beyond the region to understanding the timing, spread, and effects of epidemic diseases in native space. In sum, "occasional epidemics of varying severity"—including a virulent outbreak of smallpox (the "remarkable sickness" of Hackett's title) that ran through the western and southwestern parts of the Petit Nord and killed up to two-thirds of some of the groups that it afflicted in 1737–1738, measles, and acute respiratory diseases—affected this area before 1780.

In 1781–1783 smallpox reached into the region again, with devastating consequences. Entire bands succumbed, and estimates place Woodland and Lowland Cree losses at three-quarters of their population. Ojibway groups northwest of Lake Superior probably lost half to three-quarters of their number as well. The northern manifestation of a continent-wide pandemic that originated in Mexico City, this outbreak quickly surged north along the horse-trading network that linked Santa Fe and San Antonio with the Canadian Plains (see Fig. 9). When Assiniboine, Blackfoot, Cree, and Ojibway warriors attacked their Shoshone and Hidatsa neighbors who had been ravaged by the disease, they contracted the infection and carried it from the equestrian tribes of the plains to the hunter-gatherers of the parkland. Through the next two years, the same dreadful reports came from across the Petit Nord: "all the Indians on and near Lake Sturgeon are Dead"; "most of the Indians in and near the raney Lake is dead . . . the assineybols country is almost Depopulated" (quoted in Hackett 2002, 110, 107).

In the years that followed, the people of the Petit Nord were exposed to epidemic diseases with greater frequency, as the numbers of fur traders increased, and the growth of large cities on the American seaboard, the spread of white settlement into the Midwest, and improvements in transportation both created American hearths in which crowd diseases could become endemic and brought temporary disease pools and disease redistribution centers to the very edges of the region. To be sure, low population densities among native peoples limited the return rate of several diseases (because once they swept through a small community it required several years to reestablish sufficient numbers of "susceptibles" to allow an epidemic). But this was no defense against colds and influenza, which were recurrent. Measles and whooping cough both struck out of the east in 1819–1820, and produced very high mortality in some areas (estimates, and probably ratios, vary from one in four to one in two), and losses of 10 percent in other communities while bypassing some settlements entirely.

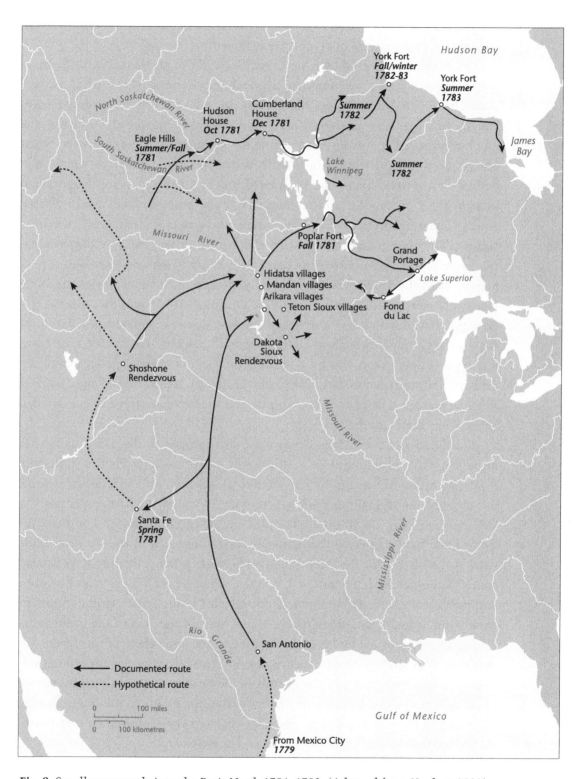

Fig. 9. Smallpox spreads into the Petit Nord, 1781–1783. (Adapted from Hackett 2002)

Although both outbreaks were so-called virgin-soil occurrences, neither disease affected the entire region or all of the people within it.

As Arthur Ray has shown in a parallel study of the spread of influenza, scarlet fever, measles, and smallpox through the western plains in the twenty years after 1830, diseases spread most often and effectively along trading routes, but the timing of any disease outbreak was crucial to its spread. Diffusion was much more rapid and widespread during the summer than in the winter, when native peoples had little contact with fur-trading posts and were widely dispersed. By the 1830s, however, improved linkages and rising settlements had brought the Petit Nord from the periphery to the hinterland of eastern disease pools. A greater variety of infections entered the area more frequently than ever before. Only the incidence of smallpox declined, driven back by the vaccinations that Hudson's Bay Company employees began to administer to native peoples in 1813. By 1839 they had probably treated most residents of the Petit Nord, and smallpox did not return to the area until 1876. Yet other disease epidemics, each of which left a different imprint upon the human population of the area, continued through the late nineteenth century. Tuberculosis, hitherto relatively unimportant, was by no means the least of these. By the turn of the century, and until it was beaten back by antibiotic treatments in the 1940s, it infected perhaps a third of those who lived in native communities across large parts of northern Ontario.

Inhabitants of the Pacific Coast were also ravaged by disease. Smallpox is known to have broken out in a number of locations between the Columbia River and Sitka, Alaska, in the late eighteenth century, although debate continues over the number, timing, and geographical extent of the first epidemics. On the evidence of a major smallpox epidemic in the Kamchatka Peninsula in 1768, some have suggested that Russian vessels involved in the sea otter trade carried the disease to the Aleutian Islands, from whence it may have spread south to the northern Tlingit by 1769. The evidence for this is thin, however, and many contemporary explorers of the North Pacific Coast attributed the introduction of smallpox among native populations of the area to the Spanish exploring expeditions led by Pérez, Bodega y Quadra, and others in the 1770s. Yet there are difficulties with this theory, too, not least that there is no explicit reference to smallpox in any of the records of the Spanish journeys. A third school of thought traces the diffusion of smallpox from the plains to the intermontane plateau and then to the coast, probably through the Columbia River basin. Again, the finer points of this argument are questioned.

Similar uncertainties surround the timing of the outbreak—half a dozen historical accounts yield dates between 1769 and 1780—and its geography, with some seeing discrete epidemics in northern and southern parts of the coast, and others arguing for a single outbreak that went unrecorded on the central coast because there were few European/American contacts with this area during the crucial period. The two most recent protagonists in the complicated debates over these issues, the geographer Cole Harris and the anthropologist Robert Boyd, also differ over the occurrence of a second smallpox epidemic in this area in the first decade of the nineteenth century. In the end, disputation over these finer details pales beside the simple and accepted fact that smallpox was "the first and by far the most devastating effect of Euro-American contact" on the indigenous peoples of this coast (Harris 1994; Boyd 1996; Boyd 1999, 21). On its initial occurrence the disease probably killed more than one in three of those among whom it appeared. It afflicted the largest and most densely settled indigenous populations in northern North America. Occupying coastal and riverbank sites and exploiting the enormously rich marine and other resources of this area in complex seasonal rounds that brought groups from different local areas into regular contact, the population of the geographically fragmented but not inhospitable territory between Puget Sound and the Aleutian Islands probably numbered 200,000 before it was thinned by disease.

Quicker and more frequent links with other parts of the world during the nineteenth century further broke down the "epidemiological isolation" of British Columbia and Alaska. Late in 1835 smallpox appeared in the Russian post at Sitka. By early January it had spread into the neighboring Aleut and Tlingit populations. Trading vessels shipped into and out of the port as the epidemic continued, and soon "rumor ha[d] it that the pox [was] doing its work in the inlets" (Boyd 1999, 116, 120). Russian-American Company efforts (initiated in 1808) to vaccinate native populations were stepped up, but often met resistance. Into 1837 smallpox continued to decimate native peoples: One estimate suggests one in four of the Sitka and Stikine Tlingit died of the disease. By September 1836 the disease had been introduced to the Hudson's Bay Company post at Fort Simpson by a vessel from Sitka, and by mid-October the pace of its spread (said a journal kept at the fort) was "getting rather alarming" (Boyd 1999, 124). It killed large numbers in the Skeena and the Nass basins.

All in all, concluded Governor Douglas of the Hudson's Bay Company, the epidemic probably carried off a third of the indigenous inhabitants of the coast between the Aleutians and central British Columbia. Mumps and influenza afflicted native populations in the vicinity of Sitka in 1843–1844 and 1845–1846, and measles spread northward from Puget Sound via the Hudson's Bay Company trading network in 1848. It affected the natives near Fort Simpson "with frightful rapidity," producing levels of "destitution, wretchedness and mortality" that Governor Douglas found "perfectly heart-rending" (Boyd 1999,

156–157). Perhaps 10 percent of affected native populations died, with a high proportion of young children among them.

Then, in 1862–1863, smallpox reappeared in British Columbia, to deal yet another powerful blow to native people in the colony. Estimates suggest that almost 14,000 died on the coast and as many as 20,000 overall. The initial outbreak, in Victoria, was mishandled, as infected native people were forced to leave town rather than being quarantined. Vaccination programs carried out by missionaries spared the Halkomelem peoples of the lower Fraser River and Strait of Georgia, and isolation protected the Nuu-chah-nulth. But north of this, into the territories of the Haida, Tsimshian, Gitxsan, and Tlinglit, death rates were high. A quarter century after the devastation, ethnographer George Emmons left a moving account of its impact with his description of a former Tlingit village off Prince of Wales Island: "It was a veritable city of the dead from which the survivors fled en masse after a visitation of smallpox. The houses with their interior carvings and totem poles still stood intact, left to weather and decay, but passing Canoes avoided that shore, while the occupants propitiated evil spirits with an offering of tobacco or food" (quoted in Boyd 1999, 201).

• • •

Vaccinations and evil spirits; missionaries and traditional beliefs: These juxtapositions open yet another set of windows onto the complexities involved in the contact process and offer a reminder that, for all their unsettling effects, the changes in native societies produced by the ravages of disease and the demands (and acquisitions) of trade were only a few of the many consequences produced by encounters between natives and newcomers. Contact is, in the final analysis, a cultural experience. It is shaped, resisted, and driven by clusters of attitudes, ideologies, and belief systems, each of which is subject in turn to adjustment and amendment according to shifting circumstances. In the end, the consequences of contact ramify and reverberate in a thousand directions. Its full and ultimate effects are beyond enumeration, and many of them lie beyond the compass of firm knowledge. But they cannot be discounted. As pioneer Canadian ethnohistorian Alfred Bailey recognized many decades ago, material displacements, economic disruptions, even the spread of drunkenness and disease precipitated changes in indigenous customs and manners—perhaps "especially conspicuous[ly] . . . in those appertaining to love and war"—that "reflected internal psychological turmoil" and "dissolved the social solidarity of groups" (Bailey 1937, 96). They were sufficiently powerful, he concluded, that they often destabilized native societies to the point of threatening their capacity to survive.

This is not to portray indigenous societies as passive recipients of European ideas or as inevitable victims of more powerful newcomers. For many years, indeed, newcomers depended in countless ways upon their indigenous hosts. Native peoples were also adept at borrowing and translating and adapting elements of newly introduced cultures for their own purposes. Customary practices were blended with beliefs promulgated by newcomers, and it was not always easy for Europeans to establish how firmly their attitudes and expectations permeated indigenous views. "You must know," a Huron told one of the French missionaries who came among his people, "that we have a 'yes' that means 'no'" (quoted in Grant 1984, 250). In religion, certainly, elements of Christian belief were incorporated into native practices to make them more acceptable to missionaries, and some professions of Christian belief may have obscured polite and subtle but nonetheless deliberate rejection of its basic principles. Through the earliest years of contact, natives and newcomers probably exchanged more than they knew, and syncretism—the fusion or reconciliation of diverse beliefs and practices—was widespread.

But such compromises were often undone by time and circumstances. When traditional remedies for illness proved less effective than those provided by Europeans, when age-old practices to propitiate traditional deities no longer produced anticipated results, the powers of elders and authority figures fell into doubt. There is perhaps no better encapsulation of these complex processes than that which John Webster Grant finds in the lines of the "Huron Carol" attributed to the Jesuit Jean de Brébeuf: "'Twas in the Moon of winter time, / when all the birds had fled, / that mighty Gitchi Manitou / sent angel choirs instead." Gitchi Manitou was an Algonquian, not Huron, deity, but Grant moves beyond this displacement to suggest that the angel choirs of Christianity came, in effect, to replace the guiding spirits of animate nature to which native peoples once looked for inspiration (Grant 1984, preface; see also Morantz 2001). As such strands of traditional belief unraveled, one by one, so the fabric of traditional societies was weakened and they lost coherence.

In the end, then, European contact had complicated, diverse, and deeply destructive consequences for the native peoples of northern North America. Population decline was undoubtedly foremost among these, and introduced diseases were instrumental in bringing it about. But devastating as they were for native populations, the impacts of introduced diseases should not overshadow other important consequences of contact. The coming of Europeans displaced native peoples, encouraged or forced their relocation, and undermined more or less directly and dramatically the ecological foundations of traditional ways of living on the land.

Like disease epidemics, each of these developments produced a cascade of social disruptions. Dietary changes; the loss of men and women in raids and military clashes attributable to trading rivalries and associated strategic concerns; the substitution of inadequate reserves for once expansive territories; the sense of disempowerment consequent upon the erosion of traditional beliefs and values; and the disenchantment, despair, and other turmoils to which native peoples were subject through the colonial period, all worked their effects into a virtually irresistible cumulative force for change. Not least, its consequences almost certainly included a reduction in the fecundity and fertility of indigenous peoples. Corresponding declines in birthrates would, in turn, have limited the capacity of native societies to recover from periods of heightened mortality. In sum, then, it is fair to say that colonialism was ultimately more insidious and more effective than introduced epidemic diseases in disrupting and destabilizing the delicate symbioses of land and life that characterized the native societies of northern North America before they felt the influence of Europeans.

PART 3

Settlers in a Wooden World

Early in February 1833, William Cooper wrote to his brother Christopher, of Graffham, near Petworth in England, encouraging members of his family to follow him across the Atlantic. William himself had left Sussex the previous year. Seven weeks to cross the Atlantic and five more to travel from Montréal into the interior had brought him to Adelaide Township, west of London in Upper Canada, where he had quickly acquired a hundred acres of land. Enthusiastic about his new surroundings from the first, he now sought to tell his relatives something of his new circumstances. "Here," he wrote, "is foxes, wolves and bears." But these were resources rather than threats. "You need not be afraid of them," he assured his English readers, for whom wolves and bears, at least, lived only in legend and fairy tale, for "they will shy off. We shoot them or catch them in steel traps."

There were also "plenty of deer, rabbits, black squirrels, rac[c]oons, porcupines [and] ground hogs" in this new land, reported William, although his claim that "they are all good for food" may have led some of his family to doubt the attractiveness of the place. No matter. "Birds of prey, eagles, two kinds of hawks, ravens, owls, turkeys, ducks, large partridges [and] wood pigeons" were plentiful. There was more snow than in England and nights were colder, but "we do not mind it; we have plenty of wood." His farm was coming along. He had built "a log house, 16 feet by 22," and planned to erect a barn, "20 feet by 30," in the summer. There was "plenty of keep in the woods" for stock. Two acres were sown to winter wheat,

97

and Cooper expected to have about four more acres cleared for cultivation in the spring.

To prepare the land, small trees were cut close to the ground, but "the big timber," he reported, "we cut about 2 feet high, and cut them into lengths, and draw them with oxen, to burn them." The soil was sandy and produced melons, cucumbers, and pumpkins as well as wheat. Yet this was only the beginning. "Bring garden seed of all sorts," Cooper urged his relatives, and "if any of you are coming out in the autumn, bring some apple pips, and pears, plum stones of all kinds, cherry stones, nectarines, peaches, gooseberry and white currant seed. I think," he added as a final enticement, "you will do better here than in England," as wages were high and "you may all draw land" (Cameron et al. 2000a, 97–99).

Cooper was only one among tens, indeed hundreds of thousands who came across the Atlantic or north from the Thirteen Colonies/United States to settle in northeastern North America—in the French colonies of Acadia and New France and the colonies of British North America—before these territories were brought together to form Canada in 1867. In many respects, his story is atypical. He was one of about 1,800 men, women, and children of the landless laboring class whose transatlantic movement was organized and underwritten by the remarkable Petworth Emigration Committee, which coordinated parish-aided emigration from Sussex between 1832 and 1837. In other ways, however, Cooper's account is thoroughly representative of attitudes held and revealing of processes repeated through successive generations of migration into this northern realm.

If not all newcomers were as fervently optimistic as Cooper about their novel circumstances, the vast majority of those who wrote of their experiences found promise in their new world. Resources of all sorts—land, timber, game—were abundant; access to them was relatively open. Settlement entailed hard work—almost everywhere forest had to be cleared before cultivation could begin—but yields were good, at least initially. Signs of improvement were palpable. Year upon year the edge of the forest was pushed back, by individuals and by the efforts of their neighbors. Useful introduced crops—wheat, oats, barley, apples, turnips, grasses, and clovers— replaced the dark and seemingly endless canopy of trees that shrouded the land.

A bush farm near Chatham, 1795. *This is what it meant to be a settler in a wooden world. A small log cabin is the center of a tiny clearing in the surrounding, and enveloping, forest. A wood rail fence marks the bounds of what may be a modest garden between forest and stream. A wooden bridge crosses the water, and oxen draw a wooden cart across the bridge. Beyond, a stump-strewn "field" suggests both the labor invested and the work yet to be done to establish a viable farm in the "woods" of Upper Canada. (National Archives of Canada)*

Decade by decade the land was transformed. Roads were built, mills were erected, towns sprang up, and cities grew. Fueled by immigration and natural increase during the first half of the nineteenth century, the rise in population and the conversion of "waste land" to sown proceeded so swiftly that some took the history of Upper Canada, in particular, to epitomize the doctrine of "progress" that seized the imagination of the Victorian age. For all that, reflected Thomas Adsett, another migrant from the long-settled, intensely cultivated Sussex countryside where brick and stone were the building materials of choice, people in this new territory lived "in quite a different way to what they do in the old country . . . this is truly the wooden world" (Cameron et al. 2000a, 45).

8

POSSESSING AND (RE)PEOPLING THE LAND

Exploration and trade worked in tandem to foster European engagement with northern North America (Hornsby 2004). Early staple trades in fish and fur brought Europeans to the northeastern corner of the continent and spawned toeholds of settlement there, and these encounters had conceptual and perceptual consequences. As new territories were explored, often to find routes to riches elsewhere or to discover and catalogue useful resources in (and information about) the places encountered, lands and waters were described, named, and mapped—and thus claimed in symbolic and practical ways. In this sense, exploration was not simply a matter of going where none of one's own kind had been before. It was part of the process of possessing.

Whether newly applied names were redolent of the old world ("Halifax") or the new ("Canada"—said to derive from the native *kanata*, meaning village or settlement), they were assigned and/or adopted by colonizers. They were expressions of authority and power. Mapping similarly converted and transcribed space into registers familiar to Europeans and unknown (at least initially) to native peoples. Names helped to domesticate territories. Maps brought them into the orbit of European authority (Barkham 2001; Clayton 2000; Harley 1988; LaTour 1987; MacMillan 2001; Zeller, 1997). By ordering and organizing, both contributed to the sense that land was there for the taking. They made space available. They initiated and eased its conversion from wild to tame, they laid the foundations for improvement, and they paved the way to transformation.

Economic and strategic considerations and the extension of European geopolitical rivalries to New World territories also hastened this process. From tiny strategic outposts in Tadoussac (1600) and Québec City (1608) the French encouraged agricultural colonization along the St. Lawrence. From 1627 the Compagnie des Cent-Associés was required to bring settlers to the region in return for its monopoly trading rights. By 1663, when the Compagnie's charter was revoked, there were some 2,500 people of French descent along the lower St. Lawrence. In the next decade or so, the French government encouraged military personnel to remain in Canada and sent contingents of settlers (including approximately a thousand women, three times the number who had arrived before that) to the colony. These levels of in-migration were not long sustained, but with high birthrates and low death rates, natural increase was a powerful expansionary force.

Although fewer than 10,000 Europeans (almost all of them French) came to settle in Canada in the 150 years after 1608, there were, by 1760, approximately 70,000 French-speaking settlers along both banks of the St. Lawrence River, from just above Montréal to a considerable distance below the city of Québec (Courville 1996). In the Bay of Fundy, small fur-trading ventures also evolved into agricultural settlements. From the Port Royal district descendants of a few dozen founding families colonized adjacent areas. By 1700 there were approximately 1,400 people, known collectively as Acadians, in the Bay of Fundy. Half a century later their number topped 12,000 and they had cultivated land in Ile Saint-Jean (later Prince Edward Island) and Ile Royale (Cape Breton Island). In sum, there were some 80,000 to 85,000 French settlers, almost all of them farmers, in Canada, Acadia, and the Gulf of St. Lawrence in 1748. British numbers were minuscule by comparison. When the Treaty of Utrecht established British control of Newfoundland in 1713, settlement there proceeded, but the island had barely 5,000 residents in the 1740s. In Nova Scotia—the peninsula section of the larger French territory of Acadia also transferred to British control in 1713—there were even fewer British settlers (numerous plates in the *Historical Atlas of Canada*, vol. 1, provide more detail about population numbers and locations).

These circumstances were radically altered by Anglo-French hostilities in the 1740s and 1750s, and the Seven Years War of 1756–1763. Louisbourg, the imposing fortress built on Ile Royale by the French after 1717, was captured by New Englanders in 1745 and returned to France in 1748. To balance its power and provide a more significant British presence in Nova Scotia, Halifax was established in 1749. In 1755 the British deported thousands of Acadians from their lands on the Bay of Fundy; most others fled. Louisbourg was put to the torch by British troops in 1758. British forces entered the citadel of Québec in 1759 and Montréal in 1760.

French fur-trading posts south of the Great Lakes were abandoned; there were battles between local native groups and the British and Americans who replaced the French. By the Treaty of Paris in 1763, France ceded all of its North American territories, save for St. Pierre and Miquelon and fishing rights on part of the Newfoundland shore. Britain claimed all of North America east of the Mississippi, as well as all of the territory draining into Hudson Bay. In October 1763 the Royal Proclamation defined the boundaries of the new colony of Québec; enlarged Nova Scotia to include present-day New Brunswick, Prince

Edward Island, and Cape Breton; and set aside all remaining lands west and north of those draining into the Atlantic, south of those draining into Hudson Bay and east of Louisiana as Indian Territory.

Just over a decade later, the Québec Act guaranteed Roman Catholics the right to practice their religion at the same time as it truncated Indian Territory and extended the boundaries of Québec south of the Great Lakes. This concession to papal religion was one of the "Intolerable Acts" that precipitated the American War of Independence and, with the Treaty of Paris, another redrawing of the North American map. In 1783 Québec encompassed the St. Lawrence and the north shore of the Great Lakes; to the south the newly independent United States extended to the Mississippi (see also Demeritt 1997).

In these years of political turmoil, British efforts to consolidate their territorial claims brought new settlers to Nova Scotia from Europe, New England, old England, and Scotland. By 1775 there were approximately 20,000 non-native settlers in this region, and that number almost tripled early in the 1780s, when Loyalist refugees from the American Revolution fled north to British territory. With the addition of a few thousand Scots in the last decade of the century, the population of the seaboard colonies of British North America numbered 75,000 to 80,000 people in 1800 (see Fig. 10). On the St. Lawrence at this time there were approximately 215,000 people, most of them rural and of French descent. Loyalist refugees and land seekers formed the majority of the 30,000 to 35,000 non-indigenous people who lived in that area of the Great Lakes peninsula designated Upper Canada by the Constitutional Act of 1791.

With approximately 350,000 non-indigenous inhabitants scattered across seven colonies, British North America was a string of underdeveloped territories claimed by, but only tenuously linked to, Britain. Well over half its population could not speak English and, insofar as they recognized their official status as British subjects, it was probably with regret, if not resentment. People generally garnered a very modest subsistence from the land, but beyond Lower Canada (which exported 400,000 bushels of wheat, flour, and biscuit in 1800) the rural economy was weak. By and large, there were few surpluses, and the limited trade that there was remained essentially local. In the Atlantic colonies where many still lived in rockbound settlements tied to the fishery, the population as a whole could not feed itself; agricultural imports were substantial. Chance rather than design had formed this necklace of scattered settlements in a land of bounded possibilities (see *Historical Atlas of Canada* 1: plates 41–53 and 68 with accompanying text for specific details).

Preoccupied with war against France and still substantially wedded to the view that the power and wealth of any country was directly related to the size of its population, British politicians saw little utility in their North American

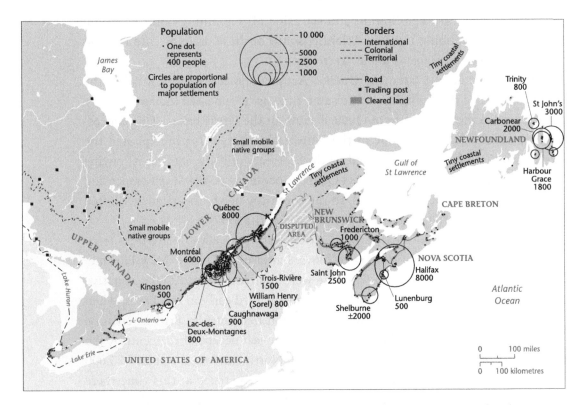

Fig. 10. *British North America in 1800. (Adapted from* Historical Atlas of Canada *1: pl. 68)*

colonies. These territories were regarded, in effect, as necessary encumbrances rather than as valuable assets. Only with the pressing strategic need to find new sources of wood, provoked by Napoleon's continental blockade, which cut off British access to Baltic supplies, did this begin to change. The transatlantic trade in timber that emerged as British wood prices skyrocketed after 1806 conferred new and sudden importance on New Brunswick and the Canadas.

When the Battle of Waterloo brought the war with France to an end (leaving thousands in Britain unemployed), and cheap westward passages in the fleet of timber ships that furrowed the Atlantic opened the way to exodus, migration quickened. In the fifty years after 1815, 1,250,000 people left Britain for British North America (Cowan 1928). This influx had an immense effect upon the colonies. Most newcomers avoided Québec, which is to say that they settled in the most thinly populated territories. Most of those who came were young and married young, which is to say that fertility rates were typically high and mortality rates relatively low in immigrant settlements.

Upper Canada, which attracted most of the migrants, increased its population from about 60,000 in 1812 to 158,000 in 1825, and topped the million mark in the 1850s. Growth was less spectacular—but still remarkable—in New

Brunswick, Nova Scotia, and Prince Edward Island. Together they grew from 80,000 at the turn of the century to approach 200,000 in the mid-1820s and to exceed half a million in 1851. In Newfoundland, the population quadrupled in the three decades after 1805, to reach 75,000 in the mid 1830s. In Lower Canada high rates of natural increase continued to drive population totals upward at an extraordinary rate. From 215,000 in 1800, the population of this colony climbed to 480,000 by 1825, approached 900,000 in 1851, and exceeded a million later in the decade (*Historical Atlas of Canada* 2: plates 4, 7–10).

In all of these territories, the number of indigenous people declined in real terms and their relative importance fell dramatically. In Newfoundland the last of the Beothuk people died in the 1820s, and there were probably no more than 150 Miq'mak in the colony in the 1850s. There were some 1,400 Miq'mak in Nova Scotia, a similar number of Miq'mak and Maliseets in New Brunswick, and about 300 Miq'mak in Prince Edward Island in the 1840s, but this was considerably less than 1 percent of the population in each jurisdiction. Estimates place the native population of Lower Canada at 16,000 in the 1780s and 15,000 a half century later. In Upper Canada a native population estimated at 14,000 in 1770 had fallen to 9,300 by 1835. Measured against non-native populations, these were minuscule fractions.

Beyond the Great Lakes, however, relative proportions were reversed. Although introduced diseases significantly reduced indigenous numbers in the eighteenth and nineteenth centuries, native peoples dominated the western interior, the far west, and the north throughout the nineteenth century. All estimates of their numbers are uncertain, but if there were 125,000 native peoples in the northern part of the continent in 1867, as often claimed, four-fifths lived west of the Great Lakes lowland occupied by newcomers. In this enormous area, the European presence was limited to modest numbers in far-flung fur-trading posts and to the Red River colony founded in 1812 to provision the fur trade (Ross 1970). Neither agriculture nor the continuing recruitment of European settlers to that colony was especially successful, however, and over the years, as European men married native women, a new people, the Metis, came to dominate the area south of Lake Winnipeg. Their number increased from 600 families in 1835, to 2,500 (approximately 12,000 people) by 1870 (Brown 1980; van Kirk 1980). Only across the cordillera was this pattern of minuscule European numbers in predominantly native space disrupted when in 1858 some 30,000 gold seekers flocked into the Fraser River valley.

Claiming territory was one thing, administering it entirely another. Early in the process of colonial expansion, imperial authorities struggled to find ways of utilizing enormous expanses of New World land effectively. Although strategic considerations often ordained territorial occupation, this was not always readily accomplished. In the sixteenth and seventeenth centuries, monarchical power was almost absolute, but it was not easily implemented beyond the orbit of the court, and its capacity to reach across oceans was sorely limited. For these reasons, responsibilities and opportunities for colonization were often delegated to individuals or companies of private adventurers. But most of these initiatives gained scant success in northern North America. Predicated on the establishment of particular concepts of property relations—and of systems for their administration—developed in the Old World and radically different from those adhered to by the indigenous inhabitants of these territories, transplanted property systems had to be adjusted to changed circumstances, even as they substantiated the notion that the land was a commodity—something that could be claimed and owned and utilized and transferred to their benefit by specific individuals or institutions—and undermined indigenous concepts of and rights to territory (Weaver 2003; see also Fig. 11).

In New France, as in most New World locations, the immediate and practical challenge of advancing settlement was to establish people on the land. Because the Compagnie des Cent-Associés received and complied to some extent with a feudal charter that required it to grant lands along the St. Lawrence to seigneurs—who could in turn grant subseigneuries or concede or lease their land as farms to individuals—this property system became the basis of land allocation along the St. Lawrence, even though only ten of the seventy seigneuries granted by the Compagnie had any settlers in 1663. By 1760 approximately one hundred seigneuries had been established, most bounded by the river at their front and the Precambrian Shield or the Appalachian Highlands to the rear. As the population increased, these seigneurial lands were taken up, typically in long, relatively narrow farms (rotures) approximately ten times as deep as they were broad (see Fig. 11).

Initially, rotures fronted on the river. With time, in places, a second line, or range, of similar properties might be developed along a road line, back from the river. This was not an entirely regular system, however. Rotures varied in size, with parcels of 60 to 120 arpents (20–40 hectares) the statistical standard. Where practical, rotures or concessions were laid out along tributary streams, and because river courses are not straight lines, interior boundaries sometimes ran into one another. The habitants who occupied the concessions were subject to various seigneurial charges and obligations, but so long as they met these they had security of tenure and effectual rights to their property, which they were allowed to sell or pass on to their heirs (*Historical Atlas of Canada* 1: plates 51, 52).

Elsewhere, radically different arrangements were implemented. In Newfoundland outports, formalities of land title and administrative surveillance

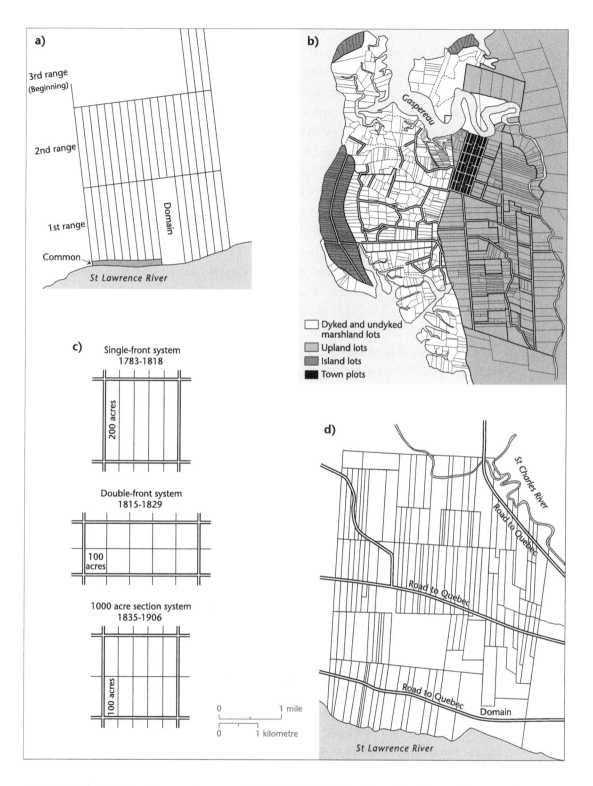

Fig. 11. *Cadastral (land survey) patterns in British North America: the seigneurial system in theory (a) and practice (d); Horton Township, Nova Scotia (b); survey systems of Upper Canada (c). (Adapted from Harris 1966; McNabb, 1986; Harris and Warkentin 1974)*

were largely absent. There settlements grew organically. Families who had claimed land by occupying it spawned new households whose members took up land (and were allocated beach room essential to the fishery) by local agreement. No official deeds or formal records marked the ownership or (often irregular) boundaries of these properties but they were lodged in the collective memory of the community (Faris 1972; Mannion 1976). For the New Englanders who came to Nova Scotia in the 1760s, the most sensible system of land allocation was the one they were familiar with, itself a transplant from an earlier England. Typically, these newcomers first classified the land according to type and location (to establish several "divisions" or sections), and then divided these areas into lots of different sizes. Decisions were then made about the amount of land to be included in each "share" of the township, about the composition of a share, and about the amount of land (the number of shares or fraction thereof, from 0.5 through 1.0 and 1.5 to 2.0) that each full member of the community—each proprietor—should receive.

In the township of Horton, a single share was deemed to be 500 acres. This included specified amounts of land in each division, including a town lot (0.5 acre), a dike lot (4 acres), an island lot (4–8 acres), and upland lots, each different in area, drawn from three distinct divisions of the township (see Fig. 11). The particular lots that constituted each share were selected by ballot. Recognizing that the quality of land in each division varied, the proprietors also set aside a certain amount of land (so-called size land) for deliberate allocation, as they considered necessary, to render the overall value of each share more or less equal (McNabb 1988).

By the beginning of the nineteenth century, such intricate, locally defined and geographically alert methods of land allocation were essentially things of the past. The challenge of settling thousands of Loyalists on the land in short order was met, in part, by implementing a quadrangular survey system. Equipped with relatively rudimentary equipment (compasses and measuring chains), government and military surveyors divided larger areas known as parishes or townships geometrically, into row upon row of uniformly sized lots. Privileged individuals might obtain two or more lots, but the basic grid became the norm. As implemented in Township 1 Cataraqui (later Kingston) in 1783, twenty-five quadrangular lots of 120 acres each were laid out along a straight baseline on the shore of Lake Ontario. Then the process was repeated on their inland boundary, and repeated again inland until seven lines (or concessions), and 175 lots, had been marked out.

Like the inhabitants of Orwell's *Animal Farm*, however, some of these uniform lots were more equal than others. Fewer than half of the baseline lots had water frontage on the lake; lots on both sides of the Cataraqui River where it flowed through the first two concessions were mostly marsh and swampy ground; on the western edge of the township, where the survey extended across an embayment of the lake, three lots had considerably less than 120 acres of land. No matter. The survey could proceed quickly; every parcel of land could be identified by simple coordinates; people could be assigned to properties—and vice versa (*Historical Atlas of Canada* 2: plate 7).

The fine grain of this survey pattern was subject to many amendments over the years. But some basic principles prevailed. In Upper Canada land had to be surveyed before it was granted. After 1789 townships generally conformed to one of two standards. Along navigable waterways they were nine miles by twelve miles; elsewhere they were ten miles square. Within these frames, different arrangements were implemented. Lot sizes—and arrangements for acquiring title to the land—varied considerably. But the broad patterns are clear. Before 1815 most surveys generally followed the Cataraqui model with rectangular-shaped lots (generally of 200 acres and an approximate breadth to depth ratio of 1:5) separated by road allowances. Between 1815 and 1830 approximately square, 100-acre lots shared a boundary between road lines (see Fig. 11). Thereafter, the survey configured 100-acre lots in 1,000-acre blocks, so that they were broadly two and a half times as deep as they were broad (Harris and Warkentin 1974). Similar survey patterns were implemented in the non-seigneurial lands of Lower Canada (such as the Eastern Townships), and less widely and less uniformly in the Maritime provinces during the nineteenth century. They would also, in time, define the geometry of the western prairie landscape.

All of these survey systems opened areas for agricultural settlement, bounded properties, eased the transfer of land to individuals, turned the earth into a commodity (because land deeds and official records smoothed operation of the land market), and imprinted a particular, usually lasting, order on the landscape. Still, their differences are important. At one level, the shift from the customary family-centered arrangements of the Newfoundland outport, through the locally notarized allocation of seigneurial rotures and the community-orchestrated patterns of property division in Horton to the colonially administered land system of Upper Canada, represents an evolution from dispersed and informal to centralized and authoritarian management of the colonial land base. At another, it also marks a loss of environmental sensitivity in the process of land allocation. As the state tightened its grip over the colonial estate, it became a great deal easier to "put people on the land"—but there was far less flexibility for individuals to define the parameters of the land they received.

Although the long-lot became the de facto standard of seigneurial Lower Canada, its proportions were never officially defined or imposed. There was nothing in the seigneurial system that dictated its form. It was, rather, a response to local conditions—a socially, economically, and environmentally sensible adaptation to pioneering circumstances that linked habitant families to, and via, the river while ensuring their access to the resources provided by a range of ecological niches upon which their survival depended: fish and eels in the river, marsh hay on its edges, meadow land, flat fertile soils for agriculture, and forests for firewood, forage, and the opportunity to hunt for game (Harris 1966; Courville 1990b). By much the same token, the proprietors of Horton devoted time and resources to defining the divisions of their township and took pains to ensure that all members of the community received appropriate access to the different environments it encompassed. There were costs involved in this: The survey took years to complete, and individual holdings were made up of widely dispersed lots. But the entire design reflected a commitment to community ideals and to ensuring that all members of that community possessed the means to realize a "modest competence."

Time and experience revealed that these principles were often honored in the breach, as lands were traded and settlement dispersed from the "town lots" intended to concentrate people in a village. Like the attention paid to the implications of environmental differences, however, they were central to the design. The grid system, by contrast, met the immediate needs of families and individuals more effectively than it fostered community, and paid scant attention to environmental variability. The grid survey increased the legibility of the landscape, but it did so only by simplifying it—by subjugating its intricate ecological complexity to a rigid geometric order. The consequences were profound. As the grid paid no heed to drainage patterns, relief, vegetation, or soil, it marched indiscriminately across ecological (or bioregional) boundaries and created individual lots that varied considerably in quality.

To be sure, many of those who settled British North America were able to select the land that they would occupy. Common lore (country wisdom or local ecological knowledge), which used vegetation as an indicator of land quality, helped in this: In broad terms much qualified by local circumstances, mixed hardwood was taken as a sign of good wheat land; pine, cedar, and hemlock betokened poor land (Kelly 1970). Surveyors also assessed and categorized the land as they moved through it. Some spoke of "warm" and "cold" soils. But many settlers lacked means to access this information, opportunity to inspect prospective locations, or the knowledge to exercise wise choices if they did. Indeed, this new world sometimes seemed determined to perplex: "The land varies very much in difference of soil in the space of half a mile," reported one recent arrival. "In some places nothing but sand—then light, rich mould—and then fine, deep, black mould; and, all of a sudden you come to stiff, hard clay"

(Cameron et al. 2000a, 189). Perhaps those who were simply assigned properties by officials or by drawing lots suffered relatively little disadvantage.

Besides, the utility of a place depended upon more than the quality of its soils. Access to water, markets, neighbors, and services were all important, and these were often imponderable; in its social and economic dimensions, location was a shifting and uncertain concept on the developing edge of the ecumene. For all of its effectiveness as an administrative instrument, the grid was essentially silent on these other matters. Even with title deed in hand, settlers faced the age-old challenge of making the land their own.

9

DOMESTICATING THE LAND

Breaking and taming the New World environment, shaping it to the newcomers' needs and desires, was an enormous task. Almost everywhere across the northeastern foreland of the American continent, settlers had to turn forested acres into farmed fields; erect houses, barns, and fences; and build roads and bridges. Mills, churches, villages, towns, and, less tangibly but equally importantly, communities had to be brought into being. Successful settlement required massive material transformations of the environment. These depended, almost exclusively, upon the muscle power of human and animal bodies, the kinetic energy of running water, the consuming power of fire, the release of heat (stored chemical energy) from organic material such as wood and coal, and to lesser extent the harnessing of the wind.

Years ago, Lewis Mumford described the long period of human history in which such circumstances were characteristic as the Eotechnic era, or more prosaically as the age of "wood, wind, and water." More recently, scholars interested in energy consumption rather than technological development have taken such heavy reliance upon animate sources of energy for the generation of mechanical power to define "the somatic energy regime" (McNeill 2000). By their estimates, human muscles generated more than 70 percent—and plant, animal, and human energy combined more than 80 percent—of all the mechanical energy used in such preindustrial circumstances (with the remainder coming from water and wind power).

Before 1850 virtually all inhabitants of the territory that became Canada lived within these approximate constraints. In very practical terms, this meant two things. First, that the harvests of agriculture and of "nature" (in the form of fish, birds, wild berries, and game) were absolutely crucial to the development of early Canada because they fueled humans to convert the chemical energy stored in plants and animals into the mechanical energy required to do work. And second, that the creation of farms from forest, or the more general transformation of the material world, was accomplished in very large part by the muscle-burning, physically tiring, seemingly almost unremitting labor of men and women toiling individually or in groups.

Chopping, burning, hauling, digging, lifting, piling, plowing, spreading, beating, scrubbing: In one form or another, these were tasks that filled settlers' days, from the seventeenth through the early nineteenth centuries. Varied as they were, their impacts were enormous. Cumulatively they substantially remade the landscape. But they were not the only instruments of change. Fire and flood were harnessed to human ends. And the effects of human disturbance, whether direct or indirect, were not always those intended. In seeking to change the face of the colonial earth to their purposes, settlers often, and often unknowingly, triggered cascading consequences that subverted their hopes and designs and had long-term, frequently deleterious, impacts upon environments and ecologies. Moreover, because individuals generally acted locally, and because the accumulating effects of local actions were often manifest beyond these immediate spheres, particular cases must exemplify larger themes and the cumulative effects of trivialities must be generalized into significant broader patterns to understand the worlds that settlers made.

In the 140 years before their expulsion from Nova Scotia in 1755, the Acadians transformed the littoral landscape in several parts of the Bay of Fundy by reclaiming rich marshland from the surge of the tides. This was both a remarkable and an exceptional accomplishment. In recent years, Matthew Hatvany and others have shown that salt marshes were reclaimed and used by settlers elsewhere on the northeastern seaboard, in Prince Edward Island and along the St. Lawrence, but nowhere were the reclamations as extensive or as central to the developing economy as in Acadia. Facing the world's highest semidiurnal tides (with an amplitude of 50 feet or 15 meters), as well as the complications implied by very high ("spring") tides twice a month, and even greater long-term cyclical tidal fluctuations (every 207 days, 4.52 years, and 18.03 years), a small population, working collectively, built and extended a complex network of dikes. Decade after decade, as Sherman Bleakney has demonstrated for the extensive salt marsh meadow on the Basin of Minas known as Grand Pré, incremental additions were made to the area of reclaimed marshland. By the 1750s Acadian reclamation works encompassed thousands of acres of land (Clark 1968; Griffiths 2005; *Historical Atlas of Canada* 1: plate 29).

On these fields domesticated from the sea, the Acadians grew hay and a range of other food crops, once salts were leached from the fine, organically rich soils by two or three years of rain. Indeed, Acadian homes and villages skirted the dikelands, and they made few inroads on the upland forests that settlers elsewhere worked so long and hard to remove. Great skill, as well as tremendous labor, went into this reclamation effort. The basic techniques of reclamation were likely transferred from Europe, but their implementation on the

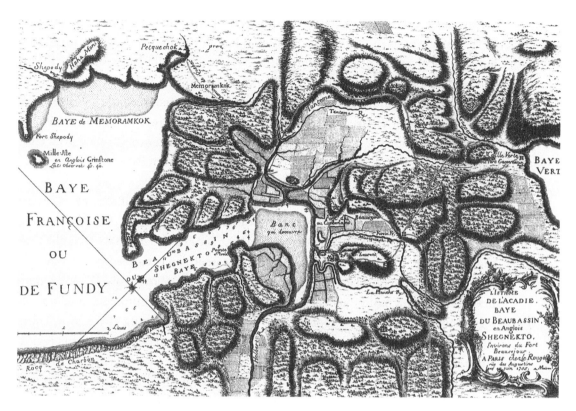

Acadian lands at the head of the Bay of Fundy, 1755. *This beautiful map illustrates something of the intricate patterns of Acadian settlement produced by the reclamation of salt marsh lands on the Bay of Fundy. Here at Beaubassin, at the head of the bay, Acadian fields protected by dikes and drained by aboiteaux (sluice gates with one-way clapper-gate valves) formed the productive hearth of a distinctive agricultural society disrupted and displaced by the Grand Dérangement, or expulsion of the Acadians from their ancestral lands, by the British in 1755. (National Archives of Canada)*

shores of the Bay of Fundy required enormous ingenuity and a remarkable level of local knowledge and adaptation (Butzer 2002).

Dikes were constructed with sods of marsh grass turf, accumulated through centuries of rising sea levels at a rate of about a meter per hundred years. Each sod was cut, carried, piled, and beaten into place on dikes that were probably at least 6 or 7 feet high and 15 feet wide at the base. Reinforced as necessary with horizontal and vertical timbers, these dikes were built near the upland margins of tidal flats. Their utility was crucially dependent upon the dense, wiry, persistent root-mat of the marsh grasses *Spartina patens* (salt meadow hay) and *Juncus gerardi* (black grass), which rendered *Juncus* sods, in particular, highly resistant to erosion.

Because dikes were necessarily impervious in order to hold back the sea, fresh water would soon have accumulated behind them, flooding the reclaimed land. To deal with this, Acadians built *aboiteaux* in the bases of their dikes. These drainage sluices included one-way, clapper-gate valves through which fresh water could drain at low tide. Substantially isolated from the main lines of seventeenth- and eighteenth-century commercial interest in the New World, and largely ignored by officialdom until the eighteenth century, this distinctive, rapidly increasing population created an extraordinary landscape in this corner of the New World.

Traveling along the St. Lawrence in the middle of the nineteenth century, the American author, naturalist, and transcendentalist Henry David Thoreau, who described himself on this occasion as a "Yankee in Canada," thought that the landscape "appeared as old as Normandy itself, and realized much that . . . [he] had heard of Europe and the Middle Ages" (Thoreau 1866, 75, 36). There may have been a malicious edge to the implication that "humble Canadian villages" were medieval in appearance, but, in characterizing the region thus, Thoreau also pointed to the scale of the changes that had occurred in this region since Europeans first settled the banks of the river. The Swedish botanist Per (or Peter) Kalm, who preceded Thoreau along this way by a century or so, noted even in 1749 that there were cultivated fields on both sides of the St. Lawrence below the settlement of Québec, "but more on the west than on the east side."

The hills, he continued, "on both shores are steep and high" and a "number of fine elevations separated from each other, large fields which looked quite white with the grain that covered them, and excellent woods of deciduous trees, made the country round us look very pleasant. Now and then," Kalm's party "saw a stone church" and noted that habitants had "erected sawmills and grist mills" on some of the large tributary streams (Kalm 1770, 2: 479–480). By contrast, Cartier, Champlain, and others who sailed the St. Lawrence until the 1660s generally found only an abundance of trees, sandbanks, meadows, and "very beautiful" forests to draw their attention.

The continually repeated processes that brought about these changes (and so misled Thoreau) have been well sketched by Colin Coates for two seigneuries on the north bank of the St. Lawrence between Québec City and Trois Rivières. Batiscan and Sainte-Anne de la Pérade were granted by the Compagnie des Cent-Associés and the Crown in the seventeenth century. Within a generation of the first habitant settlement of the area, an observer noted that "the first concessions lack wood. They are obliged to procure it in the depths of the seigniory and on the south side of the river" (quoted in Coates 2000, 36). Early in the eighteenth century, seigneurs and habitants disputed ownership of timber and accused individuals of pilfering what was not theirs.

Although estimates suggest that only a quarter to a third of the land along the St. Lawrence was under actual cultivation by this time, the clearing process was neither neat nor orderly, and large tracts of forest were quickly destroyed. "The new habitant sets fire to as many trees as must be uprooted for arable land," wrote a European military man resident in the area in 1775–1776, and the forest was soon a "quarter-league" distant from their dwellings. The result, he continued, was that "the woodland . . . appears *scandaleux;* and one thinks that fire must have fallen from the heavens into the woods, when one sees half-burned, half barren, and entirely barren trees therein" (quoted in Coates 2000, 37). In time, these scarred and dead trees would fall or be removed, until, as Joseph Sansom noted in his general *Sketches of Lower Canada* in 1817, "The trees are all cut away around Canadian settlements, and the unvarying habitations, stand in endless rows, at equal distances, like so many sentry boxes or soldiers' tents without a tree, or even a fence of any kind to shelter them" (52).

Cleared of its trees, the land was open to cultivation. This usually began in untidy fashion. Fields were replete with stumps, and unwanted species competed with crops. These included both indigenous plants seeded from nearby forests and members of the ever-active platoon of invasive weeds. Indeed, *portulaca* appeared in the grain fields of Canada as early as 1632. Some fields were cropped to exhaustion, then allowed to revert to bush; others were sown to cereals for several years before being turned to pasture or fallow (both of which resembled fields of weeds). Despite the power of fire, "indigenous nature"—the plants of the forest and its margins—was difficult to subdue. Labor was scarce, and there was much work to be done.

Before long, however, the land was won over. Its production was "Europeanized." Native species gave way to imports. Wheat, oats, peas, barley, rye, buckwheat, and flax were quickly and consistently far more important, across the colony, than maize (which by Coates's calculations never amounted to more than 8 percent, by volume, of wheat production in Batiscan and Sainte-Anne). Small quantities of maple sugar were made, and some tobacco was grown; squash and beans found their place in habitant gardens. But unless one counts potatoes (which had long been a staple of European diets) as indigenous to North America, most of the crops grown and consumed by habitants even in the seventeenth century were Old World staples. Much the same was true with regard to meat consumption. Although indigenous birds and game were significant elements in seventeenth-century diets along the St. Lawrence, the meat of imported cattle, sheep, and pigs soon exceeded them in importance. Even early in the eighteenth century, Coates concludes, habitants ate little moose, deer, or bear flesh. Pigeons, ducks, geese, eels, and tommy cod remained significant, but by the 1770s there was little wild game near the St. Lawrence settlements.

Along the shore of the river, just below Québec City, the gains of generations of cultivation of the land were evident in the pictures drawn in the 1760s by military men after the fall of Québec to British troops. The amount of land under cultivation was not large. Forest clothed the thin soils on the southern slopes of the Shield. But between water and woods, labor had been invested in fields, gardens, and houses. Here families sustained themselves and most likely had every expectation of continuing to do so. There were limits to prosperity and possibilities. But they were more evident to later generations than to those of the early eighteenth century. In very general terms, early eighteenth-century rural Canada was not the creation of people in pursuit of profit so much as a niche for the reproduction of families, a place laboriously shaped and cared for by its inhabitants.

In historical geographer Serge Courville's terms, the countryside of New France in the seventeenth and eighteenth centuries is best understood as a distinctive cultural space created by the French state's desire to "subordinate the development of the colony to metropolitan needs" and the settlers' "desire to take advantage of new-world opportunities." Two logics interacted: *"urbanite,* of French origin and the source of macro-forms and macro-structures," and *"territorialite* originating in the intimate local places shaped by habitants" (Courville 1990b, 165–166). None of this is to suggest that society and countryside in early New France were egalitarian and undifferentiated, or that all early settlers along the St. Lawrence were subsistence farmers with no interests in or links to markets. Although many earlier historians implied as much when they wrote of the farmers' "attachment to family and subsistence" in "the one continued village" of early eighteenth-century Canada, sparse data and intense local variations in circumstances, attributable to environmental and demographic conditions among others, make such claims hard to prove. More recent work has at least pointed to the existence of significant wealth disparities among the *censitaires* (peasantry) of Canada and suggested that some produced significant surpluses while others struggled (Desbarats 1992, 18).

As families increased and young men and women began families of their own, the seigneuries filled with people. As in many agricultural societies, including early modern England, this demographic shift threatened to undermine the foundations of society. Finely attuned to local conditions and depending upon the delicately balanced exploitation of the resources of wood, water, and land by people dependent upon the power generated from the muscles of humans and animals and the movement of air and water, the agricultural economy of the old, settled part of the St. Lawrence was unable to generate sufficient economic growth to sustain a rapidly rising population. For some time, population pressures were dissipated by the availability of new land. Thus, the cycle of clearing and domesticating the land that had marked the earliest years of settlement along the St. Lawrence was repeated back from the river. Then, established farms were divided to accommodate children. As families sought to sustain themselves on smaller and smaller properties, larger proportions of each holding were devoted entirely to cultivation, at the expense of other traditionally important land uses. Eventually limits were reached. Ranges ran back into marginal upland. Lowland farms could not be subdivided and remain viable.

Historian Allan Greer has traced the implications of these developments in three parishes along the Richelieu River. Some 1,750 people lived in Sorel, St-Ours, and St-Denis in 1765. Sixty years later these parishes counted 11,000 residents. By then, all of the cultivable land had been taken up, and tiny villages had grown into small towns. On many farms all land was being cultivated in rotation and in some areas the landscape was almost bare of trees. Where, in 1760, essentially similar farms sustained the families who occupied them in a fairly uniform, if modest, comfort, by 1830 society had differentiated. Hamlets had grown into villages and villages into towns. By 1815 the village of St-Ours had a handsome church, a home for the priest, and a manor house, as well as "about sixty houses, many of them substantially and well constructed of stone." A quarter century later St-Denis had 123 houses and people of "considerable property" among its 600 residents.

On poor land where farms had been subdivided and habitants had taken off-farm work in the fur trade in the late eighteenth and early nineteenth centuries, many had been impoverished by the exhaustion of soils due to continuous cropping and the decline in commercial opportunities after 1821. Early in the nineteenth century, commentators remarked that grain from Lower Canada was "frequently composed of one half weed," and described meadows "much overrun with thistles and golden rod" (quoted in Evans 2002, 55). On the better soils of St-Denis, where many had resisted the tendency to subdivide and farms were twice as large on average as in Sorel, those with land had prospered as rising urban markets for wheat had yielded them handsome returns. But here, too, the effects of rising population were reflected in growing numbers of tenants and day laborers. Habitant proprietors, who headed nine of every ten Richelieu households in 1765, were a minority among family heads in St-Denis by 1831. Up and down the valley and in the colony more generally, many suffered considerable hardship when repeated failures of the wheat crop, attributed to the wheat midge introduced into the colony in the late 1820s, placed many in economic difficulty and left others completely indigent.

Elsewhere, circumstances were even more grim, as the sons and daughters of families long established in the lowlands moved in to the agriculturally marginal territory of the Shield fringe. Historical geographer Cole Harris has written evocatively of one such instance, centered on the Ottawa Valley seigneurie of Petite-Nation east of Ottawa. To this tract of forested, knobby land, most of which was "too rough, too swampy" or covered in soils "too thin and acidic" for successful cultivation, a quickening tide of migrants came after 1830. Most of them were from the St. Lawrence lowlands. They came, for the most part, as individuals, to a settlement-in-the-making where they had few kin and there was little institutional support. By 1860 French Canadians accounted for approximately 80 percent of the population. But few held positions of power and influence in the settlement. The seigneur exacted his dues, and New Englanders, who also moved into the area in the second quarter of the century, dominated commercial life and held most of the important local offices in the seigneurie.

By midcentury, almost all the habitants of Petite-Nation scraped meager livings from a combination of work in the forests and local sawmills with hardscrabble farming. With few resources, little institutional or societal support, and meager agricultural prospects, they "stood almost alone," in poverty and vulnerable to the forces of change coursing through their narrow world (Harris 1971, 25, 50). Here, as in many other parts of this northern realm, Harris argues, a niggardly environment limited people's prospects of success and helped shape the form of an emergent society.

As the countryside of the St. Lawrence valley filled with people, and the domestication of the landscape proceeded, visitors began to describe parts of it, at least, by reference to the conventions of the European picturesque. "Pleasing prospects" and "delightful valleys" appeared in many reports. According to surveyor Joseph Bouchette the "agreeably situated" manor house of Sainte-Anne was surrounded by "elegant gardens and many fine groups of beautiful trees," in 1815 (Bouchette 1832, n.p.). Thirty years earlier, a young merchant from London thought it reasonable to imagine that the famous English landscape gardener "Capability" Brown might have planned some of the "charming plantations" along this part of the river (quoted in Coates 2000, 145). Much of this was rhetorical whimsy; commentators reached for familiar phrases to make the most of particular sensibilities in describing the landscape.

Yet it was not entirely fanciful. In 1820 the well-connected English couple John and Elizabeth Hale acquired the sixty square miles of the seigneury of Saint-Anne. Elizabeth thought the village "remarkably pretty" and found that the local scenery reminded her of "the Thames toward Putney" (quoted in Coates 2000, 151). But her New World estate was not all that it might be. Regrettably, there was "not an Oak to be seen here or in the neighbourhood." The Hales immediately set about "planting & neatifying." Deciduous trees were placed in the ground, hedges seeded, and garden plants imported. Much effort went into improving the demesne farm, by "opening & deepening drains" and by "cleaning the land which had been suffered to get full of weeds" (ibid., 158).

There was even talk of introducing English farmers who would bring with them English livestock and English farming techniques—and dutifully form the congregation of a Protestant church. When this failed, John Hale attempted to teach his French Canadian tenants how to farm according to the principles of English "improvers." Ignoring the skillful adaptation of early livestock husbandry and tillage along the St. Lawrence to local soils and climatic and economic conditions, Hale encouraged his tenants to collect manure to spread on the fields, to cultivate certain crops in rotation and so on. The Hales, and others of their ilk, sought to transform the material and the social landscape of the St. Lawrence lowlands to accord with a particular aesthetic and to conform with precepts developed in response to changing demographic, economic, and ecological circumstances in eighteenth-century England.

On the evidence of Elizabeth's skillful watercolors—replete with picturesque motifs—landscapes near the manor house quickly bore a remarkable resemblance to the countryside of her homeland. But this transformation was effected more readily with brush on paper than by labor on the ground. Re-creation was possible on a small scale and to some degree. Viewed from afar, the composition of water, clearings, clumps of trees, and clusters of dwellings about a manor house and church might evoke memories of English scenes. But this was not the old country—of either medieval France or nineteenth-century England. Despite Yankee perceptions and the efforts and the ambitions of improving English landlords, the St. Lawrence lowlands were a singular space (see also Ross 1991).

The work of Brian Donahue on Concord, Massachusetts, has demonstrated that farming in seventeenth- and eighteenth-century North America was not exactly the wasteful, ill-conceived, poorly adapted, ground-scourging practice portrayed by early nineteenth-century commentators (and many later historians). Although we still know too little of the agricultural practices of Acadians and the New Englanders who succeeded them on the borders of the Bay of Fundy, both there and along the St. Lawrence (as in early New England), it seems evident enough in light of Donahue's work that settlers adapted traditional European practices (calculated to integrate livestock husbandry and grain cultivation) to local environmental circumstances.

Migrants from New England to Nova Scotia carried knowledge of New England farming practices as well as of New England land survey and township government systems northward with them. Along the St. John River, for example, they immediately recognized the value of the extensive meadows (or interval land) where they established their settlement of Maugerville. With their counterparts in Horton, Port Royal, Chignecto, and old Concord itself, they knew that "the amount of hay that could be cut determined the number of cattle that could be kept," and that "cattle were the key to both subsistence and wealth" in the system of mixed husbandry that they practiced. Plowlands were typically established on relatively light, well-drained soils, gardens near the house and barn (from whence the soil could be enriched) and orchards on steeper, stonier or more marginal ground (Donahue 2004, 166; McNabb 1988).

In most other parts of British North America, settled after 1780 by people of European descent, the timing and circumstances of the colonists' arrival were such that Thoreau's mid-nineteenth-century conceit—that he confronted an ancient human landscape along the St. Lawrence—would have had no credibility. In these areas the great drama of landscape transformation staged by settlers in the act of making their wooden worlds was written in a different context and played out in less than a hundred years. Between the American Revolution and the middle decades of the nineteenth century, newcomers engaged and remade landscapes largely unaltered by human interference, from the narrow valleys of Cape Breton through the township lands opened up by colonization companies in Lower Canada to the sand and till plains of Essex County near Lake St. Clair (Little 1989, 1991, 1999; Clarke 2001) . Across this vast expanse of territory, settlement proceeded differently in detail. Those who occupied the land came from different situations, faced different circumstances, and produced different outcomes.

But everywhere the recentness and rapidity of the transformation were apparent. This was the "Age of Progress," said contemporaries, and its signs were all about. Not least were they evident in the radically altered landscapes of settlement. The creation of farms from forest not only brought "individual success and happiness," it also served to increase "the material resources of the country." These were "times of progress in which it is our privilege to live," observed one settler whose convictions were widely shared. Those who sought election to political office sensed the priorities of the population and almost invariably declared themselves to be instruments or friends of progress. British North Americans were reminded, more than once, that no country could "with safety be stationary" and that they should "advance with the advancing [and] keep pace with the foremost" (Wynn 1979b, 53; Fallis 1960, passim). Indeed, botanist and enthusiast Henry Youle Hind revealed the spirit of his age when he chose, in 1863, to title a book *Eighty Years Progress of British North America: Showing the Wonderful Development of its Natural Resources by the Unbounded Energy and Enterprise of Its Inhabitants.* Such were the effects of these progress-orientated settlers upon the land that they might be described, collectively, in the terms David

A settler's farm in the Canadas. *This is what many an immigrant family aspired to and substantial numbers eventually achieved in mid or late nineteenth-century British North America/Canada. This illustration warrants comparison with that on page 99 and that on page 154. Looked at thus it captures the march of material progress that formed a leitmotif of colonial discourse, while suggesting, at the same time, the self-congratulatory satisfaction that lay behind the images in typical "county atlases." Here settlers have extended their pioneer clearings, planted extensive (and stump-free) fields of wheat, invested in stock, and prepared some square timber from the nearby forest, possibly to float downstream for sale into the transatlantic staple economy, but their surroundings (and lives) are neither as ordered nor as comfortable as suggested by the sketch of Robert Fitzsimons's Prince Edward Island setting. (National Archives of Canada)*

Wood applied to those who colonized Upper Canada, as *"ecological revolutionaries* who changed the face of the earth" (Wood 2000, xvii).

In broad terms, these revolutionaries marched inland along, and upslope from, the waterways that provided the main arteries of movement through British North American space during the years of their advance. In the Maritime colonies, in particular, where good land was limited and slight differences in elevation and aspect could render agriculturally marginal climates that much less forgiving, those who acquired fertile lowland were sometimes referred to as "frontland settlers"; those less fortunate were "backlanders." By and large, the frontlands were occupied first. They spread along lakes and waterways, included fertile (periodically flooded) intervale (or meadow) lands, and benefited from relative ease of access. Backlands climbed the hills back of the first range of "frontland lots." Thin stony soils, ill-drained hollows, shorter growing seasons, and in some cases access limited to footpaths placed their occupants at an immediate disadvantage.

In the more expansive agricultural ecumene of Upper Canada, entire lakefront townships passed quickly into private or institutional hands, so that those who came later either bought into these areas or sought land from the Crown in more remote (and before long, as settlement pushed inland onto moraines and drumlins and sand plains and Shield fringe, less propitious) settings. All of this complicated patterns of settlement advance and landscape change. As one measure of this, consider that by 1820 there was no Crown land available in Hamilton Township, on the north shore of Lake Ontario midway between Kingston and Toronto (see Fig. 12). Fewer than 200 families lived in the township at this time. School and clergy reserves had been leased and speculators held considerable areas. Even in 1810, when 92 households (632 people) lived in Hamilton, almost three-quarters of the township's land had been alienated.

Speculation drove up prices (uncleared lots in Hamilton Township were about fifteen times more expensive than Crown grant land elsewhere in the early 1820s) and held up the tide of "improvement," as reserve and speculators' lands remained unoccupied and uncleared (Ennals 1978). Yet settlement proceeded. In 1831 the two front concessions were pretty much fully occupied, perhaps half the lots on the third and fourth concessions had been taken up, and there were a number of small clearings in the forest along the sixth concession. Twenty years later, almost all the land back to and including the eighth concession was occupied, and there was considerably more cleared than uncleared land through the front half of the township.

In the newly surveyed township of Essa (near present-day Cookstown), by contrast, special claims exercised by soldiers active in the War of 1812 or children of Loyalist settlers (most of whom were absentees) accounted for over half of the land in the township. Slightly more than a quarter was Crown or clergy reserve and the surveyor took almost one in every twenty lots as payment for his efforts. In effect, then, only one-eighth of the entire township was actually available to incoming settlers as government grant land (Wood 2000, 94–96). Perhaps as a result, settlement proceeded relatively slowly; isolated lots, likely to be surrounded by forest for many years, were less than desirable. Yet if this was extreme, it was not entirely unusual.

Some have suggested that over 60 percent of the land in Upper Canada was held for speculative purposes in 1825. Given the definition of a speculator as someone holding more than 400 acres, this likely exaggerates the extent of

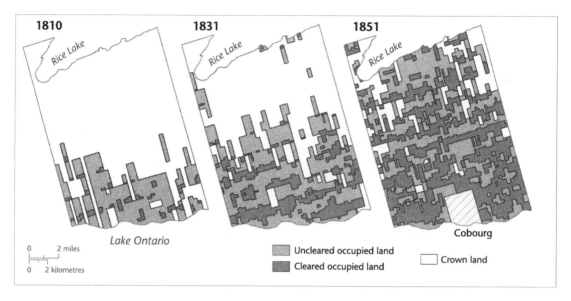

Fig. 12. *Settling and clearing the land in Hamilton Township, Upper Canada, to 1851. (Adapted from Ennals 1978 and* Historical Atlas of Canada 2: pl. 14)

"speculation for profit" at the cost of downplaying individuals' desires to provide farms for their children, but it emphasizes the patchwork quality of settlement and the persistence of uncleared land in many areas. There is no escaping the import of geographer John Clarke's calculations that between 35 and 40 percent of all the patented land in Essex County was unoccupied in both the mid-1820s and the early 1850s. A decade or so before Confederation, good Crown land was hard to find, but there were probably two million forested acres on the market.

By 1850 every colony but Newfoundland had its "rural settlement heartlands." In Canada East these were the seigneurial lands along the St. Lawrence. Elsewhere they were, mostly, relatively old-settled areas (say two generations or forty-plus years) where rural population densities were highest, the proportion of cleared land was greatest, and the productivity of farming was at its maximum. In Nova Scotia, the heartland regions included the Annapolis-Cornwallis Valley, the southern side of the Basin of Minas, parts of the Northumberland shore, and, shared with New Brunswick, the Chignecto region at the head of the Bay of Fundy. In New Brunswick several parishes flanking the middle reaches of the St. John River, parts of the Kennebecasis valley, and the Sussex-Petitcodiac district stood apart from other locales. In Prince Edward Island the central region, roughly coincident with Queens County, had the best soils and the highest population densities. In Canada West, a Y-shaped cluster of townships

Settlement growth in early Upper Canada. *Even as tiny clearings grew into farms, and the forest was pushed back, the hamlets and villages that formed the tiny nodes of an emerging settlement network across the edges of British North American territory were unprepossessing places—at least to modern-day eyes. But even here, in early Chatham, there are signs of the many roles that nascent towns played. They were "power containers," centers of commercial influence and official authority. Notice the stores along the river, an artery of movement, and the military presence— the soldiers, the blockhouse—that spelled both protection and control, as well as the continuing dependence upon wood for building and fencing. (National Archives of Canada)*

around the head of Lake Ontario and extending toward London, as well as a number of townships along the north shore of the lake and the St. Lawrence River, had population densities in excess of thirty-six per square mile.

Equally, each colony had its frontiers or fringes of settlement. In these areas, farms had relatively small amounts of cleared land, and there were few social amenities and scant evidence of prosperity. In Canada West several recently occupied areas had few people (say one family per square mile), and formed an outer fringe of settlement, skirting Lake Huron, the southern edge of Georgian Bay, the shield fringe, and the Ottawa River. In New Brunswick and Nova Scotia the "frontiers" were the extreme backlands, places similar to and including the upper slopes of the Middle River valley in Cape Breton, interior settlements such as New Ross along the rough roads cut through the forest between the Fundy and Atlantic shores of Nova Scotia, and the upper reaches of several tributaries of New Brunswick's major rivers. Those who lived in such locales generally faced a hardscrabble existence, dependent upon the produce of stumpy patches of cleared land and such off-farm work as they could obtain. Between these "heartlands" and "fringes" lay an intermediate zone, settled for a generation or more, but with farms still in the process of active development and marked by significant amounts of unoccupied, uncultivated land.

Settlers on the fringe lived in quintessential wooden worlds. "[T]he landscape is unvaried and exceedingly confined," wrote one immigrant, "nothing but trees, trees, trees continually" (quoted in Baskerville 2002, 65). Many pioneer reminiscences included a story about first settler families entering the township from different directions and living "in isolation" for months or years before realizing that others were in similar circumstances only a mile or so away (Harris et al. 1975). Typically the homes of early settlers were cabins or shanties, made almost entirely of logs. Most of their furniture was constructed of wood. The forest stood virtually at their doorsteps, forming the walls of what some at least—cowed and confined by its immensity—began to regard as a rural prison house. The fences, if any, that separated the settlers' stump-strewn fields from the forest were constructed of split wooden rails. The rough road (probably more accurately called a track) that linked small clearings with the outside world was no more than a jagged line through the forest; wooden bridges carried it over streams and where it crossed swampy ground, logs were laid down across the route (to form what was known as a corduroy road).

As clearings were extended and the edge of the forest was pushed back, the trees seemed less encompassing, and the "branches, roots [and] tendrils" that Margaret Atwood envisaged as "the dark side of light" no doubt appeared less likely to storm and break in upon the settler (Atwood 1970). Yet wood remained the *sine qua non* of existence. Wooden boxes and wooden barrels were universally employed for shipping and storage. Barns were built of wood. So, too, were the houses that replaced the shanties—those first homes of which Catharine Parr Traill said darkly, "nothing can be more comfortless . . . , reeking with smoke and dirt, the common receptacle for children, pigs, fowls" (quoted in Wynn 1987b, 250).

As farms were extended, the forest margin retreated until, in some districts, only tiny remnants of the original vegetation remained. Still, wooden fences divided fields, wooden churches stood at crossroads, and wooden buildings marked each farmstead. As prosperity increased, iron nails might replace wooden dowels in some forms of building. Hewn pine benches and tables might give way to more finely manufactured products. But rural settlers lived their days in wooden worlds. Before 1830 most of the sawmills in the colonies—and there were more than 400 in Upper Canada in the mid-1820s—catered to local needs, turning logs into the boards and planks that had dozens of uses in the construction of a new world. In many districts they were more numerous than grist mills. Later, their relative importance declined and larger mills oriented to export markets gained prominence. Yet local demand for sawn lumber and other wood products remained high. Even in the heartland districts of Canada West, dwellings built of materials other than wood were uncommon before midcentury. Most of the now-familiar brick farmhouses of rural Ontario date from the latter part of the nineteenth century (McIlwraith 1997).

Despite—or perhaps because of—the overwhelming presence of and dependence on wood, British North American settlers often appeared to be engaged in a war on trees. "The Canadian settler," wrote Anna Jameson, ". . . *hates* a tree, regards it as his natural enemy, as something to be destroyed, eradicated, annihilated by all and any means" (Jameson 1839, 72). Here, as in the eastern colonies, the axe and the flame were the main weapons of war. Settlers enrolled both in their efforts to make farms. This was a serious engagement. In Catharine Parr Traill's account of it, the Canadian "spare[d] neither the young sapling in its greenness nor the ancient trunk in its lofty pride; he wage[d] war against the forest with fire and steel" (Traill 1836, 162). In an immediate sense, settlers' lives hung in large part upon the rate at which they cleared the forest and produced crops from their land. In the longer term, family fortunes turned on this process, because clearing was "capitalization." Applying labor to "waste land" (as the forested domain was often called) to convert the raw resource into farms increased its value, drove the expansion of colonial economies, and provided an "escalator" that could carry relatively poor persons into better circumstances. Little wonder, then, that the forest was attacked with fervor.

Two main strategies were employed: "girdling" (also known as "ringbarking") and felling (or chopping). Girdling allowed land to be planted with smaller inputs of labor than required by felling, but it was probably most effective in dealing with large trees and deciduous species; on balance it probably postponed, rather than reduced, labor demands, and it was less common in British North America than in other areas of the eastern seaboard. To open up land by girdling, settlers first cleared it of brushwood and small trees, then cut a ring of bark from the trunk of each standing tree. As John Howison noted in his *Sketches of Upper Canada* published in 1821, "if this is done in autumn, the trees will be dead and destitute of foliage the ensuing spring; at which time the land is sown, without receiving any culture whatsoever except a little harrowing" (263–265). Written for British readers and prospective immigrants, this glosses the process to suggest how easily New World land could be made available. Trees killed by girdling still had to be removed in later years because they endangered humans, beasts, and hard-won improvements (such as dwellings and fences) as limbs or trunks gave way in the wind, and many species grew more resistant to the axe as they dried.

Felling entailed the chopping down of all trees (except perhaps the largest), the chopping of the fallen trees and branches into "manageable" lengths (say 10 to 14 feet long), the hauling of these into large piles (using oxen), and the firing of the wood piles a year or two later when they were sufficiently dry to burn. This was an arduous and messy business. A skilled axeman working hard might clear an acre a month during the winter. Theoretically, at least, a settler taking up a forested lot might clear five or even seven acres in the first year after erecting a shanty. But this rate could not be sustained. After a few years on the land, observed Samuel Strickland, the immigrant "has so many other things to attend to, such as increase in stock, barn and house-building, thrashing, ploughing, *etc.* which . . . give him every year less time for chopping, particularly if his family be small" (Strickland 1853, 167).

Indeed, the slowdown was significant. From an analysis of tax assessments from 140 Upper Canadian townships, Peter Russell has suggested an average long-term clearing rate of about one-and-a-half acres per adult male per year. At this rate a dedicated, energetic settler, working alone and spared illness and injury, might spend almost a decade establishing a twenty-acre clearing. In the process, he would surely have resembled those of whom John Muir wrote in the 1860s: "so many acres chopped is their motto, so they grub away amid the smoke of magnificent forest trees, black as demons and material as the soil they move upon" (Muir quoted in Wood 2000, 9).

Even then, those twenty acres would not be "clear" in any conventional sense of the term. Stumps remained rooted in place. They were enormously difficult to dislodge with the technology available in the first half of the nineteenth century, and even after twenty years of rotting in the ground their removal strained the sinews of men and beasts. Taking local definitions and census takers' understandings of "cleared" or "improved" land as a measure, however, the advance of the settler army can be charted in broad terms. In Nova Scotia 275,000 inhabitants had 840,000 improved acres. In New Brunswick there were 640,000 acres of cleared farmland in the middle of the nineteenth century. Here, in turning forests into clearings, the ecological revolutionaries had their first, and perhaps most consequential, victory.

Yet it was no more than the opening of a long campaign. Environmental change did not stop when the forest came down. Newly opened fields were quickly invaded by species that settlers generally considered undesirable—from fireweed to native blueberry, from chokecherry to wild raspberries, and from young hardwood to poplar. To keep such invaders down and to produce a "cash" crop that could be sold or exchanged to meet the cost of other requirements, most settlers on newly opened Upper Canadian land alternated cropping with fallow, thus essentially emulating in broad form a mode of cultivation prevalent in sixteenth- and seventeenth-century Britain, although the old English practice of "folding" stock (especially sheep) on the arable land—holding them overnight in movable pens erected on the fallow plow lands—largely disappeared in these New World settings, where time and labor were at a premium and sheep were relatively few.

Stock browsed in the forest, on floodplain meadows and patches of rough pasture where these were available, and in the cultivated fields after harvest (thus adding at least some nutrients to the soil). Because winters were relatively long and harsh (compared with the British Isles), keeping stock throughout the year was difficult: Shelter and fodder were both at a premium in the early years, and on most farms a significant proportion of beasts were slaughtered in the fall, both to feed farm families and to reduce the cost of sustaining the animals.

This also reduced the accumulation of manure that might otherwise have been mucked out onto the fields as well as the kitchen garden. Bare fallow was usually ploughed two or three times to prevent the establishment of weeds and saplings. Even this often seemed insufficient. In Upper as in Lower Canada, critics quipped that the agricultural system alternated "wheat and thistle pasture" (Evans 2002, 60). Especially in the first years of farm development, wheat was sown by most settlers in Upper Canada. It was relatively nonperishable, it supplied the settlers' need for flour, and could usually be exchanged at the local store or mill. It could also be sown and harvested by hand in stump-strewn fields. This reliance on wheat, and the displacement of stock from the center of the system of husbandry, have led scholars to the general conclusion that two fields (allowing a wheat-fallow-wheat rotation) and a kitchen garden were the basic components of a new farm in Upper Canada.

From the first, however, conditions on the ground were more complicated than their representations on paper. Although wheat flour was an important colonial export, especially after 1840, the farmers of early nineteenth-century British North America diversified their production, both to spread risk and to meet various and sundry needs. Oats, barley, and potatoes were added into the production mix and were particularly important in the eastern part of Upper Canada. Buckwheat, corn, rye, and peas were also grown. Township by township, wheat accounted for between one-fifth and two-thirds of field crop production (by caloric value) in 1851. By some recent estimates, wheat grew on less than a quarter of the average farmer's cultivated land by this time.

Indeed, the economic historian Douglas McCalla has insisted that most Upper Canadian farms are best thought of as mixed enterprises, and notes the variety of products that farmers brought to trade at country stores as justification. In many instances, beef, mutton, and pork were added to the aforementioned mix of crops in trade, and forests products from lumber to pot and pearl ashes further diversified the picture. One farmer brought thirty-nine different crops to market in a four-year period. Measured by what left their farm gates rather than by what they sent down the St. Lawrence, Upper Canadian farms were far from specialized. Local or internal markets were far more diverse and complex than export markets. In the Maritime colonies wheat was never as important as in the Canadas; production was significant only around the Gulf of St. Lawrence and even there it rarely accounted for more than a quarter of field crop production. Climate, soils, and cultural preferences made oats, buckwheat, and potatoes the most important crops through most of this region, and most farm surpluses went into local markets (see also Roy and Verdon 2003). The Maritime region was a net importer of agricultural produce, particularly wheat and flour.

In the short term, few who inhabited this wooden world doubted that creating farms from the forest, providing the means for families to toil and prosper, converting the wild into the sown, was beneficial. Their assessment was reflected in the very words that they used to describe their surroundings: forests were "waste lands"; areas from which the trees had been removed were "improved." Initially, at least, there was some justification for this usage. Burning raised the pH of, and added nutrients to, the generally acidic soils of eastern Canada. Calcium, magnesium, phosphorus, and potassium and potash leached from the ashes of the destroyed vegetation and pH levels may have climbed from less than 6.0 to more than 8.0 as a result. By promoting bacterial activity in the soil, this increased nitrification.

In the longer term, however, the equation resolved differently. Newly added nutrients were soon dissipated. Later experimental studies conducted elsewhere suggest that pH levels might have remained slightly elevated a decade or so after burning, that elevated nitrogen levels likely persisted somewhat longer, and that the other nutrients infused by burning were leached from the soil on a similar time frame (Ahlgren and Ahlgren 1960). All of this meant that yields from soils regarded as abundantly fertile on first cropping began to decline within years, not decades, of initial cultivation. If farmers burned the trees and collected the potash for sale as pot and pearl ashes, as many in Upper Canada did, they deprived their land of much of even this first boost of nutrients. At least one immigrant from England in the 1830s came quickly to some partial understanding of the fact that the soil was being mined. After assessing his new situation, he made a deliberate decision to purchase uncleared land near Guelph from the Canada Company, "because almost all the land that is to be bought cleared, having been first taken by people with little or no capital, is generally exhausted, and must be left fallow for three years to recover itself" (Cameron et al. 2000a, 19).

This was unduly optimistic. In the enthusiasm for development, steep slopes and shallow soils were cleared and cropped. With the removal of vegetation, a much higher proportion of precipitation was lost as runoff. Topsoil loosened to depths of 5 to 10 inches by ploughing slid almost imperceptibly downslope in a process known as sheet erosion. Left fallow and exposed to rain and surface runoff, hard clay suffered rill and gully erosion; fine sandy morainic materials were also blown away by the wind. In the 1950s the Ontario Department of Agriculture found that erosion carried off almost 100 tons of silt loam per acre from relatively gentle, one-in-six slopes that were not ploughed on the contour, and these findings have been matched in broad terms in roughly analogous situations elsewhere.

Some of the most revealing investigations of these processes have been undertaken by the geographer Stanley Trimble in Coon County, Wisconsin. His work and that undertaken by others elsewhere in the United States show that soil losses are considerable, even on low slopes planted with corn. Yields fall as topsoil is carried away, and in some places several inches were lost in a few years. Ultimately, the eroded material would be deposited downslope, but not as fertile soil. Sorted by gravity and the mechanics of fluvial sediment transport, soils long in the making were relocated as sand and silt beds on gently sloping lowlands, on lake margins, and in streambeds. Sometimes the problems of topsoil loss upslope were exacerbated by the deposition of unproductive sediment, stripped of the organic material in the original soil, atop fertile, humus-rich lowlands.

Mono Township, located approximately thirty miles northwest of Toronto, neatly exemplifies the consequences of what historical geographer Cole Harris once called the convergence of industrializing Europe with the land of North America. A rough, knobbly area dominated by postglacial sands, silts, gravels, and till, where an interlobate moraine laid down during the last glaciation meets the Niagara escarpment, the area that became Mono Township had relatively modest potential as a site for agricultural settlement. There were extensive swamps of organic muck in which cedar and tamarack grew in a nigh impenetrable tangle. There were light, sandy, and very dry soils on the edge of a former glacial spillway where hemlocks grew but farming was extremely difficult. And there were gray-brown podzols beneath maple-beech forests where hemlocks intermixed on drier sites and oaks, elm, and basswood marked the slightly moister, better soils.

European settlement of this territory began in the 1820s and gathered momentum in the 1830s when the area became an important destination of emigrants from Ulster. By the 1840s newcomers were occupying standard 100-acre lots described by the man who surveyed them as "hilly and swampy and not fit for settlement" (quoted in Harris 1975, 10). By 1850 there were 2,300 people in the township. In the years that followed, early arrivals began to note the "considerable differences in the appearance of this neighbourhood" since their arrival, remarking that "places which were then mere spots in the great forest begin to assume the appearance of farms. . . . Clearances are beginning to join each other; roads are being opened up" (ibid., 11).

In the next couple of decades, a few farms in the township were being offered for sale as "excellent" and desirable" properties with seventy-five cleared acres "nearly free from stumps, well fenced with cedar . . . and in a high state of cultivation" (ibid., 10–11). For some at least it seemed that the virtues extolled in local newspapers—"Toil, Time, and Trust" (in God) perhaps foremost among them—had borne just fruit. But the soil profiles of Mono's undulating terrain told a somewhat different story. As Harris and his collaborators discovered, on sloping land sheet erosion and gullying accompanied clearing and washed both "A and B horizons into hollows or streams, while exposed sands and silts began to blow as they had not since their consolidation under a post-glacial forest" (Harris 1975, 15).

Replacing forests with farms also had biological consequences. Removing the native vegetation cover reduced the habitat of indigenous animals and birds. This alone hardly threatened most species—broad expanses of similar habitat remained elsewhere—but it did affect their distribution, or their abundance, in particular locales. Exacerbated by hunting pressures, habitat changes could reduce local faunal and avifaunal populations dramatically. By 1820 there were almost no moose in New Brunswick. They could also lead in extreme cases, such as those of the great auk and the passenger pigeon, to extinction.

Compiling a Canadian "gazetteer" (or inventory of statistical and general information) in the 1840s, William Smith noted the variety of "living breathing denizens of the forest" but also observed that "their numbers are fast diminishing before the destructive progress of civilization" (237). Once, he continued, they had been plentiful "but as the country became settled up they were gradually destroyed, or were obliged to retreat before the advancing footsteps of their common foe." No longer need emigrants from Britain proceed, as their predecessors purportedly did, "glancing from side to side as they walk up the streets, expecting every minute to see a bear or a wolf dart out from the doorways."

This cameo was exaggerated, for effect as well as reassurance, but early in the 1840s the legislature of Canada West (Upper Canada) had moved to impose a closed season on the hunting of various game birds (ibid., 237–239). Aquatic organisms also suffered from the advance of settlement and the progress of development. Streams were dammed for mill sites; browsing stock broke down their banks; and refuse and effluent from mills, villages, and towns found its way into their waters. Where salmon had spawned in the streams of Mono Township in the 1820s, fish were no longer found in the 1860s.

In the Maritime colonies and Upper Canada, as along the St. Lawrence, agricultural settlement also meant the wholesale replacement of indigenous vegetation by imported species. Seeds crossed the Atlantic early and often. Writing to encourage his English fiancée Ann Gill to join him in late eighteenth-century Nova Scotia, Yorkshireman James Metcalf seemed almost as anxious for certain strains of wheat as for his betrothed: "if you come [he wrote] pray be so good as to bring about a bushel of wheat if you can of four different kinds for seed let yellow kent be one and Hampshire brown another for it will be of great servis hear be careful to keep it from salt water you may if you please lay it like a pillow in your bed" (quoted in Wynn 1969, 63). Writing from Upper Canada half a century later, William Phillips expressed the hope that his father would "come and help" him the following summer "and bring me all sorts of seeds that grows in England . . . and bring some cuttings of gooseberries, apples and grapes, that I may have some English fruit; you can bring them in a tub" (quoted in Cameron et al. 2000a, 23; see also Macpherson 2005).

Yet if Old World seeds (and Old World stock) came to the New World in countless ways, they rarely came clean. Weeds—thistles, wild mustard (charlock), ox-eye daisy, couch grass, shepherd's purse, lamb's quarters, burdock, and ambrosia—crossed the Atlantic in company with the seeds of treasured commercial and ornamental plants. In the eighteenth century, Upper Canada statutes relating to road labor required those engaged in it to "destroy as much as may be in their power all burrs, thistles and other weeds" (quoted in Evans 2002, 63). In Nova Scotia legislators passed a law against thistles early in the nineteenth century, and weed eradication bylaws were fairly widespread in Upper Canada after 1850. Still, botanist H. Y. Hind expressed concern, in a lecture published in *The Canadian Agriculturalist* in December 1850, that "the growth of weeds among cultivated crops is an increasing and serious evil."

Reviewing the history of inadvertent introductions to North America in 1877 and 1878, E. W. Claypole of Antioch College extended his purview from the obvious to include the "parasites which have crossed in the train of man to prey upon him and the domestic animals" (Claypole 1878, 1879). The "great army of fleas, bed-bugs, lice, sheep, cattle and hog ticks, sheep, horse or ox bot-flies, with the infamous muscle-worm of the pig (*trichina spiralis*)," he pointed out, was entirely made up of "foreigners naturalized here" (Claypole 1879, 78). Consider, he continued, that for two hundred years North American cabbages had been spared the plague of insects. Then "into this cabbage paradise, . . . the Evil One . . . gained entrance in the form of the English cabbage butterfly (*Pieris rapoe*). Crossing the Atlantic unperceived, he landed at Québec about 1857" (ibid.), and within a few years had severely damaged cabbage crops along the length of the St Lawrence and around the Great Lakes.

Further, Claypole found "one of the most striking illustrations of the truth that small causes can produce great effects if sufficiently multiplied" in the introduction of the wheat midge (*Cecidommyia tritici*)—sometimes incorrectly known as a "weevil"—into North America. "Québec [upriver of which it was noticed in 1828] appears to have been its port of landing, and its progress over the country appears to have followed the St. Lawrence and Sorel rivers," he wrote. "About 1849, it began to destroy the crops on the north shore of Lake Ontario" (ibid., 76, 75). By the 1850s enormous losses were being reported: $15 million in one year in New York State, $2.5 million in another in Upper Canada, and unspecified amounts in Lower Canada and Nova Scotia. Writing in 1865, Dr. Fitch, the leading American authority on the ravages of this inadvertently introduced beetle, claimed that "in consequence of the presence of this insect [in much of the northeast of the continent] . . . wheat has wholly ceased to be a staple product. Wool growing and dairying have become the leading pursuits. These require fewer laborers. . . . [Thus] this insect has done much toward reducing our population" (quoted in ibid., 75).

The fungus *Phytophthora infestans* or potato blight had similar, ramifying effects when it appeared in British North America in the 1840s. The self-same plant disease that swept through Ireland and precipitated the great famine migrations of the late 1840s, it turned potatoes to mush in the ground or in the root cellars in which they were stored after harvest. In the backlands of Cape Breton, where poor families on poor land were heavily reliant on the potato and where the blight struck four years in succession, "poverty, wretchedness and misery" were evident "to an alarming degree" (Nova Scotia, House of Assembly, *Journals* 1847, App. 67). Hundreds of families faced starvation, and many were left destitute. Some compared the situation in Cape Breton with that in Ireland. But it was far less dire. A smaller population, timely government aid, and the availability of fish, game, shellfish, birds' eggs, and grains from front-land farms meant that few people died. Still, the social and economic consequences that followed upon the introduction of this fungus to British North

American soil were considerable. Even in the short term, as Stephen Hornsby has shown for Cape Breton, "the blight . . . frequently wiped out the work of 10–20 years of pioneering. Livestock had died, land had been mortgaged, and many settlers struggled to pay off loans from merchants" (Hornsby 1992, 119).

Removing the forest also triggered hydrological and microclimatic changes, most of which were only vaguely recognized by those who precipitated them. When scientists began to investigate some of these processes in the twentieth century, they spoke of shifting run-off:percolation ratios, of altered hydrographs, and of changes in the surface albedo. For eighteenth- and early nineteenth-century farmers, these things were manifest as drier soils, lower groundwater tables, more rapid erosion of the surface, wider fluctuations in stream levels, and greater diurnal temperature ranges. Observing these changes, many tried to fit them within the grid of contemporary conviction about the effects of clearing on climate.

Some, persuaded that temperature should correspond with latitude, believed that removing the forest would ameliorate the extreme cold of eastern North America and bring its winter lows to correspond with those of southern and central France. Others held that rain followed the plough, and that clearing and cultivation would water the good earth. By midcentury a few had begun to suspect that deforestation caused desiccation. But cause and effect were not always easily established. There were few datum points, it was difficult to separate the effects of human interference from the natural annual variation of rainfall totals and sunshine amounts, and it was hard to calibrate and understand the significance of many of these changes.

Most farmers were assiduous watchers and recorders of the weather. Their farm diaries faithfully document dates of first frosts, the cloudiness or brightness of the days, the directions of the wind, the dates on which the river ice broke, the days on which the sap and the salmon began to run, and so on. But what to do, in 1819, 1820, or even 1827, with the information entered on January 12, 1818: "From the last Date [January 5th] there has Been Three Cold Still Nights but the Days fine Winter Weather—There has also Been Two Snows"? (Dibblee 1818). To ask this question is to confront the veil of incomprehension that shrouded lay efforts to understand the natural world and limited identification of the ways in which some of the natural processes set in motion by human action changed the face of the earth—at least until their consequences were writ so large as to be literally self-evident (and substantially irreversible).

In Ops Township, Upper Canada, residents came face-to-face with such issues as they began to develop the infrastructure of settlement in the 1830s. A gently undulating, swampy area with predominantly clayey soils, Ops was opened for occupation in the late 1820s. Its early settlers were beset by ague (also known as swamp fever) or malaria, and forced to confront the inaccessibility of their location. Given the difficulties of reaching neighboring towns, it was clear that local development would depend upon the construction of saw- and gristmills. The most suitable site for waterpower generation was identified and set aside by the township surveyors, and in 1830 an experienced miller undertook to develop it. The location, now known as Lindsay, was named Purdy's Mills after its proprietor and soon became known, to locals at least, as the site of "the largest mill dam in the world" (Forkey 2003, 50).

This was an exaggeration. Purdy's dam was 14 feet high, and it raised the water level back of the rapids on which it stood by some 7 to 10 feet. Its like might have been found on many a river in North America. Indeed Purdy's previous establishment north of York (Toronto), built some twenty years earlier, had a dam that was 10 feet high and 200 feet wide. Still, in the drainage basin known to native people as the Scugog area (literally "submerged or flooded lands"), this dam had a significant impact. When this became a cause of local contention, an experienced field surveyor and engineer investigating the matter concluded that neither Purdy nor "anyone else, had the most distant idea of the ultimate result," when they embarked on the "arduous undertaking" of building a dam in "a back county."

It was "the general opinion of the country around," reported the investigating surveyor, that "if he [Purdy] raised it [his dam] as high as a house he would do no damage." In any event, however, the "extent of overflowing" caused by the dam was "upon a much more extended and destructive scale than . . . can be found from [a similar sized dam] in these Provinces" (quoted in Forkey 2003, 61, 62). Water levels were raised some thirty miles back from the dam around the shores of Lake Scugog, and for considerable distances up two of the major tributaries of the Scugog River. Along the nine miles or so between the outlet of Lake Scugog and the dam, meadow, hay, and crop lands lay submerged several feet below the spreading millpond and the valley presented a "continuous scene of drowned and decayed timber," interspersed with "the former residences of the settlers, showing parts of the roofs out of the water, from which the inmates had to make their escape" (quoted in Forkey 2003, 61).

Although all knew, in principle, that damming the flow of a river meant increasing water levels upstream, newcomers to Ops Township, lacking both detailed local knowledge and the time or capacity to conduct scientific surveys of their territory, were evidently unable to calibrate the environmental effects of Purdy's initiative with precision. Set against dimly perceived consequences, the imperatives of convenience and economic development overrode such caution with regard to the size of the dam (and such concern for property and riparian rights) as there may have been among local residents. Not until early in the 1840s were the dam's environmental effects ameliorated. Then, a virulent outbreak of ague attributed to the greatly increased quantity of swampland in the township led an angry mob to destroy it.

10

EXPLOITING THE WATERS

For those who came to the coastal bays and inlets that encircled the Gulf of St. Lawrence, faced the Atlantic from the long shoreline of Nova Scotia, and formed the coast of Newfoundland, it was soon evident that the sea rather than the land would form the enduring basis of their livelihoods. Here climate and soils made farming a dubious proposition. Some concluded that "Newfoundland promise[d] nothing" but "Herculean labour & pains" to those interested in agriculture (quoted in Cadigan 1995, 51). But the seas were rich in fish, and in the late eighteenth and nineteenth centuries they provided the essential support of a growing population connected more or less directly to international markets by merchants who exchanged imported provisions for the harvest of local waters, by traders in the larger ports of the region, and by European importers who were the remote links of a "metropolitan-satellite chain" that "moulded a 'long-swing' in capital accumulation" based on a family-centered system of production caught up in an extended web of credit and debt (Apostle and Barrett 1992, 45).

As historical geographer John Mannion has shown, the seventeenth-century fishery depended upon commercial links with Ireland, and in the years that followed growing settlements were largely sustained by imported provisions (Mannion 1982, 2000). Indeed, life in these scattered fishing settlements was hard and dangerous. Anthropologist Philip E. L. Smith has found evidence that residents of coastal villages dispersed, in family units, into the interior of the colony to subsist on hunting and trapping during the winter. This, he argues, was a response to harsh environmental conditions and the peculiar social and political circumstances in the fishing settlements of seventeenth and eighteenth centuries (Smith 1987, 1995). Even by the standards of the time, fishing families struggled to survive: "Whoever saw a fisherman thrive?" asked one late eighteenth-century commentator on the settlement of Nova Scotia (quoted in Wynn 1987b, 227).

Settlement and the development of fishing communities began to gather momentum early in the eighteenth century in Newfoundland and in the 1760s in Nova Scotia, although the continuing close connections between British ports, British mercantile interests, and Newfoundland have led Shannon Ryan to argue that the island did not make the transition from fishery to colony until the turn of the nineteenth century (Ryan 1983). In both Newfoundland and Nova Scotia, newcomers moved first into locations chosen for their good shelter, favorable shore, proximity to good fishing, and the availability of wood and water. Typically settlements grew to accommodate kin until local resources of land and sea could support no more. Then new settlements would be formed in more remote or less well-favored locations, filling suitable coves and coasts incrementally and extending the reach of such settlements across space.

Most of these settlements were small, utilitarian, and strikingly egalitarian places. Houses were modest, and kitchen gardens and fields ran back irregularly from the shore, where simple docks, sheds, stages, and the flakes upon which fish were dried were the most prominent features of the human landscape. A few cows on rough pastures squeezed between strand, forest, and rock; pigs in roughly constructed pens; and a few hens and dogs revealed both the possibilities and limits of husbandry in these circumstances and pointed to the plural nature of subsistence in this household economy. In the larger places that emerged as the nodes of smaller settlements dispersed around an "arm" or bay, churches (in some cases one Catholic and one Protestant, reflecting prevailing beliefs in the areas from which migrants had come), merchants' stores, and a few larger dwellings occupied by commercial or religious leaders marked a level of social differentiation (Macpherson 1972; Handcock 1985).

Precisely how many people participated in and depended upon this spreading, expanding fishery through the eighteenth and nineteenth centuries is impossible to know. Certainly there were some 80,000 residents in Newfoundland by 1840, most of whom lived beyond St. John's and almost all of whom depended upon the fishery (*Historical Atlas of Canada* 2: plate 8). Numbers were smaller elsewhere, where the fishery took its place alongside other economic activities. "The bulk of our people are farmers" asserted Joseph Howe, the provincial secretary of Nova Scotia in 1854; a "large body living on the seacoast are fishermen, but not fishermen only" (quoted in Innis 1940, 334). Contemporary estimates claimed that 10,000 fishermen worked the inshore waters of Nova Scotia in the early 1840s. This would suggest that the lives of 30,000 to 40,000 people—say 10 to 15 percent of those who lived in the colony—probably depended directly on the fishery (ibid., 347). Pretty much everywhere through the first half of the nineteenth century, the organization of settlements and production around the twinned axes of household and kin networks rendered the production process resilient to the perturbations of fluctuating markets, variable catches, and competition among merchants.

In the eighteenth and early nineteenth centuries the Atlantic fishery was a diverse and predominantly small-scale affair, conducted, for the most part, in small boats close to the shore (see also Brière 1997). The 10,000 fishermen of Nova Scotia worked no more than 250 shallops (light wooden boats, usually about 10 meters long, propelled by oars and/or sails) and some 3,400 boats (each of which was crewed by two or three men). Similar patterns prevailed in much of Newfoundland. Crews took fish of many kinds. Cod and pollock, herring and mackerel were caught at different times of the year. Yields varied greatly. According to Simeon Perkins, a merchant of Liverpool in Nova Scotia, boats from his settlement made 60 quintals of fish one season at the turn of the nineteenth century, whereas boats from nearby Shelburne averaged only 20 quintals (a quintal was 300 pounds of fish "fresh from the knife"). The catch was variously salted and dried (in different ways) or pickled in brine. Oil was an important by-product and fishing families also took salmon, eels, and alewives from rivers as well as other species from the sea, as opportunities offered.

Still, this was neither the sum of the industry nor a fair measure of the pressure that the fishery exerted on the resource. New Englanders were an important presence on the southern banks and in the near shore waters of Nova Scotia in the eighteenth century, and by the second quarter of the nineteenth they were fishing off the coast of Labrador. Traders from the Channel Islands established operations at several fishing stations around the Gulf of St Lawrence, and the leading firm, Charles Robin and Company, ranked among the leading fishing enterprises in British North America by 1850 (Ommer 1991). From its Gaspé station alone it exported approximately 25,000 quintals a year through the second quarter of the century.

In these years, too, New Brunswick, Nova Scotia, and Newfoundland interests developed more and more large vessels to fish the offshore. The trend was evident soon after 1800, when almost fifty schooners (and over 425 crewmen) mainly from the Conception Bay area fished the North Shore of Newfoundland. By 1815 there were 100 ships and 1,000 men involved in this activity. When rights to fish the Petit Nord were returned to France, the fishermen "betook themselves to Labrador" (quoted in Ryan 1986, 48; see also Cadigan and Hutchings 2002). By midcentury, over 400 vessels of over 30 tons went to Labrador to participate in an industry increasingly melded (through the use of the same men and ships) with the exploitation of a new marine resource, the so-called seal fishery (Sanger 1977; Ryan 1994).

Measured by its exports, the Atlantic fishery grew during the nineteenth century. Shipments of saltfish, the main product, from Newfoundland rose from some 600,000 cwt in 1800 to 1.6 million in 1874. This expansion was far from steady. There were peaks (1815—1.1 million; 1857—1.4 million) and valleys (1834—675,000; 1854—774,000) in the export trajectory. In broad terms, annual exports averaged about 900,000 cwt between 1811 and 1825 and about

1.15 million between 1855 and 1870. Similar patterns were evident in Nova Scotia. Dried fish exports from that colony were some 20,000 quintals in 1789 and at least 65,000 quintals (the exports of Halifax alone) in 1807. By 1841 they exceeded 325,000 quintals and in the early 1870s they averaged some 580,000 cwt. In Newfoundland, it has been pointed out, saltfish exports increased at a slower rate than did the population, suggesting that the growing number of people dependent upon the fishery received ever-decreasing returns for their effort. But the picture is complicated by the fact that much of the catch was consumed locally (Ryan 1986, passim). According to an estimate by Harold Innis, the Atlantic fishery's greatest historian, only a third of the Nova Scotia catch entered international trade. This seems low, but whatever the actual figure, it is clear that local consumption constituted a significant additional demand on fish stocks.

For Innis, the fishery was an "inherently divisive" pursuit. In describing it thus he sought to contrast the cod fishery, centered on the oceanic banks and open to exploitation from dozens of coves and harbors around the Atlantic periphery, with the linear structure of the Montréal fur trade, subject to organization and control by powerful interests located in its most important node. His characterization is conceptually arresting and not without a good deal of merit in identifying important differences in the earliest northern staple trades. In recent years it has been linked with Garrett Hardin's influential arguments about the "tragedy of the commons" to suggest that the fishery is inescapably subject to diminution due to its status as a common property, open-access resource subject to competitive pressures from rival interests.

Recent research has suggested that increased fishing pressure might well have begun to harm Newfoundland's coastal fishery as early as the late eighteenth century. An analysis of historical catch rates has led fisheries biologists Jeffrey Hutchings and Ransom Myers to suggest that if discrete substocks of the northern cod population existed in coastal bays, unsustainable demands may have been exerted on some bay stocks before 1800—a possibility pointed to earlier by Grant Head using historical evidence. Sean Cadigan has also discovered that interested observers of the fishery, such as merchants, who recorded the quality of the fishing in their bay, noted more bad days than good ones, and gradually seemed to lose interest in charting glum assessments. Newspaper reports from some of the longest-settled and most intensely fished areas—Conception, Trinity, and Bonavista bays—likewise give credence to the view that the fishery was failing, locally, in the third quarter of the century. Sensing this, and perhaps "aware of a growing imbalance between people and the primary marine resource they depended on for their livelihoods," some fishers agitated for restrictions on the introduction of new, more efficient fishing gear (Cadigan 1999a, 148).

Yet this was a hard sell. The introduction of cod seines, gillnets, and trawl lines (known as bultows) in place of traditional hand lines in the 1840s made fishing effort more efficient and staved off potential disaster when there was, as the editor of the Harbour Grace newspaper reported in 1864, "little or nothing doing with the hook and line but a fair catch with seines" (quoted in Cadigan 1999a, 153). Clearly, however, the new gear was more expensive than the old, and its deployment raised questions about equity and access to the resource. Moreover, when opponents of the new gear argued against it on broadly ecological grounds—that it was frightening fish from their accustomed migration routes or that the mesh size in the ever-larger cod seines was so small as to capture and kill young fish and other species—they were dismissed with the argument that it was French and American overfishing of the offshore that was the cause of the decline. No one knew with certainty, though many might have agreed with the assertion of the Liberal government in 1856 that, "It is intended by nature that . . . all . . . productions of the sea and land should be used for the extension of human food" (quoted in Cadigan 1999a, 155).

The issues seemed both imponderable and avoidable. If foreign exploitation of the offshore was depleting resources, what could be done about it? The sea was open to all. Revealingly, the Reciprocity Treaty between the British North American colonies and the United States, signed in 1854, effectively abrogated traditional rights of control over the seas within three miles of shore by allowing American vessels to fish in Canadian waters and vice versa. When the authority of Maritime law was reinstated, with the end of Reciprocity in 1866, it was strongly defended by Nova Scotia interests but quickly bargained away yet again in return for a guarantee that Canadian fish would be permitted free entry to the U.S. market.

In 1873 (1874 in Newfoundland) Americans were permitted to fish within the Canadian three-mile zone. Canadians had the reciprocal right to fish American shores, but it was widely agreed that this was an asymmetrical concession, and there was also to be a cash payment in compensation for the "excess value" of the Canadian inshore fisheries. For those facing shrinking returns from their local waters, the obvious solution was to fish elsewhere. Many of Newfoundland's inshore fishermen responded by "fitting out large craft to prosecute . . . [the fishery] at a distance" (quoted in Cadigan 1999a, 153). Fishing the headlands, the more remote waters of the North Shore, and then Labrador, required greater capital investment and changed the structure of the fishery, but by sustaining and even allowing a modest increase in total production, it made up for declining local returns and masked the decline of the inshore fishery.

Still, there were concerns about the decline of the resource. Fisheries across the North Atlantic seemed to be in difficulty, and coincident with the emergence of interest in resource management and fisheries science, British authorities called into question the traditional belief that fish stocks were inexhaustible. Reported in the British press and reprinted in Newfoundland newspapers, these discussions called for careful scientific investigations of the life expectancy and reproductive cycles of fish on the fishing grounds of the northeast Atlantic. "Exhausted shoals and inferior fish," read Newfoundlanders concerned about declining yields from their traditional fishing grounds, "tell us but too plainly that there is reason for alarm and that we have in all probability broken upon our capital stock" (quoted in Cadigan 1999a, 157).

In 1862 the Newfoundland government established a Select Committee to examine the colony's fisheries. After hearing a great deal of evidence—much of it revealing of deep concern about the state of the fishery and the effects of new gear; some of it exposing sharp conflict within communities over, for example, whether bultows "destroyed the mother fish"; and yet more defending the right of people to catch fish any way they wished—the committee released its proposals. They would have banned the use of caplin (on which cod fed) for manure, abolished bultows, limited the period through which herring seines could be used, limited the size of cod seines, restricted the use of jiggers, and ensured that users of different types of gear did not interfere with each other's endeavors. These were sound, practical, and conservationist recommendations. But they failed to win widespread support. In 1866 a British Royal Commission had dismissed the possibility that fish stocks in the northeastern Atlantic were declining, pointing mistakenly to the large catches taken by new vessels using new gear in new waters as evidence for its argument.

In the years that followed, occasional skirmishes and protest from outlying settlements suggested that not all Newfoundland fishers were as sanguine, and that some rejected the view that nothing could be done to regulate exploitation of fish because they were by definition an "open-access" resource. Scattered and lacking any greater authority than experience, their voices failed to carry the day, although some of considerable renown—such as the geologist, naturalist, and inventor Abraham Gesner of New Brunswick—joined them in recognizing the "startling fact that the fisheries are rapidly falling off along the entire North American coast . . . and this decline has evidently resulted, in a great degree, from the modern destructive modes of fishing and the practices of foreigners" (quoted in Cadigan 1999a, 160).

11

COMMERCIALIZING THE FOREST

Some of the effort that settlers poured into clearing the forest served a dual purpose, yielding fire and fuel wood as it opened the land to cultivation. Immigrants, accustomed to bundles of sticks and twigs or slabs of peat burning in homeland fires, were usually delighted at the abundance and low cost of fuel wood. "There is plenty of very fine wood here without any expense," Ann Cosens informed her English relatives from "Waterloo, America," in 1833. "We can get it anywhere by cutting it. We need not get cold. I often think of poor Jane, she told me that I would freeze to death in Canada. There is no danger in that" (quoted in Cameron et al. 2000a, 113).

Still, the winters were quite as cold as Jane had feared, and every household in the colonies consumed vast quantities of firewood. Using the standard cord (a stack of wood four feet wide, four feet high, and eight feet long) as a measure, some suggest thirty or forty cords per household; others consider this high. Of course, much depends on the size of the dwelling, and of the type of wood used (because maple and alder release more heat per unit of volume than pine), and these things cannot be known with precision. Assuming that settlers burned as much hardwood as possible (although it took more effort to chop), that most dwellings were small, and that colonists practiced a parsimony uncharacteristic of people who believe themselves amid plenty, annual consumption cannot have been less than twenty cords per household throughout this period. In broad and too simple terms, this amounts to an acre of forest. By this simple metric, the population of Canada West would have required the yield of at least 80,000 acres of forest to supply its fuel wood needs in 1842. For New Brunswick the equivalent figure is 30,000 to 35,000; for Nova Scotia, 40,000 to 45,000. Over the decade of the 1840s, then, a million acres of forest would have been used to meet the demand for domestic firewood generated by Canada West's growing population.

Commercial, metropolitan demands further heightened the early nineteenth-century assault on British North America's forests. When Napoleon closed the Baltic to British shipping early in the nineteenth century, wood from New Brunswick and the St. Lawrence became the great staple of British North America. Shipments increased (from modest beginnings and with notable annual fluctuations) through the pre-Confederation years. Initially, transatlantic shipments were made up, almost entirely, of square timber. This was wood in relatively unprocessed form, produced by axemen who felled and "squared" trees into great balks, at least 10 feet long and of 10 inches a side (most were considerably larger), before they were shipped to the British Isles to be sawn into lumber (such as boards and planks). After 1830 the trade expanded and diversified, as colonial producers invested in and developed mills capable of producing sawn lumber of the quality required for the British market. Taken as a whole, the output of timber and lumber from British North American forests climbed dramatically in the years before 1867 (Wynn 1981, 1985; Lower 1973).

In 1825, when timber exports from New Brunswick peaked, they exceeded 400,000 tons. Calculated as 40 cubic feet, each ton might be thought of as equivalent to a single piece of timber 12 inches on each side and 40 feet long. Given the size of the pine trees that were the lumbermen's main target in the New Brunswick forest, it might then be suggested, in very approximate terms, that some 200,000 forest giants were toppled to produce the timber exports of this year. Twenty years later, when the trees that remained in the forest were undoubtedly smaller, on average, and square timber exports from the whole of British North America spiked at 40 million cubic feet, one might guess that 600,000 or more red and white pines were felled to meet the square timber demand.

At the same time, sawn lumber exports had risen appreciably. From New Brunswick alone they increased more than eightfold in the second quarter of the century. By 1867 sawn deal shipments topped pine timber shipments from the Canadas to Great Britain in value, and recently developed markets for planks and boards in the United States began to rival those across the Atlantic. Again, the implications for the forest are arresting. In Québec alone over one million pine sawlogs were cut in 1867; in Ontario the sawlog harvest was approximately 200 million board feet (say, 24 million cubic feet), which, allowing for the loss of wood to the saw kerf and a further decline in the size of trees, probably amounted to another million trees in the same year. Adding spruce, a still minor but increasingly important species for lumber, and square timber production to the tally produces the bold guesstimate that some three million trees were probably cut for commercial export from the five British North American colonies in the year before Confederation.

Initially, lumbermen logged selectively. In the mixed hemlock–white pine–northern hardwood forests found in most of the colonies, they were interested only in pine, and because the work entailed in getting these trees to market was daunting, they tended to fell only the best specimens located close to rivers of at least modest size. This meant that they generally proceeded rapidly inland. Throughout this period, logging was a wintertime activity. Trees were more easily felled when the sap had ceased to run, and it was immeasurably easier to haul them from stump to riverbank on frozen ground and compacted snow/ice "roads." Skill and muscle were required for these tasks. Once a tree had been brought down, branches had to be removed from its trunk before it was roughly squared and hauled out by oxen or horses. Although logging was selective, much damage was done to the forest by falling trees, by the clearing of hauling roads, and potentially by the accumulation of debris on the forest floor. Once at the frozen riverbank, logs were piled to await the spring freshet, when they could be floated downstream, either individually or assembled into rafts, for delivery to the ports from which they would be shipped across the Atlantic (Wynn 2003).

Thousands of colonial residents were employed, directly and indirectly, by this expanding activity. It provided farmers with winter work and the prospect of earnings sufficient to buy an ox or hire summer labor. It offered casual employment to young men anxious to build a "stake" with which to acquire their own farms. It was opportunity for the swelling band of men without land. We will never know exactly how many people worked in the woods during these years. By some contemporary accounts, written by people clearly antagonistic toward the timber trade, it appears as though almost every able-bodied individual in the vicinity flocked to participate in the adventurous (and critics said dissolute) life of the lumberman. Such claims are clearly exaggerated (Wynn 1980). Even in relatively small districts heavily involved in lumbering, it was rare for more than half of the adult male population to engage in woods work during the winter. About 8,000 men, employed pretty much year-round in preparing roadways, felling trees, hauling out the cut, and bringing it downriver, could probably have produced and delivered the timber brought to Québec from the Ottawa Valley in 1845–1846.

Clearly many other economic consequences flowed from the commercial exploitation of the forest. Lumbering created new markets and improved prices for certain types of agricultural produce, especially oats and hay and pork and peas (to feed the men and beasts employed in lumbering camps). It encouraged investment in sawmills and the development of shipbuilding and other industries, all of which provided jobs and put money into local communities. As one measure of this, note that the New Brunswick census enumerated 4,300 mill hands in the province in 1851. For these reasons, many regarded the timber business as almost beyond criticism. One midcentury commentator effectively described the New Brunswick trade as the classic staple that it was, long before economists connected the term and the concept. This activity, he argued, "has brought foreign produce and foreign capital into the Province, and has been the chief source of the money by means of which the country has been opened up and improved; by which its roads, bridges and public buildings have been completed; its rivers and harbours made accessible; its natural resources discovered and made available; its Provincial Institutions kept up and its functionaries paid" (quoted in Wynn 1981, 33–35).

More difficult to measure, not least because contemporaries rarely remarked upon them, were the ecological changes set in motion by nineteenth-century lumbering. Over the years, selective exploitation effectively weeded prime species from the forest. We have no way of knowing how much pine grew in British North America in 1800, or 1860 for that matter. Yet it is clear that the forests of eastern British North America were substantially changed by logging and fire. Large pine, once so plentiful, had largely disappeared from most southern areas by the second half of the nineteenth century. As the demand for wood increased and the best trees were removed from the most accessible locations, patterns of exploitation changed. Smaller trees were taken out, and lumberers used ever-smaller streams to "drive" their timber out of the woods. More and more areas of the forest felt the lumberers' axes. Logging took place further and further from settlements.

Decade by decade, operations grew larger and more specialized. Although farmers, alone or in small groups of friends or relatives, continued to work forests near their dwellings and to trade small quantities of logs or timber for credit at a local mill or store, larger lumbering parties worked more winter days in remote locations, often on contract to large mill owners or entrepreneurs. Soon such groups, moving ever further into the hinterlands, felt impelled to build better access roads to supply their camps and to "improve" the streams and smaller rivers on which their operations concentrated. Rough road lines were cleared into the backcountry. Explosives were applied to remove rocks and rapids that complicated and "jammed" the drive. Flumes were built to carry logs and timber around irremovable obstacles. And dams were built to flash accumulated water downstream to lift and float the cut through otherwise difficult sections of the streams across which they stood.

All of this had environmental consequences. As capital was invested in dams or flumes, it became costlier for lumberers to move on and more likely that more and smaller trees would be removed from the forest above their "improvements." Flash floods and flumes altered the hydrological regimes and water levels of streams or particular reaches thereof and enhanced scour along banks and beds. Dams backed up water and flooded hitherto dry ground. More insidiously, logs sank or came ashore on islands and meadows, and sawmills habitually dumped sawdust, bark edgings, and off-cuts into the streams on which they stood. When sawmills were built near the mouths of rivers, their dams typically obstructed the spawning migrations of salmon and other species upstream.

By and large, even these fairly obvious and immediate environmental consequences were of little concern to contemporaries. Many of them went unnoticed, and others were accepted as necessary costs of development, even as signs of progress. In this vein, it is revealing that the English Prince of Wales took a much celebrated ride down a large flume built to carry timber around the Chaudière Falls on the Ottawa River during his visit to the colonies in 1860.

Here and there, however, disquiet surfaced. Initial concerns over the dumping of mill refuse in streams reflected anxieties about the interruption of navigation. In New Brunswick this was a particular issue on the St. Croix River and in the harbor of Saint John, where there were several large mills. Late in the 1840s a report to the provincial assembly claimed that so much sawdust was being dumped into the river and harbor of Saint John "that it is impossible for the whole of it to float on the surface till carried away by the tide," and that "a large proportion necessarily sinks to the bottom and is daily rendering the slips and floats around the shores more and more shallow, and also, most probably, proportionately filling up the main Harbour itself" (quoted in Wynn 1981, 94; see also Gillis 1986).

Similar effects on nonnavigable streams were ignored. Concerted efforts to avoid the fouling of rivers with mill waste did not come until late in the century. Although the earliest settlers of Charlotte County in New Brunswick ensured that fishways were built around milldams there, the legal provisions they established were ignored when the large Union Mill Dam was built to meet the needs of an expanding milling industry in 1826. Elsewhere mills barricaded streams near the head of tide with impunity. Noting this, the New Brunswick lawyer, naturalist, colonial office holder, and prolific author Moses Perley expressed indignation but not surprise when his midcentury survey of the province's river fisheries revealed that catches had declined almost everywhere in the previous two or three decades. Where economic progress was the overarching goal; where anxiety for short-term gains trumped concern for long-term consequences; where legislators lacked the knowledge, courage, and constituency to stand against the tide, there was little ground in which environmental concerns might grow. The natural world was essentially regarded as an input into the development process; for the most part, its qualities and its forces were ignored (Allardyce 1972).

Fire was another matter. It was the frequent ally and occasional scourge of settlers across British North America. Invaluable as a source of heat, it was indispensable in the making of farms. Subjugated and controlled, it was an integral part of the settlers' technological arsenal. But liberated by the wind and fueled by nature, it often ran decisively and disastrously beyond the limits of human management. Wildfires were endemic during these years. Every summer, clearing fires escaped the limits within which they were put to work. Acres of standing forest were burned. Wherever this occurred, and depending upon the intensity of the fire, the ground would be left "almost destitute of verdure."

Fireweed, raspberry bushes, Labrador tea, and associated species would appear within a season or two; five years after the fire, young shoots of light-tolerant species—from alder and aspen through birch to pine—would be evident; ten to twenty years on, young birch trees "with their silver stems and fresh green leaves" grew amid charred trunks and fallen trees in what contemporaries sometimes described as "snarls" (Alexander 1849, 181). Sixty or seventy years after this, and depending upon the soil and drainage conditions of the site, a grove of straight young pines might occupy the land. Most of the blazes that initiated these changes were small, confined in both space and time by natural circumstances. An open clearing, a swamp, a shift in the wind, a change in the weather, each and all might limit the extent of destruction. There were losses, to be sure. Flames consumed dwellings, stock, and human lives, as well as trees. Local newspapers noted these things, but treated them for the most part as unfortunate setbacks and inescapable tragedies. Every so often, however, fires consumed larger areas and the flames jumped into the headlines.

The most notorious, and tragic, example of this occurred in New Brunswick in 1825. After a dry summer, fires were reported early in October in the forest near Fredericton. Quickly, they raced toward the city, parts of which were burned. Elsewhere the damage was even more extensive. On the banks of the Miramichi River, the prosperous town of Newcastle was reduced to a smoking ruin; not one in ten of the settlement's two hundred houses and stores survived intact. The destruction was just as severe in smaller, neighboring settlements. Beyond, fire—fueled by the debris accumulated through twenty years of timber making—destroyed enormous tracts of forest. Contemporary accounts referred to the "lengthened and sullen roar" that "came booming through the forest," late on October 7, "driving a thousand massive and devouring flames before it" (Cooney 1832, 69; Wynn 1985; Ganong 1906). Entire parishes were laid waste, and in the cold light of the morning after the fire, former settlements exhibited a "vast and cheerless panorama of desolation and despair" (Cooney 1832, 76).

Two weeks after the fire, a New Brunswick newspaper reported that burning debris falling into rivers had led salmon to abandon them for terra firma! But oft-repeated claims that the fire ravaged six thousand square miles of New Brunswick are probably exaggerated. So, too, were contemporary attempts to link the New Brunswick fires with other, roughly contemporaneous, burns in Maine and Upper Canada into accounts of the "ungovernable rapidity" with which the flames had wrought their "extensive range of mischief" (ibid.; Wynn 1985). All forest fires trace seemingly idiosyncratic paths through space, shaped by local winds, terrain, and other conditions; some areas are incinerated, others left largely unscathed. What we do know, at this remove, of the "Miramichi Fire" is that there were several more or less simultaneous and highly destructive blazes across northern New Brunswick. Four hundred square miles of timberland were burned over in the vicinity of Newcastle; large areas along the Southwest and the upper Northwest Miramichi rivers were similarly destroyed. So, too, were thousands of acres north of Fredericton. Although settlers fled, and families reportedly sought refuge in the nearby streams as the flames roared about them, at least 160 people died in the fires of October 7.

"[A] greater calamity [than this "Miramichi Fire"] . . . never befell any forest country," wrote Robert Cooney, a resident of the region and, a few years later, the author of one of its first histories (Cooney 1832, 73). For Cooney, a Methodist minister, and others, there were moral lessons in these tragic events. The death and destruction visited upon the people of New Brunswick by the fire were, he insisted, a form of divine retribution for the dissipation and debauchery that moralists associated with the timber trade. Sensational comparisons sometimes likened the lumbering towns of eastern Canada to the Sodom and Gomorrah of biblical legend. But for all their lurid qualities, such critiques owed more to the deeply rooted agrarian values of English tradition than they did to everyday circumstances in the British North American colonies. Articulated by those who saw the "adventitious," opportunistic, commercial, and male-dominated characteristics of the lumber industry as a threat to their conviction that farm life, rural existence, and family values formed the bases of the ideal society, they were ideological claims rather than careful descriptions of nineteenth-century circumstances (Wynn 1980).

No one has seriously disputed Cooney's claims about the unprecedented magnitude of the Miramichi Fire, especially in light of the loss of human life in New Brunswick in 1825, but this should not blind historians to the fact that the forests of northern New Brunswick continued to yield considerable quantities of timber after 1825, or to the less notorious but still substantial cumulative effects of fire in the forest across British North America. Blazes consumed hundreds of acres every year. At least initially, when the forest seemed endless, this was rarely a cause for concern. But in the late 1840s, when lumbermen began to appreciate that there were fewer of the large pines upon which their industry had been built, and that the forest beyond the Ottawa-Huron tract (between the Ottawa River and Georgian Bay) was less rich than that to the south, the loss of timber to "careless" fires became a heated issue. As the supporters of agricultural expansion urged the extension of settlement into the Shield along roughly opened routeways that extended from older settled districts into the northern forests and that were named "Colonization Roads" in express recognition of their purpose, lumbering interests resisted their advance, protesting at the loss of timber attributable to the settlers' incursions into the forested Shield, and arguing for the establishment of a "fair line of demarcation" between lumbering and farming activities.

These were not inconsequential developments. As environmental limits began to pinch the prospects of those interested in continuing exploitation of regional resources, recognition of the pragmatic value of retaining Crown ownership and leasehold tenure of the forest may have sharpened. And as farming foundered in these northern fringes of the habitable ecumene, views of the tree- and lake-covered rock of the Shield began to shift. The "land of hope for emigrants-soon-to-become-farmers" was soon transformed, in the popular imagination, into "a harsh and lonely northern realm." Later in the nineteenth and early in the twentieth century, this perception achieved iconic status as Canadian novelists, poets, painters, and others set about framing a distinctive, northern identity for their country (Wynn 1979; Berger 1966; Bordo 1992–1993; Campbell 2005).

12

FIELDS OF EXERTION

In the 1850s and 1860s many celebrated the achievements of British North American settlement. Positive, optimistic, even euphoric assessments of the colonies abounded. Nova Scotia, wrote one essayist, had "the healthiest climate under the sun," and "inexhaustible resources" (Knight 1862, 80–81). Its soils bore comparison, in the words of another, with those on the "Banks of the Nile" (Gesner 1849, 167–171). In the neighboring jurisdiction, visiting British soil chemist and agricultural expert J. F. W. Johnston, hired by local authorities to produce a *Report on the Agricultural Capabilities of the Province,* came to the arresting conclusion that the soils of New Brunswick were considerably more fertile than those of Ohio.

And in Upper Canada many a prize essayist extolled the advancements of the age. According to J. Sheridan Hogan, for example, the colony was a "matter of wonder and instruction" for the way in which "upwards of a million of the working classes had within a short space of time, and by means hitherto unknown or unthought-of, raised themselves to comparative affluence and independence" (Hogan 1855, 1). In much the same vein, major towns were celebrated for their imposing official buildings (courthouses, legislatures, customs houses, gubernatorial residences) and described as places where "all is a whirl and a fizz and one must be in the fashion" (quoted in Wynn 1987b, 257).

Settlers had changed British North America as they occupied it. From the shores of St. John's capacious harbor (as Joyce Macpherson has shown) to the farthest and most remote frontiers of their engagement with North American space, the newcomers had set processes of environmental change in motion. The colonies of the mid-nineteenth century were significantly different from those entered and made by earlier settlers. The social and environmental verities to which settlers had clung so staunchly in the 1830s and earlier—that these were lands of opportunity and inexhaustible resources—were neither as self-evident nor as convincing as they once had been. The mantra repeated endlessly in prose and verse during the third quarter of the century, that work and self-improvement would yield economic success, rang increasingly hollow.

While contemporaries were assured that "The noblest men I know on earth/ Are men whose hands are brown with toil; / Who backed by no ancestral

The Sunny Side stock farm, Prince Edward Island, 1870s. *As the rigors of pioneering faded into memory, British North Americans-become-Canadians marked the substantial material progress they had achieved in dozens of "county atlases." Produced on a subscription basis, they offered established settlers an opportunity to memorialize themselves and their achievements with illustrations such as this depiction of Robert Fitzsimons's farm near New London, Prince Edward Island. Notice the tidy landscape, the majestic setting, the ample orchard, the substantial dwelling, the impressive cluster of farm buildings, the well-bred livestock and even the trotting track—a sure sign of wealth and leisure attained—in back of the farmstead itself. (National Archives of Canada)*

groves,/ Hew down the wood and till the soil" (quoted in Holman 2000, 20–22); opportunities for ordinary men (and women) to get ahead by getting onto the land had diminished significantly by the 1860s (Wynn 2002). Pioneers who had come with little and succeeded bequeathed material advantages to their children. As the generations turned, some were afforded a head start, and others faced what the previous generation had not: opportunities preempted by prior arrival. So long as land was readily available nearby, the implications could be avoided. But the British North American ecumene was limited. By midcentury settlers were running up against the thin soils of the Shield or the short growing seasons of the uplands. They were encountering the limits of agricultural possibility. The social and economic distances between people were both increasing and increasingly entrenched).

Conditions in the colonies' growing urban centers were also troubling, even though the largest and busiest early nineteenth-century cities were small by modern-day standards. Montréal had 40,000 residents in 1840, and Toronto 44,000 in 1860. In these places, and in their smaller counterparts scattered across the territory, however, oxen and horses pulled carts and carriages through the streets and left manure (some 15 to 30 pounds per horse per day) in their tracks. Cows and pigs and chickens lived alongside humans in most urban centers, and butchers and tanners practiced their trades alongside druggists and lawyers and bankers. Quite typically, one "delightfully situated

Toronto, Canada West, 1854. *This hand-colored lithograph by Edwin Whitefield offers a glowing, almost idyllic representation of expanding mid-nineteenth-century Toronto. Front Street separates the town from the (symbolically) busy harbor. A few churches dominate the skyline, and there is an obvious concentration of commercial buildings in the center of the picture, but the countryside is close at hand, there is a good deal of greenery within the town, and several citizens have done well enough to erect spacious and attractive dwellings within ten minutes' walk from the center of this pedestrian city. (National Archives of Canada)*

cottage Residence," with a view of the lake and harbor in the rising town of Cobourg on Lake Ontario in 1843, had the trappings of considerable refinement—five bedrooms, dining and drawing rooms, a china closet, and a garden lawn—but it also boasted a stable yard and a "three stable cowhouse" (Wynn 1987b, 263).

The urban centers of British North America were, to use the American environmental historian Ted Steinberg's term, organic places in which pigs foraged in streets, in which most human waste accumulated in cesspools and privies, and in which domestic water was generally drawn from wells. As they expanded, the problems associated with urban living increased. Human effluent overflowed from privies and slopped into streets when these were emptied. Animal manure attracted flies (that transmitted diseases), dried in the sun to be pulverized by passing traffic and blown about in the air, and lay beneath winter snows to produce a fetid stench in the spring. As populations grew and filth accumulated, typhoid and (most devastatingly) cholera, carried across the Atlantic on emigrant ships, spread through the cities.

Québec City, Canada East. *Commanding the entrance to the St. Lawrence, Québec was important strategically, but by the mid-nineteenth century, when this illlustration was made, it had been surpassed in population and commercial importance by Montréal and Toronto, in particular. Here the upper and lower sections of the town are visible as well as the busy commercial artery of the great river. Like most towns of its time and size, Québec City in 1860 was a compact and complex place where land uses and urban functions were mixed in tight and intricate proximity, and "nature"— whether the "countryside" or the horses, swine, and chickens that moved through city streets—was never far away. (Illustrated London News Picture Library)*

For several dark days in mid-June 1832, when cholera ravaged Québec and Montréal, more than 100 people a day succumbed to the disease in each city. By September, when the epidemic ended, it had claimed over 5,000 victims in these two cities, and several hundred more beyond (Bilson 1980). Two years later another epidemic exacted its lesser toll. Believing that the disease was spread by miasma in the atmosphere, contemporaries were at a loss to prevent its transmission. Not until 1855, when Dr. John Snow famously attributed a cholera outbreak in London, England, to contamination of the local water supply, were the correlated perils of filth, pollution, and waterborne diseases that afflicted early nineteenth-century urban dwellers more properly understood.

Prevailing spendthrift attitudes toward the environment and resources posed a more complicated set of concerns. In early Canada and British North America, as in virtually all pioneer societies, the majority of settlers had assumed with Marc Lescarbot on the shores of Acadia early in the seventeenth century that "Man was placed in this world to command all there is below" (Lescarbot 1907–1914, 3: 137). Contemplating plenitude, there seemed little need for restraint. "Wild Geese / And Cormorants," wrote Lescarbot's contemporary, the Sieur de Diereville, "roused in me / The wish to war on them" (Diereville 1933, 75–77). For two centuries, the assault on nature's New World bounty portended by such comments continued almost unabated. Although statutes and regulations served to restrict or slow resource utilization and environmental exploitation in the late eighteenth and nineteenth centuries, most such initiatives were intended to generate revenue or to prevent conflict among users rather than to conserve and protect nature. In those few instances in which there were efforts to reserve resources, settlers generally honored restrictions in the breach.

Short-term, utilitarian, spendthrift attitudes were never entirely hegemonic, of course. Farmers developed detailed knowledge of their surroundings, and in some of the earlier-settled districts of the colonies some had developed an affection for their land that led them to take a long view of its importance for the continuation of their children's and their children's children's futures in the region. As early as 1663, for example, Pierre Boucher, governor of Trois Rivières, coupled detailed, closely ordered observations with an unusually "affectionate appreciation" of Canadian nature in his *Histoire Véritable et naturelle des moeurs et productions du Pays de la Nouvelle France vulgairement dite le Canada* (Berry 2001). In this account, New France is presented as "a good country," its beavers are possessed of "wonderful skill," and the flying squirrels of the forests were "delicate and pretty" creatures.

In the nineteenth century, societies devoted to the study of natural history were formed in many towns. Their members spent much effort in studying, collecting, talking, and writing about the wonders of nature, and in making inventories of the birds, plants, fish, animals, and insects within easy radius of their homes. Others, influenced by William Paley's *Natural Theology; or, Evidences of the Existence and Attributes of the Deity* (1802) found the hand of God in "the appearances of nature." In Lower Canada in the 1830s, Philip H. Gosse, an Englishman who had come to the eastern townships via a commercial apprenticeship in Newfoundland, attempted to emulate the English country parson Gilbert White, author of the *Natural History of Selborne* (1789), by setting down a detailed description of the countryside surrounding his farm through the different seasons of the year in *The Canadian Naturalist*, published in London in 1840.

For Gosse, "Nature was a teacher," and everything that the naturalist saw tended "to enrapture and delight him" (Gosse 1840, 337, 360). At much the same time, Catharine Parr Traill, an English gentlewoman and new settler in the Newcastle District of Upper Canada, reflected that the scientific botanists of her homeland would consider her "very impertinent in bestowing names on the flowers and plants I meet with in these wild woods." But she averred in defense, "I can only say, I am glad to discover the Canadian or even the Indian names if I can, and where they fail I consider myself free to become their floral godmother and give them names of my own choosing" (Traill 1836, 102, 120).

By and large, however, most nineteenth-century commentators were quite acerbic in their purportedly practical critiques of the extensive, exploitative, land-blighting techniques employed by British North American farmers. To read these assessments is to be informed that most Upper Canadian and Maritime farmers practiced a wasteful form of agriculture in the early nineteenth century. Labor was expensive, by comparison with land, and there was correspondingly less incentive to maximize production per unit area. Manure generally went uncollected, even when animals were housed in barns over the winter. Most settlers were said to be too lazy to engage in such work—or, at best, they are counted to have shrewdly, if misguidedly, figured that the costs of processing, carting, and handling the manure to spread it on distant fields exceeded any benefits that might derive from the work. In Nova Scotia, Scots were at the center of countless stories that joined their failure to manure with their legendary thriftiness in criticisms of their farming practice. Elsewhere, critics lamented the appearance of fields full of "weeds and wild strawberries" and condemned farmers for "scourg[ing] the land most unmercifully" (Haliburton 1862, 27–28; MacNeil 1985, 1986).

On balance, there can be no question that most farmers were spending down the capital accumulated in their land. Whether they knew it or not, they were using up the wealth invested in the earth by biological processes that had been operating for centuries. But whatever reasons underlay these prevailing practices and gave some substance to criticisms of them, it is equally clear that many of the critics' claims were exaggerated. Reflecting the prevailing discourse of agricultural improvement in an age of progress, these individuals wove reports and personal observations with preconception (prejudice) and ideology into a seamless rhetorical web.

Even in the eighteenth century, colonial farmers were falling short of the expectations of their "social superiors," educated English gentlemen accustomed to the precepts of "improved agriculture" in their homeland. In Halifax in 1789 the colony's bishop and lieutenant governor joined with other gentlemen of the town to lament that colonial farming "was carried on . . . without system," and to form a "Society for Promoting Agriculture in Nova Scotia" to change things (Martell 1940). Even the promise of medals for those who raised the caliber of their farming proved ineffective, however. When "Agricola" took up the cause of improvement thirty years later, in a series of letters to a local newspaper that were full of advice for farmers, he lamented that the principles of vegetation "were grossly misconceived" in Nova Scotia, where farmers neither rotated their crops nor tended their summer fallows properly (Young 1822, introduction). In the 1820s and again in the 1840s the Colonial government funded a Board of Agriculture and made monies available for the support of local agricultural societies that were intended to import and disseminate seeds, stock, and implements and to offer prizes to encourage agricultural improvement.

Similar initiatives occurred in the Canadas and in the other Maritime provinces. Everywhere improvers sought to "excite a spirit of emulation across the countryside" (Wynn 1990a). They held exhibitions; awarded premiums; imported stud horses and sheep and oxen; distributed selected seed; and circulated texts on improved agriculture, agricultural chemistry, and stock husbandry. Here and there, small groups of farmers imbued with desire for improvement and sufficiently wealthy to bear the extra costs in time, labor, fencing, and infrastructure that even North American variants of English "high-farming" required stood as exemplars of "new and improved principles of cultivation and better methods of securing manure, making composts and a greater use of lime and new improved implements of husbandry" (Wynn 1990a, 39).

Steven Stoll has argued recently that the American counterparts of these British North American "improvers" recognized "a link between an enduring agriculture and an enduring society," and adopted the term *improvement* to signify "changes that enabled land to be cultivated in the most prosperous possible way over the longest possible time" (Stoll 2002, 20). In this view, advocates of improved agriculture embraced an ethic of permanence; they "saw the folly of land-financed expansion in the nutrient bankruptcy of common husbandry," and sought "enduring occupancy through conservation, pointing toward the creation of stable agro-environments." In Stoll's words, "Americans in the old states [who had exhausted their soils and laid waste their farms] saw the bones of the earth after the Revolution" and the question became whether farmers could "learn to prosper by generating fertility in one place, even while knowing that numberless acres lay beyond the mountains" (Stoll 2002, 31).

Whatever the merits of this argument for the United States, it has less purchase in the northern colonies. There, the timing of settlement and of the articulation of improving principles, as well as the limits of the agricultural ecumene (that made it harder to contemplate lighting out for numberless acres to the west), suggest other frames for understanding the passion for improvement.

In British North America, as indeed in the United States, improved farming was an obsession of "merchant squires" rather than of yeomen, of officeholders, merchants with country properties, and military men on half-pay pensions who could afford its costs, rather than of the majority of settlers for whom the land was the basis of existence. The enthusiasm for improved agriculture was also clearly derivative; it drew its pedigree from British high farmers, through British authorities, and appealed to colonials whose aspirations were determined by reference to British standards.

As those active in agricultural societies all too often complained, most settlers revealed a "dogged obstinacy" in their adherence to extensive rather than intensive farming practices. Only a small fraction of farmers in any district paid the subscription necessary for membership in an agricultural society, and most of those who did so seem to have had as much interest in receiving a share of the government largesse distributed by the societies as prizes and premiums as in the thoroughgoing transformation of their agricultural practice. Improvers scratched their heads at "the apathy, nay the positive indifference, of the *Plodholes* around them," but all things—and especially the expense and disappointments entailed in trying to implement English agricultural methods in colonial conditions—considered, the vast majority of settlers who kept their stock in fields rather than barns, who "turn[ed] up the sluggish soil, sow[ed] and reap[ed] in due season" and who tended as best they could to the myriad of tasks confronting them, showed an acute intuitive appreciation of their particular circumstances (Wynn 1990a, 49, 50).

There were environmental consequences to the ways in which settlement and development unfolded, and settlers were not entirely blind to them. Across the colonies, people recognized that crop yields fluctuated, that some birds and animals were no longer as readily abundant as once they were, and that time and again it might have been said that the "spot over which the Indians roved, free of control, is now a large and wide-spreading town" (Traill 1852, 209). They also knew, all too well, that fires destroyed property and took lives, that spring floods could be catastrophic, that mill dams obstructed fish runs, that mill ponds flooded stream margins, that sand and silt washed downslope, and that sawdust fouled riverbeds. Had they stopped to reflect, they might also have noted, though likely not without a certain "pride and satisfaction," that in village and town upon village and town the "sound that falls upon the ear is not the rapids of the river, but the dash of mill wheels and mill dams, worked by the waters" (Traill 1852, 210). Yet it was no easy matter, in the mid-nineteenth century, to understand these isolated events as expressions of a broader, cumulative, and ultimately injurious transformation of the earth.

For all the detailed, local environmental knowledge that newcomers built up through years of labor and a far more intimate set of connections with land and/or sea than the majority of twentieth-century inhabitants of these territories would know, most eighteenth- and early nineteenth-century settlers were unable, for want of time, inclination, or broader understanding, to translate local observations into a coherent, integrated account of the effects of human actions upon the environment. Even those who endeavored to "husband the earth" and to ensure the continuing productivity of their hard-won farms interpreted what they saw about them in light of prevailing ideas about their settings and about the challenges of colonial development.

By and large, the colonies were regarded as new and raw lands in need of the most basic form of improvement: the conversion of waste or wild land to uses that would sustain human lives. For most of those who occupied British North America in the nineteenth century, the New World was a field of exertion. They knew along with Nova Scotian T. F. Knight that "a country is what its people make it be," and they poured enormous energy, industry, and self-denial into realizing the potential of their new lands. Even in the 1850s and 1860s, it was commonly said that the colonies were still in the adolescence of their development. There was much growing to do.

Rare were those who could look beyond these considerations and see the cumulative long-term impacts of their actions. Indeed, the American George Perkins Marsh has long been celebrated as the pioneer of new ecologically sensitive attitudes toward the environment for doing just that in his 1864 book, *Man and Nature; or, Physical Geography as Modified by Human Action.* Here, he argued cogently (and some have said for the first time) that humans had the capacity to transform the face of the earth, and more specifically that deforestation and the cultivation of hill slopes quickened erosion.

Some have challenged the originality of these claims by insisting that Marsh borrowed from the wisdom of ordinary farmers, "stood on the shoulders of rural reformers to formulate his ecological ideas," and "copied conservation insights and observations already widely disseminated" (Judd 1997; quotes from Lowenthal 2000, 419–421; Wynn 2004). Others such as Richard Grove have found the origins of modern environmentalism in responses to the destructive social and ecological consequences of imperial encounters with small Atlantic islands and to colonial ventures in subtropical India and Africa. Yet Marsh's champions have remained resolute in the view that he was a lonely prophet, far ahead of his time, in "rethink[ing] the long sweep of human history," and "cautioning against the risks of careless growth" (Lowenthal 2000, x, xii–xiii).

Early in 1835 residents of Halifax were informed, in no uncertain terms, that fifty years of colonial settlement in Nova Scotia had exacted serious ecological consequences. The bearer of this message was Titus Smith Jr., known to many in the city to which he had come as a child in the 1780s as the "Dutch Village Philosopher." Commissioned in 1801 to survey the little-known and largely uninhabited interior of the colony, Smith had spent the early decades of the century farming and engaged in close observation of the flora and fauna of his adopted home. A founder of the Mechanics' Institute, he was regarded by his contemporaries as a giant intellect and was a towering figure among the relatively small population of learned individuals in early nineteenth-century Nova Scotia.

On January 14, 1835, a small audience gathered in the Halifax Mechanics' Institute heard Smith reflect upon the "Natural History of Nova Scotia." His talk, noted the *Halifax Times* of January 27 that year, abounded "with original views, [and] minute and curious information." It began in natural theology, paid obeisance to the work of "the great Cultivator," and continued to describe the intricate mechanisms that bound biotic communities together in each of Nova Scotia's two great vegetation zones, broadly characterized as the deciduous and the evergreen forests. In this it was well in advance of its time, foreshadowing the work of many twentieth-century botanists, ecologists, and foresters.

But this was not all. A substantial part of Smith's message in 1835 concerned the effects of human settlement upon the vegetation and soils of the colony. By his account, human imprints upon the flora and fauna of Nova Scotia were slight before 1783. Here he strategically ignored the settlement of New England planters and fishermen after 1760, the foundation of Halifax in 1749, and nearly 150 years of Acadian marshland reclamation to argue that the great influx of Loyalists to the colony in the 1780s had disturbed the equilibrium within which indigenous peoples had lived in this land.

Establishing settlements the length and breadth of the colony, Smith argued, the Loyalists had set clearing fires that ran out of control and destroyed extensive tracts of forest. They pastured cattle on the initial regrowth, burned the land again, and quickly exhausted the soils. By clearing, burning, and cutting trees for fuel, even the swamp-forests that might have served as seedbeds for regeneration were largely destroyed in areas close to settlements. Rains washed organic matter from newly cleared lands. Kept in clear fallow, soils lost their "fertile principles." Continued through the half-century following the Loyalists' arrival by other, equally spendthrift newcomers, these processes had serious and cumulative effects. Nova Scotians, Smith told his listeners, were neglecting the dictates of nature and were sure to suffer in consequence. They were embarked upon much the same course as earlier generations elsewhere, who had "by mismanagement, impoverished some of the finest countries on earth." Consider, he intoned, that "Ancient Syria and the neighbouring countries . . . present such an appearance of sterility, that, were it not for the magnificent ruins that remain, it would be almost impossible to credit the accounts that historians have given of their population in former ages" (Smith 1835, 656–658).

Smith was no Marsh. He lacked the Vermonter's ringing turns of phrase; after making his potent claims about the long-term consequences that flowed from the disturbance of nature's harmonies, he moved on to frame a social critique of industrial society and human avarice rather than to call, on that January evening, for greater public responsibility toward the natural world; and his insights were never published for broad public consumption. Yet Smith's sharp and solitary awareness of the effects of colonial development upon New World environments is a powerful reminder both of the magnitude of settlement's impact and of the challenges inherent in understanding the consequences of those impacts where so much was predicated on development, and most colonists were wedded to the idea of progress.

Time and again nineteenth-century British North Americans were reminded, in one way or another, that nations had to advance with the advancing and to improve with the improving. Few were able to stand aside from this stream of tendency. If some paused to find the woods "lovely, dark and deep," most kept their eyes fast on a future in which farms replaced forests, and prosperity reigned above modest competence (Frost 1923). "O wail for the forest, the proud stately forest/ . . . O wail for the forest, its glories are o'er," wrote Catharine Parr Traill (quoted in Forkey 2003, 106), whom some have called the "Canadian Gilbert White" on the strength of her love for Canadian nature.

In blind anticipation of Robert Frost, most pioneers felt that there were implied promises to keep, and miles to go before they could quit the path of development upon which their settlements were so energetically embarked. Like Dunbar Moodie in Margaret Atwood's poetic rendering of *Roughing it in the Bush*, newcomers facing the formidable task of establishing themselves in the New World inclined to "deny the ground" upon which they stood, and to "pretend this dirt is the future" to avoid being overwhelmed by circumstances (Atwood 1970, 16–17). Only thus, Atwood implies, could the unremitting toil be endured; only thus, nineteenth-century settlers might have insisted, could those who pioneered enable those who came after them to inherit productive earth.

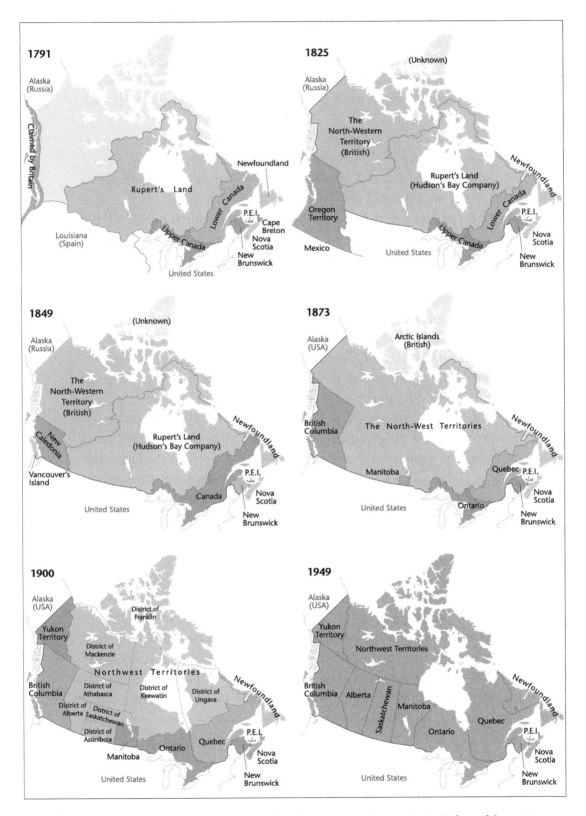

Fig. 13. *The administrative units of northern North America, 1791 to 1949. (Adapted from Wynn 1987a;* Historical Atlas of Canada *2: pl. 21)*

PART 4

Nature Subdued

W inter was a cold fact of life in British North America. None could ignore its icy grasp. In the far north, where fur traders recorded the falling mercury in their thermometers with dutiful devotion (and perhaps a soupçon of trepidation) each year, land and water began to freeze in October. By November, *la prise des glaces* had the banks of the lower St. Lawrence in its grip. By the end of the year the great river was impassable for shipping. Not till April, May, or June on the shores of Hudson Bay did old Jack Frost release his lock on the country (Catchpole, Moodie, and Kaye 1970).

Rhythms of existence—animal, vegetable, and human—were shaped by the orbit of the earth around the sun and by the progression of the seasons orchestrated by this movement. Birds and beasts responded by migrating or hibernating. Caribou, geese, ducks, passenger pigeons, and dozens of other species followed age-old passages south to north and back again the next year. Bears, wolves, and squirrels retreated to their dens. Shrubs died back, small plants disappeared from the woods and fields, deciduous trees shed their leaves, and all of the vegetable kingdom fell dormant.

Human life was also molded by climatic circumstances. Native peoples dispersed in pursuit of game or concentrated in winter villages. Lumbermen took to the forests, while farmers hauled and chopped wood and busied themselves with barn work such as mending equipment and threshing grain. Fishermen made and repaired nets, boats, and traps. Saw- and gristmills ceased operation as their millraces froze. Boats were hauled ashore, town laborers were laid off work. In the winter, wrote one fervent critic of the situation from Montréal, "an embargo which no human power can remove is laid on all our ports." Wharves and warehouses were deserted. Ships lay idle, their "naked spars . . . from which the sails have fallen like the leaves of the autumn" constituting a "blasted forest of trade." Up and down the St. Lawrence, "the great aorta of the North," the "life blood of commerce . . . [was] curdled and stagnant . . . blockaded and imprisoned by Ice and Apathy" (Keefer 1850, 3). In the towns, at least, many families were plunged into poverty by the seasonality of the labor market, the costs of winter fuel, and the rising price of provisions as February turned to March and March to April (Fingard 1974).

Yet some farmers were liberated by winter. The daily demands upon their time and energy were lessened, and ice and snow made travel by sleigh to church, neighbors, and town a good deal easier and faster than the journey by cart or carriage on rough summer roads (McIlwraith 1990). Buffalo robes, blankets, fur hats, and gloves kept travelers warm, and in the larger centers officers from the garrison, "young sporting bloods," and their ladies cut a dash as they piloted their cutters and sleighs through streets busy also with sleds hauling firewood and other commodities to urban markets. For those who had achieved a modest competence, in short, winter in the city or the countryside had its charms. It was, wrote Anna Jameson, a time "for balls in town, and dances in farmhouses and courtships and marriages"—an observation borne out by the demographic record that reflects, in the incidence of late summer and fall births, the effects of climate upon the most intimate aspects of human life (Jameson 1839, 22).

Immutable though these patterns might have seemed in 1830 or 1760, and as broadly stable as winter temperatures have been over the last few centuries, the cold, hard facts of climate are not the same as their human consequences. Climate, a statistical construct fashioned from long-term average measurements of certain phenomena (particularly temperature and precipitation), is not invariable, but it generally changes relatively gradually. By contrast, human experiences of climate—or human perceptions of the seasons and of their implications for human activities—are far from constant. Technology (to say nothing of culture) intervenes between people and nature. It shapes the ways in which humans interact with the environment and has the capacity to transform the ways in which they experience the world. In the middle decades of the nineteenth century, British North Americans began to implement, feel, and respond to a technological shift unlike any that they had experienced before.

The Eotechnic age gave way, in Lewis Mumford's terms, to a Paleotechnic era; dependence on wood, wind, and water was superseded by the rise of coal, iron, and steam. Based upon the use of fossil fuels to produce mechanical energy—the basis of what John McNeill calls the exosomatic energy regime—this transition produced a massive increase in energetic capacity (McNeill 2000; Kammen 2004). Even the small, inefficient steam engines in use in Britain by 1800 had the power of two hundred men, and there were steady improvements in efficiency thereafter. Converting chemical energy to drive machinery by burning coal to produce steam allowed British North Americans to transcend the limits of their dependence on human and animal muscles and the kinetic potential of wind and water, by which they had been largely bound through the first half of the nineteenth century. It also led them, in turn, into radically different relations with the natural world.

These changes were neither instant nor complete. Such transformations never are. The shift began gradually in the 1830s, gained momentum in the 1850s, and achieved almost tectonic proportions by the end of the century. It took many forms, but it turned upon the production of mechanical energy from coal. Settlers in parts of Nova Scotia had long burned coal from local outcrops to produce heat for cooking and warmth; by midcentury, small quantities of coal, shipped from Cape Breton, fueled the new cast-iron stoves of some Montréal and Halifax homes. Even before this, some breweries and foundries in Montréal had used British coal, carried across the Atlantic as "back-haul" cargo in the timber trade, in their production processes. But coal consumed thus produced heat, not motive force, and in this sense burning it was little different from burning wood, which continued to meet most domestic fuel needs across the country until well into the twentieth century. Steam rather than running water was used as a source of power in a small number of large, strategically located new sawmills in the northern colonies before 1840. The Royal William, launched with great fanfare in Québec in 1831, became the first Canadian vessel to cross the Atlantic under steam alone in 1833.

But the railroad was, without doubt, the most important single instrument and symbol of the Paleotechnic age. British North America's first railroad, built in 1836, provided a 14-mile, summertime portage between the St. Lawrence and Richelieu rivers. It ran on iron-capped wooden rails and was a rickety, hazardous, and yet profitable concern. By 1851 this line had been improved, extended to the American border and connected to the Vermont Central Railroad. A couple of years later another railway, the St. Lawrence and Atlantic, ran east from Montréal to ice-free Portland, Maine. In Upper Canada the first railroad ran north from Toronto to Collingwood. Three others followed, each linking Lake Huron with Lake Ontario. The orientation of these lines was no accident. They were all designed, in a sense, as mechanical portages: modern, indeed revolutionary, means of truncating circuitous summertime water routes and breaking winter's icy lock on shipping.

Railroads were transformers. Unlike oxen and horses, locomotives never tired; they ran day and night; and they devoured distance with astonishing speed, to the point that contemporaries often described these noisy, smoke-billowing machines as "frightful" and "fearsome" contraptions. Little wonder that engineer Thomas Keefer, who seethed at "Ice and Apathy," called railroads "iron civilizer[s]" and compared their potential economic and social effects, their "influence over matter," to those that "the discovery of Printing had exercised upon mind" (Keefer 1850, 9: 11). At Confederation in 1867 the new Dominion of Canada had 2,278 miles of operating track, most of them in Ontario. By 1891 there were almost 14,000 main-line miles, and railroads spanned the continent from the Atlantic to the Pacific.

Still, coal and iron did not hold complete sway. Wood burned in locomotive fireboxes and wooden trestles carried rails and rolling stock across valleys and gorges. Moreover, track-miles and the trains that ran on them were only a small part of the larger story of transformation in the northern part of the continent through the last half of the nineteenth century. Behind railroads stood factories, in which locomotives were built; behind them, the mills that produced the iron and steel from which rails and cars and engines were constructed; behind them, the mines where men toiled underground to release the ore and coal upon which the mills depended. Most of these enterprises floated on rafts of development capital, the raising of which had required far-reaching changes in colonial fiscal arrangements. Thanks in part to railroads, the reach of ideas and the influence of businesses were extended: Half the print-run of the *Toronto Globe* sold outside the city in the 1870s; two decades later, beer brewed in Ontario reached markets in British Columbia.

Yet, if Keefer was correct in his broad assessment of the effects of railroads upon economy and society, he said too little about the Paleotechnic revolution's consequences for nature and about the changing ways in which people thought about and experienced the natural world during this period. As steam engines and iron rails overcame winter's impediments to commerce, they also demonstrated humanity's growing power over nature. The telegraph laid across the Atlantic on the eve of Confederation carried much the same message, and so did dozens of other innovations in science and technology, from the rotary snowplough that kept trains running through drifts and storms to the new varieties of wheat that pushed back the northern limits of commercial agriculture. Even the sun, it seemed, was brought to order when Canadian railroad engineer Sandford Fleming developed (and in the 1880s saw implemented worldwide) the idea of standard time zones, which attenuated the link between local time and the earth's diurnal rotation to rationalize railway timetables and long-distance communication. If human power had failed to remove nature's winter embargo on Canada's ports, it had, surely, found effective ways around it and other natural obstacles to human enterprise.

Just as Canadians began to assimilate the major implications of the Paleotechnic revolution, however, its technological foundations were challenged. Improvements in the generation, and particularly the transmission, of electricity late in the nineteenth century brought a new source of power—hydroelectricity—into commercial significance. According to some Canadians impressed by the abundance of potential hydropower sites on the country's rivers, these developments signaled a "Great Divide" in the path of Canada's economic development. "Black coal," the polluting, carboniferous rock found only on the flanks of the country, in Nova Scotia and New Brunswick and British Columbia and Alberta, would be replaced by clean and potentially ubiquitous "white coal," or hydroelectricity, as the source of power for manufacturing and other purposes.

This was no small promise. Early in the twentieth century, Canada imported approximately half of its coal requirements (some fifteen million tons in the years immediately prior to the First World War) from the United States to feed the demands of central Canada (mainly Ontario). In addition, hydroelectricity held out hope of reversing some of the troubling tendencies of the Paleotechnic era, the specialization and differentiation of the spatial economy that had come with industrial concentration. Early twentieth-century Canadians were not alone in these enthusiasms. Italians seized on the potential of water flowing from the Alps to electrify and industrialize the Po Valley.

A few decades later, Lewis Mumford argued that these developments were of sufficient moment to constitute a third phase of technological evolution: the Neotechnic or age of "electricity and new alloys" (such as aluminum). In Mumford's schematic vision, electricity was a critical innovation. Because it could be transmitted across considerable distances to power machines with individual electric motors, it promised to free industry from the locational restrictions imposed by coal and the mass-production line, and allow the development of smaller widely dispersed factories, able to deliver their products where and when they were needed.

Fully three decades after the world's first large hydroelectric power plants came "on stream," however, Mumford acknowledged that electricity had yet to deliver many of the liberating effects he attributed to it. Its potential to shape new economic and societal forms and, in the process, to change human relations with nature from the exploitative, polluting mode of the Paleotechnic to a more conservationist ethos remained substantially unrealized. Neotechnic means were being used to further

Paleotechnic ends. Even in the early 1930s, wrote Mumford—drawing upon Matthew Arnold and Oswald Spengler in characteristically erudite fashion—(Western) civilization was living "between two worlds, one dead, the other powerless to be born"; like the mineralogist's crystal formed of one substance but having the shape proper to another, it was a "cultural pseudomorph" (Mumford 1934, 265).

Be this as it may, hydroelectric power made a significant contribution to the transformation of Canada in the first quarter of the twentieth century. The new energy source powered massive new industrial plants, spawned new manufacturing activity, propelled streetcars along city streets, lit the homes of thousands of Canadians, and transformed many of the chores of everyday existence by running newly invented tools and appliances. By 1930, barely sixty years after the confederation of four British North American provinces and three million people into a new nation, and three-quarters of a century after a burst of railroad construction symbolized the dawn of the Paleotechnic era in the region in the 1850s, Canadian economies, societies, and environments had been so reshaped by what Mumford rightly regarded as two revolutionary shifts in socio-technic organization as to be virtually unrecognizable to those who knew them in the Eotechnic phase.

In the half-century after Confederation, immigrants came into the country from an ever-broadening horizon of source regions. From northern and Mediterranean Europe they came to settle in eastern cities, to take up farms, and to work in the forests. From Iceland and the steppe lands of what is now Ukraine they came to people the prairies. And for a while in the late nineteenth century, migrants from Asia came to labor in western resource industries and on the construction of the transcontinental railroad (Wynn 1987a). By 1930 the four original provinces of Canada had become nine. The country had ten million residents. Canada's territory stretched from the Atlantic to the Pacific and from the American border to high Arctic islands. To the northwest the former Russian territory subsequently known as Alaska had been purchased by the United States in 1867, the same year as Canadian Confederation, and brought more closely within the orbit of North American geopolitics, development, and consciousness.

By 1915 not one but three railroad lines spanned the northern part of the continent, reducing to days, and comfort, and the mundane, a journey that had been an arduous, slow, and dangerous adventure, rarely undertaken, only fifty years before (Brown and Cook 1974). In 1930 well over 40,000 miles of mainline track crisscrossed Canadian space. Each year these lines carried, among vast quantities of other things, enormous amounts of wheat from millions of acres of farms in the western interior, where W. F. Butler, traveling in the early 1870s, had considered himself in the middle of a "great lone land" (Butler 1872).

In 1929 Canadian wheat exports averaged out at a million bushels a day, and wheat production was twenty-five times greater than it had been at Confederation. The total value of forestry production increased almost fourfold between 1890 and the early 1920s. Yet annual growth rates in both forestry and agriculture were below that for the Canadian economy as a whole (3.8 percent per annum) between 1891 and 1926. Mineral production led the way among primary industries (with a growth rate of 4.5 percent per annum over thirty-five years). But it was in manufacturing that growth was most spectacular. Before 1850 the economic growth of British North America was essentially predicated on the expansion of settlement and resource exploitation; although agriculture and natural resource extraction remained important thereafter, they accounted for a shrinking proportion of national income (Norrie and Owram 1990; Paterson and Marr 1980; Wynn 1987a). Manufacturing employment more than tripled between 1870 and 1929, and the value-added increment soared from $94 million to $1.7 billion in the same period.

Urbanization proceeded apace. Less than 20 percent of the population lived in urban centers in 1871; by 1931 over half did so. At Confederation coal (and coke) contributed less than 10 percent to the Canadian energy budget. By 1900, it was the country's leading source of energy; in 1920 it accounted for three-quarters of Canadian energy consumption, and even in 1930 its contribution exceeded 60 percent. Crude oil production, initiated with discoveries near Sarnia late in the 1850s, quadrupled between Confederation and the end of the century, although it made only a minuscule contribution to national energy needs; before 1900 most oil was used as grease or refined into kerosene to be burned in "oil lamps."

Demand for petroleum fuel increased in the twentieth century with the development of the automobile, but domestic supplies accounted for less than 10 percent of Canadian oil consumption as late as 1946. Meanwhile Canadian hydroelectricity generating capacity increased almost exponentially, from approximately 100,000 horsepower at the end of the nineteenth century to 5 million horsepower in 1929. Between 1891 and 1926, Canada's Gross Domestic Production almost quadrupled, from approximately $860 million to $3.2 billion (in 1913 dollars). In much the same period, the country's share of world trade climbed from slightly more than 1 to almost 5 percent.

13

HARNESSING BLACK
AND WHITE COAL

Coal was a familiar but elusive substance in early nineteenth-century British North America. Enthusiasts for development were convinced of its importance to the future prosperity of the colonies. After all, Britain's rise to economic pre-eminence during the early nineteenth century had rested upon the exploitation of coal and iron and, "as surely as the child becomes the man," it was each and every colony's destiny to "follow in the footsteps of her parent" (quoted in Zeller 1987, 96). But coal was hard to find, especially in the Canadas. Developing geological knowledge, largely attributable to the work of provincial geologist William Logan in the 1840s, suggested that the bedrock south of the ancient Shield predated the Carboniferous, and was therefore unlikely to contain coal. By the 1850s it was widely acknowledged that this was the case.

The British government had blithely and in ignorance ceded the coal deposits of Michigan to the United States after the War of 1812, leaving them "ready to supply American steamers with fuel on the lakes, while ours on the same waters, in case of war, must depend on wood or coal expensively transported" from elsewhere including, perhaps, the United Kingdom (ibid., 54). There were rich deposits of anthracite in Pennsylvania. There were rumors of "a vast coalfield, skirting the base of the Rocky Mountains [in Hudson's Bay Company territory] . . . and continu[ing] probably far into the Arctic Sea" (quoted in ibid., 98). But hopes of coal deposits even as far east as the Gaspé Peninsula proved ill founded. New Brunswick and Nova Scotia had the only significant coal deposits in British North America. The most accessible of these, in Pictou County and Cape Breton, had been in the hands of a British monopoly, the General Mining Association, since 1826, but were returned, after protracted negotiation and with the important exception of some 88 square kilometers of territory, to the government of Nova Scotia in 1858.

Canada, rich in copper and iron ores, "had coal upon all sides of it," and a prosperous industrial future might be assured if these resources could be combined in a "union of interests" (ibid., 83, 96). Thus, Suzanne Zeller has argued, emerged a powerful stimulus to the confederation of British North America, a stimulus heightened during the American Civil War, when the United States prohibited the export of coal to Canada. Geological science offered a new understanding "of the limitations and possibilities" of the colonies, and "seemed to provide a solid foundation on which an expanded Canada would have to be built if it were to prosper through industrialization in future" (ibid., 110).

Confederation—the political union of Nova Scotia, New Brunswick, and the two Canadas in 1867, followed by the transfer of Rupert's Land and the far northwest to the federal government in 1870 and the addition of British Columbia and Prince Edward Island to the Dominion in 1871 and 1873—had other stimuli and was not in and of itself sufficient to produce the anticipated development of a "community of interests" based on east-west trade (see Fig. 14). For the better part of a decade, the economic promise of union was stillborn. A worldwide economic downturn in the 1870s and the disposal of surplus American industrial production in the Canadian market depressed local output. Then the Canadian government implemented a system of protective duties cleverly known as the National Policy Tariffs to foster the development of a relatively self-sufficient economy based on expanded domestic production and increased interregional trade.

Between 1879 and 1887 a complex schedule of import duties was brought into effect. Levies were calibrated to make domestic manufactures competitive with imports, and to encourage the use of Canadian resources. Thus, finished consumer goods incurred the heaviest charges: Imported furniture and clothing were taxed at almost 35 percent. Manufacturing inputs such as pig iron and rolled steel were liable to imposts of 10 to 20 percent, and coal was charged 50 cents, then 60 cents (and after 1897, 53 cents) per ton. Textile machinery, essential for the establishment of manufacturing in Canada, was admitted without charge, and there were additional incentives to domestic manufacturing in the form of government requirements that railroads lay Canadian-made track and the payment of bounties to steel producers. Schedules were tinkered with, duties on particular classes of goods were adjusted periodically, and the tariffs were reduced overall in the 1890s and again thereafter, but for sixty years Canadian manufacturing enjoyed a higher level of protection than at any time before or since. Its transformation was thoroughgoing and rapid. Although it is difficult to disentangle the effects of the tariffs from other forces shaping the economy, it is clear that manufacturing output increased at an annual average rate of 4.2 percent between 1870 and 1915, well above the growth in aggregate income during that period. In general terms, coal launched this expansion and hydroelectricity sustained and extended it.

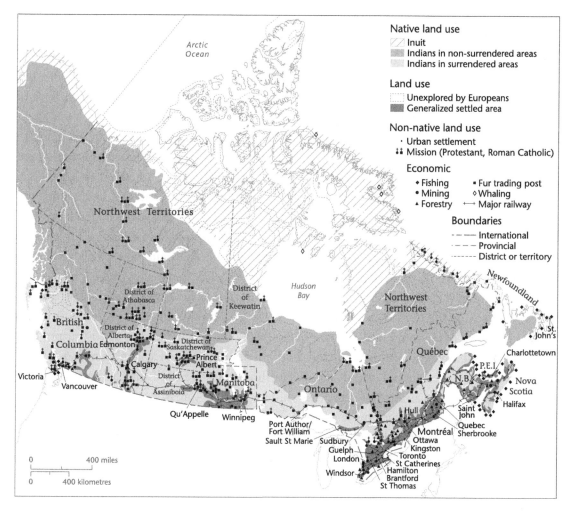

Fig. 14. *Canada in 1891. (Adapted from* Historical Atlas of Canada *3: pl. 1)*

The roots of the Paleotechnic revolution were the pits, the underground workings from which the essential fuel of the age was extracted. In Cape Breton, bituminous coal seams crop out roughly parallel to the coast for some thirty-five to forty miles between Cape Morien and Cape Dauphin and dip gently northeast (or seaward). Coastal inlets and geological structures complicate the geography of the coalfields by interrupting the continuity of the outcrops and producing significantly steeper angles of dip in certain locales. Seams vary in thickness and in purity, and thus in value. By and large, seams less than three feet thick were of little commercial utility; the best seams averaged better than six feet in thickness and were up to ten feet thick in some places. After 1870, when shallow, non-mechanized mines were superseded by larger, more expensive investments, most development concentrated on the best seams. Because exports to the United States and the St. Lawrence region accounted for a substantial share of output before 1900, mines depended upon access to a sheltered harbor. Over time, the development of a transport infrastructure (rail lines and port facilities) also fashioned the geography of exploitation. Thus, the location and growth of mines were shaped by an intricate, shifting array of forces.

The geographer Hugh Millward (1985) has found order in this complexity by suggesting a five-stage model of mine development. The first two stages, encompassing quarries and adits (stage 1) and shallow pits (stage 2), were characteristic of the pre-Confederation years; the largest of the stage 2 operations produced about 10,000 tons of coal a year. The fifth, highly mechanized stage was not reached until the 1950s. Throughout the late nineteenth and early twentieth centuries, production from the Cape Breton coalfield (and from deposits in Pictou and Cumberland counties) was massively expanded by the investment in the new technologies and the implementation of new organizational structures that distinguish stages 3 and 4 of Millward's model. In 1858 the coal mines of Nova Scotia produced slightly over 300,000 tons; early in the 1870s output was triple this amount; by century's end, production exceeded 3 million tons a year. By 1915 provincial output was 5.5 million tons (over half the Canadian total). In thirty years after 1880, the gross value of mining (mostly coal) production in the Maritime provinces more than quadrupled.

Large steam-driven pumps were crucial to expansion. Without them, mines became waterlogged. With them exploitation could proceed to depths of 500 feet or so. But the large capital investment represented by pumping engines required massive increases in output. New machinery was introduced to haul coal underground and to bring it to the surface. Production was concentrated on the larger seams, new larger mines were opened, along with rail lines and shipping piers designed to serve several mines, and output rose to about 150,000 tons per mine per year.

Yet work at the coal face remained unmechanized. Men used black powder (dynamite) and wielded picks and shovels to extract the coal. They worked long hours in difficult and dangerous circumstances to excavate as much as they could safely take from the seam, creating rooms separated by pillars (or walls) of coal in the practice of what is sometimes known as "bord and stoop" mining (Frank 1985). With time, the exhaustion of the uppermost commercial seam on each lease forced miners to go deeper underground or to push their workings "offshore" down the dip slope of the coal seams. Both were expensive. Haulage and ventilation costs increased along with the costs of sinking shafts and opening stopes. Operations were centralized and mines were amalgamated. New mining machinery and new techniques increased production efficiencies. And the typical annual output of the mines rose to some 300,000 tons.

By Millward's careful analysis, seventy-three mines worked the Cape Breton coalfield before 1980. Almost all were established before 1945. Approximately half (thirty-seven) were major mines that produced at least 500,000 tons of coal over the period of their existence. The first of these major mines opened in 1834, nine more opened between 1858 and 1871, and eighteen between 1901 and 1920. Mining is, ultimately, an ephemeral business, however. The exhaustion of accessible reserves, consolidations, and other factors led to the closure of seven major mines between 1885 and the end of the century, and to fourteen closures between 1914 and 1932. At the peak of mining activity, just before the First World War, twenty-three major mines were operating on the Sydney field. This reflected, in part, the establishment of two major primary iron and steel mills on the Cape Breton coalfield at the turn of the century. These were the main trunks of the new coal-based economy of the Paleotechnic era.

Sydney became the center of operations for the Dominion Steel Company, and across the harbor, the Nova Scotia Steel and Coal Company, which had been established in New Glasgow in 1882 and amalgamated with the Nova Scotia Forge Company in nearby Trenton a few years later, developed operations in Cape Breton to make use of iron ore from Bell Island, Newfoundland (McCann 1981; Wynn 1982). Throughout these years, corporate consolidation brought most major mines into the hands of large industrial conglomerates such as the Dominion Coal Company (merged into the Dominion Steel Corporation in 1910) and the Nova Scotia Steel and Coal Company (based in New Glasgow), both of which were subsequently merged into the British Empire Steel Corporation in 1921. In a manner entirely congruent with trends toward the economic and spatial concentration of production, and the expanding reach of business and trade during these years, Nova Scotia's coal and steel industries were thoroughly incorporated into national and international networks of financial control and corporate capitalism.

Closely associated with the primary iron and steel producers were the secondary manufacturing plants that formed the metaphorical branches of the Paleotechnic system, as they used iron and steel from Cape Breton and Pictou County to produce a broad range of commodities. Along the line of the Intercolonial Railway, Amherst became an important manufacturing center on the strength of manufacturing plants producing railway rolling stock, steam engines, and boilers, as well as on the existence of factories making furnaces, enamelware, woolen goods, boots and shoes, pianos, and furniture. In the first five years of the century, the value of manufacturing in Amherst more than quadrupled to $4.5 million. Tiny Sackville had two foundries whose cast-iron stoves warmed homes as far west as the prairies. Rolling mills were established in Halifax and Saint John. Steel shipbuilding grew into an important industry.

Following the rise, a couple of decades earlier, of textile plants, sugar refineries, and chocolate factories across the region as new railways linked ports in Nova Scotia and New Brunswick to central Canadian markets protected by the National Policy Tariffs, these developments were widely seen as portents of coming industrial maturity in the region. Even textile mills, which typically drew their motive power from running water rather than from steam, were hailed as beacons of transformation and improvement (Wynn 1982). Months before its completion, the four-story cotton factory with 34,000 spindles erected in Milltown, New Brunswick, early in the 1880s was welcomed for its potential to "convert one of the lowest, most squalid parts of the village into a neat and tidy hive of industry" (DeLottinville 1980, 106).

Urbanization proceeded in lockstep with industrial development. In Milltown the new mill created 500 jobs (most of them for women) and drew many newcomers into the village of 1,600. The St. Croix Cotton Manufacturing Company built dwellings for its managers and erected a boarding house for eighty workers. Private interests opened two more boarding houses, and several families took in lodgers (DeLottinville 1979). Between 1871 and 1911 the population of Cape Breton's industrial districts (the mining areas and the town of Sydney) increased from 12,000 to 57,000. Dwellings and towns sprang up to house the miners and their families, many of whom moved into the area from rural Cape Breton, and thus marked in a small but significant way the drift from countryside to town that was a significant feature of the population geography of eastern and central Canada during these years. The population of Sydney, static at approximately 3,660 in the 1880s, increased by 270 percent in the 1890s and jumped from under 10,000 to almost 18,000 in the first decade of the twentieth century. As in the coal towns, company housing accommodated a substantial part of the labor force. In Amherst the population doubled in the first five years of the twentieth century to reach almost 10,000. Across the region towns and cities of widely differing size experienced rapid population growth, as new industries were established and local economies expanded.

Similar patterns were evident across the country. In broad terms, Canada's expanding population—rising from 3.5 million in 1871 to 10.4 million in 1931—urbanized at a striking rate, as industrial expansion concentrated economic activity and employment, and thus people, in fast-growing towns and cities. Landscapes, ecologies, and the environments of daily life were profoundly altered by these developments. Through the last thirty years of the nineteenth century, levels of urbanization increased by 6 percent per decade; between 1901 and 1911 the advance was approximately 8 percent; for each of the next two decades it was 4 percent. In sixty years the proportion of Canadians living in urban centers rose from less than one in five to greater than one in two.

As in the Maritime Provinces, part of this transformation was accounted for by the expansion of steel towns and dozens of small manufacturing centers associated with them (Wynn 1982, 1987a; Nelles 1974). In Québec the iron forges at St. Maurice, which had operated since the French regime, were closed in 1883, as their technology—using charcoal to turn bog ore into iron—became obsolete. Henceforth, iron was imported, initially from England and subsequently from Nova Scotia, for secondary processing in rolling mills located in and around Montréal. Prominent associated industries included the construction of railway equipment; as early as the 1880s the Grand Trunk Railroad shops in Point St. Charles employed about 2,000 people. In Ontario pig iron was produced after 1895 from blast furnaces in Hamilton using ore from the Mesabi Range on the American side of the Great Lakes, coal from the Appalachians, and technical expertise from the United States. A few years later steel production began in Hamilton.

At the turn of the century the American promoter Francis H. Clergue developed the hematite ores of the Helen mine in the Precambrian Shield to serve a steel plant established alongside a cluster of other industries tied to an early hydroelectric power plant at Sault Ste. Marie. With astonishing chutzpah he won subsidies for construction of a railroad between his mine and mill and a contract to supply steel rails and other products to government-assisted railways (Nelles 1974). In southern Ontario, smaller centers, such as Guelph, and companies, such as the International Malleable Iron Company, were counterparts of the towns and metal-using establishments along the Intercolonial Railroad. So, too, Milltown, New Brunswick, had its approximate equivalents in the central Canadian textile towns of Valleyfield, Coaticooke, Chambly, Magog, Montmorency, Dundas, Stormont, Cornwall, and Merriton.

According to the census of 1891, there were more than 4,500 industrial establishments in a hundred or so small centers broadly comparable to Milltown (Baskerville 2002; Parr 1990). Together they employed 25,000 industrial workers in plants valued at more than $2.5 million. Increasingly, however, industry and population concentrated in larger centers. In 1881 Montréal had 140,000 residents and was the only Canadian city with more than 100,000 inhabitants. Half a century later, a fifth of all Canadians (2.3 million people) lived in cities of 100,000 or more. Reflecting the agglomerating tendencies inherent in the Paleotechnic regime, as well as the new economies of scale associated with factory production and reductions in the friction of distance, seven major urban centers (Montréal, Toronto, Winnipeg, Vancouver, Hamilton, Québec, and Ottawa) dominated the Canadian urban hierarchy and accounted for 45 percent of Canada's urban population in the 1920s.

These developments had substantial environmental consequences. Although almost all coal came from underground workings, mining transformed the face of the earth. Each mine had its headworks—a cluster of structures that included the winding gear and the sheds in which equipment was housed and men washed and changed. Railways sliced across the landscape, and heaps of shale and other rock brought up with the coal and discarded grew beside the mines. In time, too, the surface of the earth subsided where abandoned stoops and bords collapsed underground, disrupting drainage and imparting a disordered, even chaotic, appearance to mining sites. By attracting migrants, collieries also increased local population densities and forwarded the march of urban land uses over formerly open countryside. In the steel towns of Cape Breton, Sault Ste. Marie, and Hamilton, coke ovens, blast furnaces, open hearth furnaces, and settling ponds soon became dominant features of the landscape. Smoke, noise, and other forms of pollution were widespread.

In smaller industrial centers, workers often toiled in environments that were damaging to their health, and rare strictures against the pollution of air and water by industrial processes were poorly enforced. Even water-powered industries caused unanticipated environmental damage. In Milltown, where decades of forest exploitation and sawmilling had left surrounding hillsides scarred and significant concentrations of sodden sawdust in the St. Croix River, residents grew angry with the mill owners when the discharge of wastewater from the dyeing process killed salmon in the river. By the end of the century it seemed clear that urban growth produced environmental damage. All too often, urban expansion led, at least, to "overcharged cesspools, neglected privies, and filth-laden sewers" (Baskerville 2002, 155).

There were also less obvious dimensions to the environmental impacts of urbanization and industrialization. Factory workers in Canada as elsewhere were subject to the rhythms and routines of industrial life. Subject to what E. P. Thompson (1967) called "time and work discipline," they found their lives shaped, in greater or lesser degree, by the clock (or its surrogate, the factory whistle), rather than by the cadence of the seasons, the tempo of necessity, and the inclination of the individual. Despite their support for the new cotton factory, longtime residents of Milltown complained once its bell began to toll mill hands to labor at 6:00 AM. Those dependent upon mill wages had no such luxury. Their work lives were governed by factory routine. In time, almost everyone grew accustomed to such changes in the way they lived with nature.

But memories of older arrangements faded slowly and provided fertile seedbeds for nostalgic reminiscence about days gone by. Cape Breton Scots

Iron and steel plant, Ontario. *Such plants were integral to the industrialization and modernization of the Canadian economy. Concentrated in Cape Breton and Pictou County, Nova Scotia, Hamilton, Ontario, and at Sault Ste. Marie in northern Ontario (shown here in the 1970s), mills developed in the late nineteenth century were upgraded and enlarged in the twentieth, but obsolete equipment and new competition brought many of them into economic difficulties late in the twentieth century. (Courtesy of J. Lewis Robinson)*

who moved from farms into the industrial heart of the island (and whose predecessors had crossed the ocean in pursuit of family-centered independence only a century or so before) lamented their circumstances with particular poignancy—but they were, at base, little different from those experienced by tens of thousands of newly urban Canadians. "Oh isn't it a shame," wrote a local bard, "for a healthy Gael living in this place to be a slave from Monday to Saturday under the heels of tyrants, when he could be happy on a handsome spreading farm with milk-cows, white sheep, hens, horses, and . . . clean work on the surface of the earth, rather than in the black pit of misery" (Dunn 1953, 132). Another "wage slave" put his feelings even more succinctly: "I work in the pit. It's a terrible hole/ Getting paid by the company for hauling their coal" (quoted in Frank 1985, 213). Similar sentiments were widespread and deep rooted. They were the bedrock upon which new attitudes toward nature, both romantic and indulgent, would be built through the turn-of-the-century decades.

For all that, many celebrated the "Industrial Ascendancy" of Nova Scotia in the years immediately before the First World War. One publication, devoted to boosting expansion, took *The Busy East* as its title. Cover illustrations showed harbors crowded with massive steel ships, neighboring wharves jammed with goods and warehouses, and the landward horizon filled with enormous factories belching enormous volumes of smoke (clearly a symbol of progress and prosperity rather than pollution) across the sky. But this optimism was misplaced. Although the region's industrial economy prospered under the stimulus of wartime demand for munitions and steel, the postwar years brought depression and decline. The loss of two in every five manufacturing jobs between 1919 and 1921 spelled deindustrialization and depopulation (as the net value of manufacturing output fell by half in six years after 1919, and 150,000 people left the region in the 1920s). Increasingly difficult economic circumstances meant an unwillingness to burden such industries as remained with environmental and other regulations and a lack of capacity to address some of the worst problems of pollution and despoliation produced during the heyday of industrial expansion.

Meanwhile, the economies associated with increases in the scale of production, the advantages of industrial linkages, and the capacity of the railroad to integrate space favored central Canadian producers over those on the periphery. In an increasingly integrated national economy, the rationalization and consolidation of manufacturing capacity and financial organizations produced clear patterns of regional specialization. In agriculture, commerce, finance, and manufacturing, the agglomerating tendencies of quicker and cheaper communication, improvements in technology, modern systems of marketing, and modern managerial practices undermined relatively high levels of regional self-sufficiency in the provision of foods, goods, and services. Southern Ontario and neighboring Québec became the hub of an increasingly extensive Canadian economy; peripheral regions contributed resources (minerals, energy, food, and labor) to the national enterprise, and central Canadian manufacturers and retailers established branch businesses to provide hinterland populations with goods and services. By 1929 over 80 percent of Canadian manufacturing took place in central Canada, and Ontario accounted for over half of the national output.

The development of hydroelectricity provided an enormous impetus to the spatial reorganization of the Canadian economy. The kinetic energy of running and falling water was harnessed for the generation of electricity shortly after Thomas Edison invented the direct-current generator in the 1880s. But these initial developments were very modest. Generators were driven with belt drives or run-of-the-river turbines. They produced relatively little power, and it had to be consumed within a kilometer or two of the generator because the problems of transmission were formidable. Most of this electricity was used to light factories and streets. It was enormously useful and highly valued. Early in the twentieth century, dozens of municipalities in Ontario and Québec operated or purchased power from generators on nearby streams. But the development of high-voltage alternating current and transformers (that stepped up voltages for transmission and reduced them for use) in the 1890s transformed the potential of hydroelectric power by allowing its transmission, through copper wires, over relatively long distances without prohibitive loss or expense. This unleashed the potential of rivers remote from demand centers and encouraged the development of larger generators (see in general Froschauer 1999; Bellavance 1994; Manore, 1999).

Central Canadians were quick to tout the promise of electricity. One enthusiast envisaged the Ottawa valley as "the power heart of the world and the centre of a delightful district unsullied by coal smoke and beautified by reservoirs of unrivalled natural beauty" (quoted in Nelles 1974, 217). Development proceeded at an astonishing pace. In 1901 Québec derived approximately 150,000 horsepower, slightly more than half of its total industrial generating capacity, from hydroelectric power; Ontario produced less than 100,000 horsepower of hydroelectricity, barely a third of its industrial power consumption. By 1928 hydraulically generated electrical power accounted for 2.5 million of the 2.7 million horsepower generated for industrial purposes in Québec, and 1.8 million of 2.15 million horsepower in Ontario. Elsewhere industrial hydroelectric power generation was minuscule—except in British Columbia, where it went from zero to half a million horsepower between 1901 and 1928, and in Manitoba, where hydroelectricity accounted for about 85 percent of the 300,000 horsepower used by industry.

In Québec electricity was generated at several important sites across the province. In 1928 the Lac Saint-Jean–Saguenay, Shawinigan–Saint Maurice, and Ottawa-Gatineau regions each produced approximately 0.5 million horsepower, and the Montréal–St. Lawrence region generated slightly more than half this amount. In Ontario, by contrast, the Niagara region dominated by producing over 1 million horsepower. In all of these areas large dams and generating stations were erected on major rivers. On the Saint Maurice, the Shawinigan Power Company established its plant and transmission lines early in the century. By 1903 it was transmitting power some 90 miles to Montréal, and a decade later transmission lines ran under the St. Lawrence to Sorel and Thetford Mines.

But these markets took only a small part of the company's output. By offering long-term contracts at relatively low rates to large industrial consumers, the company fostered the industrialization of the Saint Maurice Valley. Pulp and paper mills, textile and clothing factories, electrochemical plants, and an aluminum smelter were built in the growing towns of Shawinigan, Trois Rivières, Cap de Madelaine, La Gabelle, Grand Mère, and La Tuque, and the population of the valley increased from 61,000 to 130,000 in thirty years. Similar patterns prevailed in the other major producing regions, as individual companies developed local capacity and attracted large industrial enterprises to locate nearby.

At Niagara the first generators turned on the American side of the river in 1895, a decade before production began in Ontario. Even in 1911 two-thirds of the power generated at Niagara was consumed in the United States. On the Ontario side of the border, the government-established, publicly financed Hydro Electric Power Commission of Ontario (HEPCO) was given responsibility, in 1906, for the transmission and sale of power. It sought to encourage domestic consumption. Transmission lines were extended to Toronto and London (and numerous municipalities in between) by 1911. Later in the decade, HEPCO moved into power production. In 1917 it acquired the Ontario Power Company, and by 1921 its Adam Beck #1 Generator was the largest in the world. Late in the 1920s HEPCO's transmission grid encompassed southwestern Ontario and extended to Ottawa, Kingston, and the border of Québec.

Charges for domestic electrical service fell steadily—in distant Windsor and Kingston they declined from 12 cents and 10 cents per kilowatt in 1911 to less than 2 cents and slightly more than 3 cents per kilowatt respectively in 1925. Hardly surprisingly, provincial consumption for domestic purposes soared, from an average of about 10 kilowatts a month on the eve of World War I to 90 kilowatts a month in 1925. At the same time, electrical power was supplied to industrial consumers across the province. It became, in effect, a substitute "for imported coal in a wide range of existing industries and locations" and fostered the expansion of industrial capacity in many Ontario urban centers (*Historical Atlas of Canada* 3: plate 12).

Major electrochemical and metallurgical plants sprang up on the Niagara peninsula, automobile-related production gathered prominence in Windsor and Oshawa, and many towns flourished as local entrepreneurs advanced one or another industrial specialization. In 1911 approximately half (or more) of those engaged in the labor force in Hamilton, Brantford, Berlin/Kitchener, and Peterborough worked in manufacturing. Guelph and London returned 42 percent in this category, and Toronto, Windsor, and Kingston each had approximately one in three of their workers in the manufacturing sector.

14

NEW URBANISM

Cities grew and changed almost immeasurably between 1870 and 1930. At Confederation, even the largest urban places were pedestrian cities, compact centers clustered around wharves and warehouses crowded, during the daylight hours at least, with people, draft animals, carts and carriages, and with a bewildering mix of buildings and land uses in their tight confines. Smaller centers at river crossings or crossroads were little more than large villages by modern-day standards. By 1930 Montréal and Toronto were substantial cities, even by the benchmarks of the twenty-first century, with populations in excess of 1 million and 800,000 respectively. Smaller centers such as Hamilton and Ottawa, with populations approximately equivalent to that of Montréal in 1871, bore few resemblances to their earlier counterpart. New cities in the west were radically different again. Even smaller, long-settled places such as Halifax and Saint John, with 50,000 to 60,000 residents, revealed the impress of the technological forces and new ideas that reshaped urban forms and urban lives during these years.

According to American historian Sam Bass Warner, the streetcar was a crucial instrument of urban transformation. By quickening movement—albeit modestly at first—horse-drawn and later electric-powered streetcars broke the bounds of urban space. In broad terms, pedestrian cities in which most people walked to work rarely extended more than a couple of miles from their core. Well before this perimeter was reached, people generally began to trade density for distance. Rising numbers of inhabitants meant that city spaces became increasingly closely occupied; the development of rowhouses, tenements, and three- and four-story "walk-up" apartments, as well as the subdivision of land parcels and the colonization of alleys and lanes, were some of the results. By extending the effective range of daily personal movement for substantial numbers of urbanites, streetcars changed all this. They were the communication arteries that spawned and pumped life into new suburbs, that allowed urban populations to grow to unprecedented totals, that made labor available for industrial expansion, and that thus set in motion a series of multiplier effects that further stimulated the growth of the urban economy.

All of this promoted the differentiation of urban space. As cities grew, functions and land uses that were typically clustered together in pedestrian cities expanded and began to occupy, and define, distinct districts within the urban fabric. According to the best-known ideal-type model of the modern city, formulated by scholars at the University of Chicago and substantially based on developments there, the radial extension of streetcar lines from the city center prompted linear growth, the fingers of which then coalesced to form concentric rings or land-use zones. In this conceptualization, the Central Business District (CBD) formed the core of the city. It was encircled by a "Zone-in-Transition" where older structures were converted to various relatively temporary uses, including the provision of housing for recently arrived immigrants, as property owners awaited the redevelopment of the area with its anticipated incorporation into the expanding CBD. More or less discrete housing belts surrounded these inner-city areas, with modest working-class neighborhoods closest to the center and more-affluent middle-class zones with larger houses on larger lots beyond. In big cities, the CBD was itself divided into discrete zones devoted variously to wholesale, retail, financial, and other purposes.

This is a useful way of thinking about the changes that remade urban space, but like all such representations, it presents an archetype that was rarely replicated precisely on the ground. Few cities grew as rapidly, across such relatively accommodating terrain, under the expansionary impetus of the streetcar, as did Chicago. History (earlier-established patterns of land use), topography, political decisions, speculative adventures, and a dozen other factors shaped the growth paths of each and every city in slightly different ways to produce diverse urban geographies.

Thus, no Canadian city quite matched or mirrored Chicago during these years, but basic characteristics of the Chicago model were evident in one or another Canadian center. In Montréal and Toronto the rise of corporate capitalism and the growth of its "ancillary services," such as stockbrokers, insurance agents, law firms, and accounting companies, produced clear concentrations of different types of economic activity and the construction of new, tall buildings to capitalize upon premium locations. By 1928 Toronto had more than forty buildings over six stories in height (there were none in 1890), and the Canadian Bank of Commerce Building, at thirty-two stories, was the tallest in the British Empire (*Historical Atlas of Canada*, 3: plate 15). Earlier, in 1909 and 1911, the Dominion Trust Building and the World Tower (seventeen stories) in the up-and-coming downtown of rapidly expanding Vancouver had each claimed this title in succession. Both Montréal and Toronto had inner-city neighborhoods marked by poor housing, crowding, and high mortality rates—characteristics associated with the zone-in-transition—early in the twentieth century. In Toronto "the

Ward," a densely inhabited predominantly Jewish immigrant neighborhood near the core, was the focus of a good deal of concern before the First World War.

Tellingly, philanthropist Herbert Brown Ames marked the growing social distance between rich and poor in Montréal with a spatial metaphor when he titled his investigation into conditions in the old working-class/immigrant district located between the river and the comfortable neighborhood of elegant townhouses on the slopes of Mount Royal, *The City Below the Hill* (1897). Similarly, Winnipeg had its North End, where thousands of immigrants speaking a dozen different tongues made their first Canadian homes during the massive surge of eastern and central European immigration to the western interior after 1896. In Vancouver, Chinatown, located on poorly drained and undesirable lands next to the tidally flooded False Creek flats just to the east of the downtown, persisted for decades as a tight-knit and, to the population at large, somewhat mysterious and troublesome community apart from the rest of the city, even as neighboring streets were occupied by immigrants from eastern Europe, Russia, and Italy.

Early in the twentieth century, every Canadian city also had its expanding halo of suburbs. Yet, for all their common dependence upon the streetcar (or in some cases suburban railroads), these were far from uniform creations. They ranged from the modest to the grandiose, and from the planned to the adventitious. Some were corporate developments, others the product of piecemeal individual speculations. Here they were developed around particular industries or works yards, there they lacked employment opportunities or services of any type. Geographer Richard Harris has usefully categorized this diversity into four types in his book *Creeping Conformity* (2004), a study of "how Canada became suburban." In this typology, "affluent enclaves" marked the elite end of the spectrum. Plans (not always realized) for suburbs in Victoria (The Uplands), Vancouver (Shaughnessy), Calgary (South Mount Royal), Montréal (Mount Royal, Westmount), Saint John (Coldwater), and other places were influenced by the ideas of Ebenezer Howard, Frederick Law Olmsted, and John Olmsted: curving streets, tree-lined boulevards, and green spaces, with large lots and (often) covenants or restrictions that determined the size and value of houses that could be built (as well, in some cases, as the racial or ethnic origin of those allowed to live in them).

At the other end of the spectrum were the "unplanned suburbs." Beyond the establishment of a street grid over an area of several acres (that was often unconformable with the grids developed on neighboring parcels) and the division of this grid into narrow lots (with frontages as narrow as 20 feet), the speculators who developed the land upon which these suburbs emerged generally avoided both the provision of infrastructure (sewerage and water lines) and the regulation of development. Most houses in these suburbs were built by amateurs, the "owner-builders" whom Harris identifies as constructing 40 percent of all the new single-family homes in Toronto between 1900 and 1913. Typically, they built piecemeal, perhaps starting with a shack and extending and improving as finances and family circumstances allowed. Some dismissed the results as "grotesque improvisation[s]," but for those who lived in them they were homes and footholds that gave them a stake in expanding urban economies (Harris 2004, 101).

"Middle-class" suburbs were more closely regulated and more orderly in appearance. Many of them were established on a straightforward grid pattern, especially in the west, although modest imitations of the curved streets of affluent enclaves worked their way into some designs. Zoning provisions and municipal regulations generally shaped the evolving landscape, and water—if not always piped sewerage—systems were usually put in place from the outset. In these locations speculative builders, working the narrow turf between preference and profitability in competitive markets, tended to build conservatively. By and large, they chose dwelling designs from increasingly widely available pattern books, repeated a single or a few basic patterns, and tailored their selections to the preference of local buyers. Although they might build houses in succession, bringing one to completion and using the proceeds of its sale to finance construction of the next, their work tended to replication. Suburb after suburb took shape as a conglomerate of minor variations on a very small lexicon of standard housing styles.

"Industrial suburbs" were distinguished by their development in association with outlying industrial activity as much as by their physical appearance. Initially, most lay beyond their neighboring cities, although almost all of them were subsequently amalgamated into those centers. Generally, they focused upon a railroad branch line. They also "subverted" the expectations of the "Chicago model," by creating significant enclaves of working-class housing on the far periphery of the CBD. Some, such as the Canadian Northern Railway development around its works yard in Leaside (in 1912), included areas of attractive housing laid out on lines influenced by garden city ideas. Others—Maisonneuve and Montréal Est, West Toronto Junction and Elmwood in Winnipeg— were much more workaday places, with modest houses constructed by speculative builders for industrial workers. Indeed, Maisonneuve, which began with plans for great boulevards, splendid public buildings, and a huge park, soon began to promote itself as "le Pittsburg [sic] du Canada" (Linteau 1981, 73) with pamphlets and posters portraying smoke-spewing factories with docks in front and a railway behind. Most of its residents were working-class tenants who lived in narrow unadorned row houses that lined the street grid. Their dwellings were hardly less ugly and forbidding than those in the slums of Montréal described by Gabrielle Roy in her 1947 novel, *The Tin Flute.*

Although urbanization was one of the dominant trends of the age, and its cumulative imprint was enormous, the course of urban growth in individual cities did not always run smooth. In the eyes of many contemporaries, urban expansion was a magician's rod to equal the railroad lines that knit the country together. When cities, like Topsy, "just growed," it seemed a simple thing for individuals to profit by buying cheap land on the urban fringe, sitting tight for a few years, and then selling at a premium (Wynn and Oke 1992, 87). The process could be quickened if development could be hastened, and thus speculation was quickly coupled with boosterism throughout urban Canada. Property developers jostled one another to attract streetcar service and buyers to their holdings. Streetcar companies developed their own variant on the game by purchasing outlying land, running their rails into it, and offering purchasers a pass to ride the line free for several months or a year. The result was a frenzy of activity. But economic booms were often followed by busts.

The results were felt in all cities, but in none perhaps so markedly as in those in the west where, as Peter Smith pointed out in the *Historical Atlas of Canada* (3: plate 20), "optimism . . . ran easily to excess," and led in the case of Edmonton to "repeated and unnecessary extensions of the city's boundaries, over-ambitious railway schemes, streetcar lines on unbuilt streets, a broad fringe zone of scattered development, and a still broader fringe of empty speculative subdivisions," all of which constituted a "negative legacy" of the city's frontier epoch. In city centers east and west, many deplored another negative legacy of modern urban growth: the proliferation of utility poles and wires erected to carry the electricity that lit the night, transmitted telegraph messages, powered streetcars, and (after the First World War) operated the traffic signals that sought to bring some order to increasingly chaotic patterns of traffic flow. Percy Nobbs, a professor of architecture at McGill University in Montréal, lamented that the once-beautiful streets of his city had come to resemble "a Chinese harbour after a typhoon" (quoted in van Nus 1984, 170). Others felt similar regrets at the loss of once-pleasant waterfront districts to warehouses and railroad tracks.

As cities grew, it became apparent that urban environments created challenges to life and health as well as to aesthetic sensibilities. Montréal presented a particularly troubling case. Through thirty-five years after 1896, its general and infant mortality rates were consistently and significantly above those in Toronto, and some suggested that the city's death rates were among the highest in the North Atlantic world. Between 1900 and 1910, three of every ten children born in Montréal died before their first birthdays (Copp 1974). This was a class, as well as a societal, problem. Infant mortality was extremely high in the poorest parts of the city. Inadequate housing, unsanitary conditions, and the want of sufficient income (later studies have shown that the majority of Montréal's working class lived below the poverty line during these years) took a dreadful toll. And it was not only children who were affected. Between 1900 and 1908 two of every thousand residents of the city succumbed to tuberculosis, more than in any other large American city. Typhoid and other diseases were also recurrent killers. In twentieth-century Canada as in nineteenth-century England, the rapid growth of cities strained society's capacity to cope with the environmental consequences of the growing concentration of humans, animals, and industries in small areas.

Waste disposal and the provision of clean drinking water were significant issues. Late in the nineteenth century boards of health in the larger urban centers began to recognize some dimensions of the emerging problems they faced. They collected data, condemned a few of the worst dwellings amid dozens of inferior structures, and required—more or less effectively—efforts at garbage disposal. Their endeavors were supplemented by those of various sanitary and philanthropic reformers, whose efforts to improve the cleanliness and thus the healthfulness of working people led historian Marianne Valverde to characterize this "as the age of light, soap and water."

But there were enormous obstacles to overcome. Neither piped water nor sewerage systems were widespread in the 1880s. Horses and oxen polluted city streets, and dogs and pigs also ran freely through many areas. People kept chickens in their backyards, and garbage was dumped on vacant land, on roadways, and in rivers. In this respect, electric streetcars were a boon, as they replaced draft animals and reduced the amount of manure in the city. Cesspits and privies were built, but many were poorly constructed, with improper understanding of local soil and topographic conditions, and their sheer proliferation almost invariably overloaded the environment's capacity to absorb domestic waste. Wells were polluted periodically, and afflictions such as dysentery were probably almost endemic. The creation of parks and green space, which some reformers advocated on the grounds that they would give residents of crowded neighborhoods access to fresh air and open space, did no harm, but was hardly a solution to the progressive pollution of intricate and still poorly understood urban ecologies (Kaika and Swyngedouw 2001).

When municipalities pursued more effective strategies, such as the extension of piped water and the development of sewerage systems, they often met resistance. These were costly projects, especially in earlier-settled areas. Yet even in new centers, where the timing of development allowed the provision of water and sewerage infrastructure almost from the first, concerns about cost long and often worked to reduce the effectiveness of the utility system. Thus, Vancouver councillors presented with a state-of-the-art plan for the development of their utility system in 1911 opted to save money by building an inferior, combined sewerage/drainage system that continued, on occasion, to pollute local beaches into the third quarter of the twentieth century, despite expensive investment in its improvement.

In many cities, moreover, sewerage systems could not be extended fast enough to keep pace with the spread of suburban development. Especially—but not only—in unplanned suburbs, many dwellings relied on privies and cesspools. In Vancouver, city inspector Robert Marrion voiced the frustrations of harried officials when he noted in 1912 that "nearly every householder demands an up-to-date water closet and every water closet requires a septic tank, this needs an overflow which usually discharges the excrement in solution into the channel of the nearest street or lane, thus causing complaints to be made from the people in the locality who are generally creating nuisances themselves" (quoted in Keeling 2004b, 76). Even when sewerage systems were built, they relied upon the "assimilative capacity" of local waters to dispose of waste. In practice this meant that they piped untreated sewage into rivers or lakes, in the generally false belief that problems would be dissipated by dispersion and dilution.

As early as 1878 a Select Committee of the Ontario legislature discovered that water supplies were polluted by human waste in sixty of the eighty municipalities that responded to their inquiries. A few years later a Toronto newspaper described the city's water as "drinkable sewage." As Peter Baskerville has pointed out, little was done with this knowledge because "provincial legislation allowed for the collection of information and inspection but provided little in the way of enforcement" (2002, 154–155). Even the establishment of a Provincial Board of Health did little to change these circumstances, as the local authorities responsible for the provision of infrastructure were unwilling or unable to raise the revenues required for construction. When they did, their efforts were as often misguided as they were inadequate. In 1893 the town of Waterville was indicted (in phrases shaped by legal considerations but redolent of the mounting frustration of all those affected), on the grounds that it "wilfully and injuriously did construct, make, build and maintain certain sewers and drains" that "greatly filled, impregnated, polluted and fouled" the Detroit River "with . . . refuse matters and substances" and rendered its waters "corrupted, fouled, offensive and unhealthy to the great damage and common nuisance" (quoted in ibid., 155) of residents. The cumulative effects of such actions were revealed almost twenty years later, when the Canadian Commission of Conservation pointed out that only one among nine industrialized nations reported a higher incidence of typhoid infection than Canada. For Dr. Helen McMurchy, the author of a report on infant mortality, it seemed quite evident that "the Canadian city is still essentially uncivilized—it is neither properly paved nor drained, nor supplied with water fit to drink, nor equipped with any adequate public health organization" (quoted in Cook 1987, 395).

Some have argued that improvement waited upon catastrophe—that a major fire or epidemic was necessary before taxes could be increased to acquire control of and/or improve water and sewerage systems. Such disasters were undoubted stimuli, but local circumstances were generally more complex than this simple equation allows: Political, commercial, and personal interests were refracted through imperfect understandings and imprecise perceptions of the mechanisms and risks involved in particular cases. Even in Toronto, where elaborate plans for improvements in the water supply were drawn up in 1896, where much of the downtown was destroyed by a great fire in 1904, and where extremely high infant mortality rates (increasingly associated with polluted water) were an ongoing concern, the city did not effectively separate its sewage outfall from its water intake in Lake Ontario until 1910. Then, the water supply was filtered and chlorinated, and sedimentation tanks (the most basic form of primary sewage treatment) were constructed to reduce the input of raw effluent into the lake. Still it took ten more years and a major investment in urban infrastructure to bring running water and a sewer connection to most of the homes in the city. And as novelist Michael Ondaatje reminds readers in his evocation of life in Toronto, *In the Skin of a Lion* (1987), construction of the splendid Art Deco–style R. C. Harris Filtration Plant, the treatment works built to improve the quality and quantity of the domestic water supply and known to contemporaries as "The Palace of Purification," did not begin until the 1930s (Benidickson 2006).

15

COUNTRYSIDES
IN TRANSITION

Rural worlds were also remade during these years. In earlier-settled agricultural districts from eastern Nova Scotia to the western limits of Ontario, significant changes influenced farm families, communities, and the countryside in various, often seemingly paradoxical, ways. In the most general terms, transportation improvements, mechanization, and market shifts differentiated regional patterns of agricultural production, while competition, science, and technology reshaped agricultural practices. Yet these influences played unevenly across the landscape. They hardly touched some parts of the countryside until well into the twentieth century; in other areas they were felt in the 1860s (broadly see Wynn and MacKinnon, n.d.).

By and large, however, changes in transportation destroyed local monopolies and opened prospects of new markets. Urban and industrial growth placed new demands upon and yielded new opportunities for farmers. Demographic pressures put people in motion. New machinery squeezed labor from the countryside. Scientific advances increased returns from land and labor. And experimentation and experience demonstrated the economic advantages of specialization in agriculture. Professor A. B. Balcom, an authority in economic science at Acadia University in Wolfville, Nova Scotia, saw the fundamental forces at work here in the 1920s, when he urged Canadian farmers to recognize "the consumers of the world" as their customers and to accept that "the producers of the world . . . [were] their competitors" (Balcom 1928, 42).

With the opening of the Intercolonial Railroad in 1876, farmers in Canada's eastern provinces were immediately subjected to competition from grain (and later meat) producers in central Canada. In New Brunswick and Nova Scotia the wheat acreage fell by almost two-thirds during the 1880s. Regional production of barley and potatoes also fell (although that of oats increased) and the area in field crops never again exceeded the level of 1880. Competition from imported meat also produced a decline in sheep and cattle (other than milch cow) numbers in the region.

On the other side of the ledger improved access to markets favored specialization. The Annapolis Valley became a prominent supplier of apples to Britain, and American demand spawned the development of egg and bacon production in Nova Scotia. In Québec, dairying came to dominate a mixed farming economy in the last quarter of the nineteenth century, but local specializations also emerged in response to local markets or particular ecological conditions: market gardening near Montréal; livestock in the Eastern Townships and along the south shore of the St. Lawrence below Québec City; hay for sale in lumber camps and New England in the Beauce, the St. Maurice, and the Ottawa valleys; and horses in Chambly County.

In Ontario the proportion of land sown to wheat fell from 18 to 5 percent between 1880 and 1900. Mixed farming with an emphasis on livestock became the norm. Dairying expanded. Feed grains, hay, and root crops assumed a larger place in the production mix, as cattle raising for British markets increased in importance. Here as elsewhere, agricultural production was gradually adjusted to make the most of local soil and climate conditions and to take advantage of market opportunities. Economic historians have debated the reasons for these adjustments at some length. Using econometric techniques and hindsight unavailable to the farmers whose decisions they seek to understand, they have concluded that, in Ontario at least, land use reflected variations in the comparative prices and comparative yields of the various field crops, and that risk—the variability of prices and yields—was not a major determinant of farmers' decision making (Ankli and Duncan 1984).

Be that as it may, these changes, and the shift to production for exchange rather than use value that underpinned them, led to the rationalization and reorganization of farming operations. In Ontario the number of occupied farms rose until 1891 and then fell back; in 1941 there were only a few thousand more farms than in 1871. The average farm increased in size, from slightly more than 90 acres in the 1870s to 125 or so in 1941. Overall, the cultivated area rose from 6.5 million acres in 1871 to 9.6 million acres in 1911 and 10.4 million acres in 1926, and the amount of improved land per farm increased, on average, from 51 to 65 acres between 1871 and 1901.

All of this suggests complex patterns of adjustment—including the abandonment of marginal holdings, the consolidation of farms, and out-migration from agriculture. Each of these developments had ecological consequences: Forests began to recolonize abandoned fields, and in the early stages of this process the increased availability of browse for deer likely encouraged an increase in their numbers; at the same time, the conversion of forest and rough pasture to cropland made soils more vulnerable to erosion and changed local hydrological regimes; larger, more specialized farms simplified cropping patterns and generally reduced the mix of plants under cultivation; and machines were substituted for the labor of humans and animals in many stages of the production process, a change that altered ratios of pasture to cropland and helped precipitate the shift toward the use of artificial fertilizers in place of manure.

Although immigrants came to Canada and cities grew, rural populations in each of the four original provinces increased more slowly than they should have according to prevailing rates of natural increase. People were leaving the land. In the half century after 1881, seventy counties in some of the least-productive agricultural areas of the Maritime Provinces, southeastern Québec, and Ontario lost more than 300,000 people (Conrad and Hiller 2001). In New Brunswick members of the 1909 Agricultural Commission were given a simple reason for this: "Stories of big crops and big wages take the men West" (New Brunswick Royal Commission on Agriculture 1909, quoted in Weiger 1990, 228). As individuals and families departed, communities declined. Because most of those who left were young, rates of natural increase fell. Before long, "that feeling of discouragement which comes from seeing desertion and decay" (ibid., 232) led others to move as well. Aging couples would find their migrant children unwilling to return to the farm. Where they had labored to carve fields from the forest and to plant lives as well as crops, invading species of weeds and shrubs and trees began to obliterate their hard-won "gains." The forest, forced to retreat before the settlers' advance, reclaimed its ground. Human imprints faded. Buildings decayed and fell, fence lines gradually disappeared, and orchards grew gnarled and untended in the woods, forgotten until members of a later generation returned to reap (in the poignant words of poet and novelist Charles Bruce) "a crop of winter firewood" (quoted in Wynn 1988, 44).

At the same time, those who remained on better lands or who found themselves in locations that offered access to sizable markets beyond the local often adapted quickly and decisively to new circumstances (Baskerville 2002). New machinery and new techniques were critical here. Although horses remained integral to farming well into the 1930s (in 1941 two-thirds of Ontario farms had at least one horse), animal power was increasingly supplemented and then replaced by mechanical (or engine) power. Stationary steam engines were fairly commonplace by 1900. In Ontario the first steam threshing machine was operated in 1877; by the end of the century most of these machines could thresh the entire grain crop of even the largest farms in a day or two. A decade later, gasoline engines began to appear on farms. In 1921 a fifth of Ontario farmers owned one, and almost a third owned a car or truck, the engines of which could be used to power farm machinery using the "take-off" mechanism that was standard on many vehicles.

Less revolutionary but increasingly effective machinery (including horse-powered rakes, reapers, mowers, binders, and thrashers) also found increasingly widespread use. Stump pullers allowed fields to be more thoroughly cleared and opened the way to deeper and more complete working of the soil. Cream separators saved much labor after 1880. Patented, purchased steel churns also replaced homemade wooden equipment. Such improvements enabled the expansion of herds and production and helped change the mix of stock and patterns of land use on farms. Electric milking machines were labor saving and revolutionary, but in Ontario only one in fifteen farms had electricity in 1921. Even as that proportion increased to one in six by 1931, the number of milking machines climbed to 4,000. Long before, the "whack, whack" of the flail on the barn floor and the "whiz" of the spinning wheel in the kitchen had faded from the most productive farming districts; "they, along with the scythe and sickle," said one observer, "belong to the days of our grandfathers" (quoted in *Nova Scotia House of Assembly Journals* 1885, 4: 521).

Farmers also transformed their environments more directly and deliberately. Through the pre-Confederation years, settlers struggled with and avoided certain types of land. They understood that the organic and alluvial soils of swampy areas were potentially extremely fertile but balked at the cost and difficulty of draining them. And they discovered that wet loams and clay plains were not only difficult to work without some form of drainage, but that they tended to compact and retain more moisture after several years of cultivation. Because wet loams and clays were extensive, farmers tried to address the problems they presented by cutting open ditches or, more commonly, by plowing their fields into ridges and furrows. Made skillfully, furrows that ran downslope to a stream or ditch carried away standing water quite successfully—although they also less visibly transported nutrients and hastened soil erosion. Less artfully constructed, the ridge and furrow system simply transferred water from elevated to lower-lying areas, or created a microgeography in each field, where ridge tops were dry and cultivable and furrows were either flooded or sun baked, and generally left unplanted. This reduced yields per acre and provided a niche for weeds; it also impeded the use of field machinery and required a certain amount of labor each year to maintain. Tile- or under-draining techniques were known to farmers with "friends in England" from whom they "frequently received publications . . . in which the subject was treated ably," but the costs of implementing them were prohibitive before 1850 (Kelly 1975, 281).

As the drive to clear the land of its forest abated and farming became more specialized and more commercial, however, farm journals zealously advocated underdrainage. In some respects, the changing emphases of late nineteenth-century farming carried this enthusiasm onto deaf ears. Farmers moving away from wheat production to cattle raising could cultivate their well-drained fields and use those that were wet for hay or pasture. In time, though, such practice became less and less economically viable. Wet pastures were sometimes unsafe for cattle, and the growing demand for heavier stock placed a premium on the production of feed grains and root crops (which did better on well-drained land) rather than hay. In earlier-settled districts where there were good markets for various crops, from spring wheat to barley, oats, and peas, and where signs of "soil exhaustion" were accumulating, the incentive to improvement was always greater.

In southern Ontario, a million acres of uncultivated marsh and swamp also beckoned with new urgency. By 1867 the northern limits of good farmland had been met, and the difficulties of agricultural settlement on the fringe of the Shield were becoming all too evident. Agricultural opportunities were needed to sustain a rising population and many came to the view that wetlands, "our best lands[,] are . . . lying waste and worthless while our high dry lands are quite exhausted with overcropping" (Kelly 1975, 286). Suddenly both public resources and private effort were pitched into the improvement of swampland. Government and municipal surveyors limned the extent of wetlands and estimated the possibilities of reclamation by gravitation. The Ontario Drainage Act of 1869 made $200,000 available to finance the construction of main drains, and this was quickly expended on 139 miles of construction in nine townships in far southwestern Ontario.

The immediate results were gratifying. Large areas once characterized as "all swale in the bush," and "just one fever and ague swamp," were converted into "first class lands . . . fit to produce any kind of crop" (Kelly 1975, 289–291). New areas were opened to settlement, and new landscapes reticulated by large drains and straightened streams as well as the survey grid took form. But drainage in one place all too often caused flooding in another. Early reclamation schemes typically used the lowest-lying parts of the region as sinks. Later extensions of the reclaimed area, or small-scale private initiatives that utilized government drains to remove water from older, established farms, overtaxed channels designed for initial swamp drainage schemes and inundated lands once "unwatered" by them.

In the 1880s and 1890s, the use of steam-powered pumping engines and the construction of major earth works including artificial embankments became necessary to reclaim and remediate areas despoiled by the use of gravitational drainage techniques. Still, this was not the end of it. Swamps were natural reservoirs that regulated the flow of rivers and streams. When they were drained, the hydrological regime was changed. Rainfall flowed more quickly through drainage systems. Especially in the spring and fall, undrained downstream swamps and valley-bottom towns were subject to floods. Frog ponds, nesting areas used by several species of birds, and other wildlife habitats were also destroyed by swamp drainage.

By the end of the nineteenth century, new ideas about the rural landscape and new principles of land use were also remaking the countryside, especially in Ontario. Reacting to the damage produced by the wholesale assault on the forest before Confederation, an assault that had removed 80 percent or more of the forest in some townships and led to widespread erosion, flooding, and silting, officials began to imagine and encourage reforestation of substantial parts of the rural landscape. Heights of land and areas of poor soils—those parts of the province from which agriculture was particularly in retreat—were to be planted with trees "to secure favourable climatic conditions and regulate the water supply" (Kelly 1974b, 6). Ideally 20 to 25 percent of each agricultural township would be placed under permanent forest cover. In addition, farmers were encouraged to plant and maintain woodlots (the government provided advice on the species to plant and on management techniques) and bonuses were granted for the planting of shade trees along highways and farm boundaries. Although progress toward these ends was uneven, and almost always less dramatic than promotional efforts envisaged, the cumulative effects of these initiatives were to "impose a new series of planned modifications and land-use controls" that helped to "rebuild" landscapes ravaged by a century of untrammeled exploitation and development (ibid., 12).

The dairy industry exemplifies many of the changes that transformed the countryside during these years. In British North America most farmers kept a few cows to provide milk for their families. Traditionally, farm wives churned butter by hand, and small dairy surpluses would enter local trade by barter or cash exchange. On the fringes of towns and growing cities, in the days before refrigeration, some farmers developed modest specializations in dairying, and carried milk to town each day for sale in small quantities to customers along a regular delivery route. To keep their herds in milk as long as possible, periurban dairy operations tended to keep more cows than traditional mixed subsistence farms, to house them in barns at night in both summer and winter, to collect manure, to fertilize fields of fodder crops, and to purchase mill feed. But they rarely focused solely on dairying, and most of them remained relatively small: A dozen cows and half a dozen calves were normally sufficient to provide a modest living and (with the wide range of other necessary chores before them) to keep a farm family fully occupied.

As commercial horizons began to widen, however, dairying began to attract scientific interest and government expenditures. With an eye on the huge British market for dairy produce, provincial officials began to imagine rural prosperity for eastern Canadian farms newly converted to dairy production.

The centralization of dairy production. *In the course of the nineteenth century, dairying changed from a small-scale activity in which modest quantities of butter and cheese were hand-churned on farms that kept a few cows to a far more specialized activity dependent upon improved breeds of stock and central dairy plants that processed the milk of many producers. Here milk is being passed over refrigerating pipes in the separating room of an unidentified, probably early twentieth-century, dairy factory. (Library of Congress)*

Denmark, after all, had raised its agricultural sector from near-hopeless impoverishment to a progressive and happy state by producing butter and cheese for British consumers. In the 1890s various branches of the Canadian state appointed dairy commissioners, established experimental farms to conduct research into dairying, imported breeding bulls to improve blood lines, and offered bounties for the construction of creameries and cheese factories.

Such effort had its results. In the three easternmost provinces, where there were thirty small butter and cheese factories with an 1890 output of $85,000, some 144 factories produced $920,000 worth of butter and cheese in 1900. A decade later declining output puzzled and frustrated government officials, who sometimes attributed the difficulties of the industry to the gendered division of farm labor. By this account, farm women limited the growth of factory dairying because wives, whose "perquisite" the proceeds of homemade butter had been, feared that checks received for milk sent to the creamery or the cheese factory would go straight into the pocket of "his lordship" (Ruddick 1909, 110).

But the decline owed more to structural factors than it did to cultural and sociological reasons. Most of the dairy factories were shoe-string cooperative enterprises, erected in haste and underfinanced. Typically, they relied upon small herds, returning relatively low milk yields per cow. Variations in output and in the butter-fat content of milk from different farms made it difficult to sustain production at an efficient level throughout the season and undermined the foundations of trust and compromise on which cooperation rested. Skilled cheese makers were hard to find. And problems of cleanliness and quality control increased as quickly built factories aged, and butter and cheese came from more producers. An undercapitalized, small-scale, marginal industry could not compete in the British market with the highly efficient producers of Denmark, New Zealand, and even central Canada, whose advantages of location, climate, organization, scale, investment, and herd quality variously allowed them to dominate the developing trade. Indeed, dairying effectively reconnected Québec agriculture with the market in the second half of the nineteenth century, after midge and rust devastated wheat production in the colony in the 1830s. By 1900 there were almost 2,000 butter and cheese factories in Québec, and production expanded into the late 1920s, when it accounted for well over a quarter of agricultural income.

Yet success in dairying carried costs, too. In Ontario, where the industry was earlier established, larger, and more prosperous than in the Maritime Provinces, cheese and then butter production moved from farm to factory in the 1870s and 1880s. Milking machinery; pedigree breeding programs; new feeds; and new barns designed to make the handling, feeding, and milking of stock more efficient radically altered the nature of farming. For the *Farmer's Advocate* the successful Ontario farmer had been forced to move "from the rank of a strenuous toiler to the more complex status of a business proprietor" (Wynn 1987a, 391). A decade earlier, a contributor to the 1895 *Canadian Livestock Journal* had claimed that "the modern dairy cow in her best form is a highly artificial animal. The more artificial she is the better. The dairy cow has been trained and made over by the hand and brain of *man* for a perfectly natural purpose, for giving milk, yielding butter and making money" (quoted in Derry 1998, 36).

Too much can be made of this claim. Selective breeding for particular attributes had changed the appearance of bovine breeds over decades, and the dairy herds of 1890 were, undoubtedly, better milk producers than the forest-browsing stock of early nineteenth-century pioneers. Indeed, the development of dairying had prompted farmers to replace the dual-purpose breeds (Ayrshires and Shorthorns) popular early in the century with specialized producers: Holsteins for milk and Jerseys for cream. The circumstances in which prime dairy herds spent their days, moving between milking shed and tended pasture, nurtured on enriched mixes of factory-prepared feed, kept indoors through the winter, and bred to a schedule designed and determined by the farmer, were also less "natural" than conditions in the wild, or indeed than those in which most British North American cattle were raised. But these cows were not cyborgs.

Similarly, the contrast between strenuous toiler and business proprietor should not be exaggerated. Although machines had eased the burden of human labor on most farms, no early twentieth-century farmer was spared the need for arduous toil in nature. Although the business side of farming loomed larger on sizable commercial ventures than it did, or had, on small-scale subsistence operations, and although improving and ambitious farmers were enjoined to keep careful records of their expenditures, returns, and investments, few would have readily equated their lot in life with that of the urban businessman. They knew nature through their work in it, day upon day and season after season. Still, these contemporary observations point to significant dimensions of change in the countryside. Spawned by the dawning recognition that mechanization and modernization were recasting traditional farming practices and modes of life, they also reflected the seemingly inexorable transformation of rural lives and landscapes in modernizing, twentieth-century Canada.

16

THE ASSAULT ON THE FOREST

In 1938 the historian Arthur Lower published an important book with an arresting thesis. Reduced to its essence, this work argued that although British North America had been "Great Britain's Woodyard," Canada quickly came to serve the voracious demand of the United States for wood. With new technologies facilitating the conversion of trees into timber, and railroads providing new and improved links between forests north of the international boundary and consumers to the south of it, Lower argued that the assault on the Canadian forest quickened until "the sack of the largest and wealthiest of medieval cities could have been but a bagatelle" by comparison (Lower 1938, 26).

The unadorned statistical bones of this story go a long way to justify his extravagant rhetoric. Between 1850 and 1870 the value of British North America's wood exports to Britain increased more than fivefold. Yet sales to the United States grew even more dramatically. In 1850 the United States took 22 percent of British North American wood exports, by value. Twenty years later the proportion was almost 42 percent. Despite an economic depression in the 1870s, a lesser downturn in the 1890s, and a complex series of tariffs and duties imposed on trans-border wood shipments by both Canadian and American authorities in the last thirty years of the century, Canadian lumber exports to the United States climbed more than threefold (in value) from the depression-induced trough of 1879, to a nineteenth-century peak in excess of $15 million in 1896.

The consequences for the forests of central Canada were immense. The onslaught began across the southern boundary of the Shield, in the Trent watershed (north of Peterborough) and in Simcoe County. With the construction of the Northern Railroad, from Toronto to Collingwood on Georgian Bay in 1855, Simcoe County rose from obscurity to leadership among lumber producers in the province. In 1861 its rich pine forests yielded three times as much lumber (almost 208 million board feet), as its nearest competitor. As railroad lines proliferated, exploitation proceeded. In the town of Midland the first sawmill was built in 1872, with the completion of a railroad between its fine harbor and Port Hope on Lake Ontario. Ten years later the town had six mills and two sash-and-door factories, and the Midland Railroad (which served other mills than those in Midland) carried 105 million feet of lumber, most of it for shipment to Oswego

Loggers atop a "brag load." *The rhythms and technologies of logging changed little, relative to developments in sawmilling, during the nineteenth century. Felling and hauling remained essentially wintertime activities, saws replaced axes for some tasks, and refinements in the organization of logging enterprises as they grew larger led to greater investment in the preparation of ice roads and the use of horse-drawn sleds or (very late in the century) railroads to haul logs to riverbanks. Here a group of workers poses atop what was probably a "brag-load" of logs bound for a sawmill, but the cumulative effects of such extraction were to "lay waste" large tracts of forest. (Corel)*

in the United States. By 1886 the Georgian Bay catchment produced 221 million feet of lumber. At the end of the century, Midland was second only to Ottawa among Ontario sawmilling centers, and it drew logs from several river basins flowing into the north shore of Georgian Bay (Lower 1938, 172–175). Lumbermen had swept through "a region larger than many a European country" in less than two decades (ibid., 178).

Rapidly rising demand for lumber in Chicago and the American Midwest further hastened forest exploitation on the northern shores of Lake Huron after

1880, when many Michigan mills, having denuded their immediate surroundings of high-quality pine, turned to Canada for sawlogs. Enormous booms, each comprising 30,000 to 60,000 logs, were towed across the waters of lakes Huron and Michigan, and by 1886 it was said that American interests held licenses for 1,750 million board feet of timber in the Georgian Bay area. Through the next decade the two countries skirmished over the trade, until the Ontario government prohibited the export of sawlogs cut from Crown land. This dramatically reduced shipments of unmanufactured wood to the United States, but did less to limit exploitation of the forest. According to a report in the *Canada Lumberman* of March 1902, American mills built in Ontario as a result of the export prohibition were capable of producing 265 million board feet of lumber a year (ibid., 155–159).

Technological changes were vital to the quickening pace of forest exploitation. In the last half of the nineteenth century, sawmills and the woodland operations that supported them grew ever larger and more efficient. In eastern Canada the most striking changes were in milling. Steam replaced water as the source of motive power, and highly efficient multiple gang saws carved logs into boards and planks. With their enormous buildings, extensive yards or piling grounds where new-sawn lumber was stacked, and associated booming grounds to hold the supply of logs, mills dominated the communities and landscapes in which they were built. Most of them employed hundreds, and so long as timber remained in their hinterlands, they were the economic lynchpins of their communities and the engines of growth in the north country. But such was the pace of exploitation that some of them lasted only a decade or two. In the exuberant prose of W. H. Withrow, who wrote a "scenic and descriptive account" of Canada at the end of the nineteenth century, "the ravenous sawmills in this pine wilderness are not unlike the huge dragons that used in popular legend to lay waste the country; and like dragons, they die when their prey, the lordly pines, are all devoured" (Withrow 1889, 127).

These mills typically depended upon the labor of hundreds of men who worked in winter camps and on the spring/summer logdrives. In most respects, the rhythms and technologies of bush work changed little through the late nineteenth century, especially in eastern forests. Camps increased in size, and more and more of those who worked in the woods made this their main occupation. Specialization increased as cooks and clerks and sawfilers were added to managers and teamsters and fellers and "beavers" and swampers (who cut and maintained roads). But axes were used to fell and handsaws to buck logs into manageable lengths. Horses skidded logs to the landings, and hauled sleighs out of the woods. Technological improvements were refinements of an age-old set of practices. Cross-cut saws replaced axes, sleighs were improved in design and increased in size, and efficiencies were gained by better road making and new arrangements in loading areas.

Steam came late to these forests and initially its impacts were small. Mounted on barges or scows, steam engines hauled log booms across lakes. Early in the twentieth century the Lidgerwood Manufacturing Company developed a steam-powered skidder to winch logs out of the bush. It was intended to do the work of horses, and in advocating its adoption the company resorted to the classic arguments of the period. It should be used, they said: "Because it is inanimate, does not die, eats nothing when it does not work, is unaffected by the weather, disease or insects, is constant or tireless and gets cheap logs" (quoted in Radforth 1987, 78). Steam-powered tractors and logging railroads were also promoted for similar reasons. Some were used in northern Ontario and northern Québec early in the twentieth century, but by and large they were judged too costly, cumbersome, and unprofitable for widespread use in the forests of eastern Canada at this time.

On the west coast, where the trees were larger, the terrain was more difficult, and labor costs were greater, the technological equation resolved rather differently. Rapid expansion—there were twenty-five sawmills in British Columbia in 1888, and production capacity doubled in the next six years then doubled again by 1900 and continued to increase until the province had 261 mills producing almost 1,620 million board feet in 1910—carried logging away from the coastal littoral (where trees could be felled and rolled into tidewater using axes, saws, manual jacks, and gravity, or hauled to water by oxen and horses) and saw the widespread use of steam engines (or "donkeys") for yarding and hauling from the 1890s onward. The logging railroad was also adopted relatively early, in the 1890s, and by the 1920s there were 700 miles of logging railroad track operated by seventy-nine different concerns in British Columbia (Rajala 1998).

It is extremely difficult to know how much wood lumberers encountered, and how much of the forest they removed, as they "laid waste the country." Stands varied in value and data are inadequate. In eastern Canada, informed observers reported yields of 20,000 board feet of pine per acre in the Trent watershed, and of only 1,000 or 2,000 feet per acre on the north shore of Lake Huron in the late nineteenth century. As measures of high and low commercial yields, both of these figures are probably underestimates. If the cut averaged 9,000 feet of pine per acre through the forest as a whole (this is likely a high estimate), an area of 100,000 acres would have been worked over to produce the lumber shipped to the United States in 1870. Repeated at this level and more, year upon year, the effects of the industry were enormous. Bernhard Fernow, one of the leading foresters of his day, suggested that at least 100,000 million board feet of

Steam engine hauling logs. *Logging railroads were integral to the exploitation of the massive cedar and Douglas fir forests of British Columbia. They allowed loggers to move considerable distances back from tidewater and to haul out their cut efficiently. But they also required heavy capital investment. This fine image speaks volumes to the power of the logging railroad: fifty-two carloads of Douglas fir logs snaking to the horizon behind a single locomotive, quite possibly built by the Shay Engine works in Michigan and thus reflective of the growing continental integration of economic activity. But look beyond this, to the buildings on the right—company bunkhouses, part of another form of resource camp. And recognize that the logs from this train (probably on Vancouver Island) are to be assembled into rafts to be towed, by steam tugs (across the Strait of Georgia?), to gargantuan mills (likely at the mouth of the Fraser River). (National Archives of Canada)*

pine had been cut from the "original" forests of Canada by 1899. This implies the systematic removal of pine trees from at least 18,000 to 19,000 square miles of land. In practice, a much greater area was subject to the lumberers' on-slaught.

Domestic demands also took their toll on these forests, as the requirements of railroad construction, settlement, and development were satisfied. Each mile of railroad constructed—and the expansion in Canadian railroad mileage was enormous in the half century after 1880—required thousands of ties and, before they were treated to reduce rot, replacement was a constant necessity. By one estimate, 55 million railway ties (almost all of them hemlock) were cut from the forests of Ontario northwest of Sault Ste. Marie between 1875 and 1930. In British Columbia some estimates suggest that a million ties a year were needed when railroad construction was at its height in that province early in the twentieth century.

Appropriately sized cedar trees, peeled of their bark to provide the poles that carried telegraph, telephone, and electricity lines across country and through growing towns, were sold beyond British Columbia, and shingles split from the western red cedar trees so abundant in southern coastal forests were a major commercial commodity. In the 1920s four-fifths of Canadian shingles came from the province. Tamarack pilings for wharves and other construction were also in considerable demand. A conservative estimate suggests that some 2.2 billion board feet of northwestern Ontario lumber was used for these purposes alone. Wood was needed for houses and commercial buildings to accommodate the demands of a growing population and an expanding economy. And wood was still much used for fuel; consumption for this purpose was little different in 1900 than in 1871, although energy from wood accounted for less than 50 percent of total requirements at century's end.

In sum, the assault on Canada's forests seemed almost unrelenting. But much though it was reduced and scarred, the forest endured. According to one estimate, some 40,000 million board feet of pine remained at century's end: Slightly more than half of this was in Ontario; rather less than half in Québec; and possibly 2,000 to 2,500 million feet in the Maritimes. Still, this should not obscure the real damage done to forests and ecosystems by a century or more of essentially unrestrained resource exploitation. Early in the twentieth century, when the first generation of professionally trained foresters began to assess the state of the forests over which they were slowly assuming charge, they found much to lament. In Nova Scotia most of the forest had been degraded. In the Trent watershed of Ontario, once among the richest "pineries" imaginable, according to those who studied them, unimpressive second growth dominated the landscape by the end of the nineteenth century.

A careful survey in 1913 revealed that little commercially valuable forest remained, and that much of the land had been "severely culled and burnt over" before being colonized by light-loving species. In Snowden Township, Haliburton County, which was hardly atypical, only one-fifteenth of the landscape bore mature forest (much of which had been heavily logged). Young poplar and birch covered more than four-fifths of the area, much of which had been burned after earlier logging. Loggers sought cheap wood. Their interests were immediate and they paid little heed to the long-term health of the forest (Howe and White 1913; see also Gentilcore et al. 1984, 144).

Because the increasing size and technological refinement of the mills that were the critical elements of this expanding business required enormous capital investments, the capitalists who owned them exerted enormous influence in late nineteenth-century Canada. Power, prestige, and influence afforded them important roles in debates about the resource when concerns about wasteful destruction of the forest surfaced in the 1850s and 1860s. In the 1870s James Little—a lumberman who had cut through the forests of southwestern Ontario and the Georgian Bay area before moving his interests into the St. Maurice area of Québec—bemoaned the dire effects of fire and overcutting on the forests of northeastern America and offered a package of suggestions—including stricter control over the forests, prohibitions against cutting small trees, fire protection, and the exclusion of settlement from designated forest areas—to maintain the wood supply. These concerns were to the fore when the American Forestry Congress held its second meeting in Montréal in August 1882, and by 1885, both Ontario and Québec employed forest rangers to detect and fight fires, and Québec had set aside (temporarily) a large area of the Ottawa Valley solely for timber production.

On the other hand, the hubris of the lumber kings led to ongoing struggles over "the vexed question of sawdust" in Canada's rivers. The damaging environmental impacts of sawmill waste on both fisheries and navigation were well known by the 1860s. In the 1830s and 1840s both New Brunswick and the Canadas had introduced legislation penalizing the discharge of slabs, edgings, bark, and waste wood (but not sawdust) into navigable rivers, and at midcentury there were efforts to limit the dumping of sawdust into the St. John River (Wynn 1981; Perley 1852). By 1865 fisheries legislation applicable to the united Canadas added sawdust and mill rubbish to the list of deleterious substances that it was unlawful to "drift or throw" into "any stream frequented by salmon, trout, pickerel or bass."

But these prohibitions were largely ineffective. Generally, provincial legislatures lacked the resolve to act effectively against an important industry. Fines for transgression were small, and enforcement was sporadic. When efforts to implement the law were made, mill owners were quick to protest that compliance would be ruinous, that it would cost more to control the sawdust produced by their operations than to produce the lumber that they sent to market. Even the new federal fisheries act of 1868 allowed the minister to permit the discharge of sawdust and mill waste into streams if there was no clear and compelling public interest argument for preventing it (Allardyce 1972; Gillis 1986).

When J. R. Booth became the first lumberman to be fined for dumping mill waste into the Ottawa River in 1875 he was charged a mere $20. In the face of such effrontery, the industry seemed to redouble its resolve to ignore the law, and several citizens protested the hurt they would suffer if they were deprived of opportunities to haul mill waste from the river and its banks. Captains of industry were effective lobbyists. In 1880 mills on the Gatineau River and in the Ottawa-Hull area were exempted, by Order-in-Council, from an earlier act prohibiting the discharge of mill waste into streams. Five years later many of the same interests secured legislation in the province of Ontario directing judges not to grant injunctions against the dumping of sawdust and other mill waste into rivers, if "the public interest in preserving the lumber industry in a particular neighbourhood outweighed the private injury or interference caused by the waste" (McLaren 1984, 204).

Yet the industrial Goliath eventually met its David in the person of Antoine Ratté, who leased recreational rowboats from his wharf and boathouse in the city of Ottawa. In the eighteen years that he had owned his business, he claimed in 1885, the bay in which he operated had become terribly fouled by mill waste. Here, as elsewhere, the effects of waste discharge had reached alarming proportions. The "refuse accumulates in great floating masses, substantial enough occasionally for a man to walk upon . . . [.] Depositions of sawdust . . . result not only in fouling the water, making it offensive both to taste and smell, but produce from the gas generated underneath the surface frequent explosions which are disagreeable and sometimes dangerous" (ibid., 225). Denied recompense on a technicality in 1886, Ratté appealed and the case dragged on through six additional appeals until 1892, when he was finally awarded damages. Two years later he mustered over 700 signatures on a petition to the Privy Council, complaining about the continuing deterioration of the Ottawa River.

In 1894 new legislation established a blanket prohibition against the discharge of sawdust into Canadian rivers. Again the major lumber producers of the Ottawa Valley were quick to respond. By their lights, declines in fish stocks were attributable to dams without fishways, and sewage not sawdust was the cause of poor water quality. It was as well to remember, they argued, that "this is a utilitarian age and that the interests of any important industry, the success of which affects the well-being of so many people, are unavoidably held to be paramount to the gratification of more aesthetic taste, satisfactory and desirable as they may be under proper conditions" (ibid., 238).

From across the country, others—mill proprietors, municipal councils, mill workers, commercial interests, and individuals who salvaged mill waste for firewood and other purposes—offered variants on the theme. They were countered by citizens in support of action, people who lamented that a once-fine salmon river had been turned into "the most dirty stream I ever saw," and who recognized that "since the sawmills at Brompton Falls went into operation," the St. Francis River "has become depopulated of its finny tribes" (ibid., 241, 240). But "utilitarianism" won the day. Officers of the Fisheries Department were told not to enforce the law against dumping sawdust into rivers.

Rapidly rising markets for newsprint helped to deflect attention away from the pollution of rivers by sawmills. Between 1885 and 1900 the number of newspapers published in Canada grew from 644 to more than 1,200; the number of "dailies" rose from 71 to 121. In the United States expansion was even more dramatic as the circulation of larger, more frequently published newspapers grew by 80 percent in the four decades after 1870. From 452,000 tons in 1880, North American consumption of ground wood pulp rose to 2 million tons by 1900 (Roach 1996). Because technological limitations associated with the chemical sulfite process made it necessary to use high proportions of spruce (a northern species) in manufacturing newsprint from wood fiber, U.S. mills were unable to satisfy their escalating demands from domestic forests. Canada, said a new generation of boosters, was "the land of the spruce," a place where "an area equal to that of England could be cut over every year, and still the reproductive powers of the spruce would maintain the equilibrium of demand and supply" (Johnson 1900, 19). With such reserves and widespread enthusiasm for the potential of hydroelectric power generation, prospects for northern development seemed almost endless.

The new era began with developments in Sault Ste. Marie in the 1890s. Here, an American corporation headed by Francis H. Clergue established the first modern pulp and paper plant in Ontario. With its remote location close to the supply of wood and the development of its own hydroelectric power source, this mill set patterns that would become characteristic. It also defined what would soon become a common method of proceeding by negotiating a long-term agreement (in this case twenty-one years) that allowed essentially unregulated access to pulpwood over a large area of Crown forest (in this case fifty square miles), for a specified fee (in this case 20 cents per cord), in return for the construction of a mill. A few years later the Nova Scotia government followed suit in negotiating the so-called Big Lease with New England financiers who agreed to build two pulp mills and pay an annual rental of $6,000 in return for cutting rights over 620,000 acres of Cape Breton for ninety-nine years (Sandberg 1992).

Other, similar arrangements followed, but few were successful. Three firms that began building mills in Ontario were forced out of business by the economic recession of 1903. Although Clergue persisted, with difficulty, five other companies deferred negotiations with the government in hope of brighter prospects. After making various efforts to suggest good faith, and extracting a string of concessions from the government of Nova Scotia over a couple of decades, the owners of the Big Lease sold and resold their rights, at handsome profit to themselves but with no pulp mill in operation (Nelles 1974; Sandberg 1992; Clancy 1992).

These faltering first steps were attributed, in part, to American tariffs on imported paper, established in response to Canadian restrictions on the export of unprocessed wood. But as the demand for wood pulp rose and American paper prices climbed, support for these tariffs declined. In 1911 the duty on newsprint imported to the United States was reduced and in 1913 it was eliminated. At much the same time, Québec and New Brunswick prohibited the export of raw pulpwood. Thereafter, Canadian newsprint production increased dramatically, as existing plants expanded capacity and new ones were built. In Dryden, Iroquois Falls, and Smooth Rock Falls before 1918, and in Fort William, Port Arthur, Kapuskasing, and Kenora in the 1920s, pulp mills and associated clusters of housing and services appeared across northern Ontario.

In Québec, mills at Chicoutimi, La Tuque, Dolbeau, and Kenogami were, similarly, the foci of new resource towns. In 1911 alone, nineteen new companies with a capital in excess of $41 million were incorporated in Québec. In New Brunswick, Bathurst, Dalhousie, Edmundston, and Grand Falls joined the list of places with pulp mills at their center. Between 1900 and 1920, the value of Canadian pulp and paper exports rose from $1.8 million to $163.1 million. Between 1913 and 1920 newsprint production climbed from 350,000 tons to 876,000 tons (as prices tripled), and in the five years after 1924 Canadian newsprint capacity more than doubled. Toward the end of the decade, more than 75 percent of output went to U.S. markets.

Huge capital investments underwrote this expansion. By comparison with even the largest, most modern continuous-flow sawmills, pulp mills were a different order of enterprise. They were thoroughly modern businesses, dependent upon the widely scattered labor of cordwood cutters and the concentrated expertise of highly educated, highly specialized groups of employees (engineers, chemists). Both were incorporated into a complex, hierarchical managerial organization. A simple comparison suggests something of these differences. In 1926 Ontario had 45 pulp and paper mills and 676 sawmills. Yet the pulp and paper sector stood far ahead of the lumber industry in capital investment ($168 million compared with $51 million), number of employees (10,312 vs. 7,640), and value of output ($81 million vs. $31 million). It also moved decisively through the first thirty years of the century to extend and secure its control of the wood supply. In 1921 the major pulp mills of Ontario held rights to 14.3 million acres of wood; in 1930 the equivalent figure was 35.6 million acres. In New

A pulp and paper mill on the Saguenay River, Québec. *Early in the twentieth century there were dozens of mills like this one on the Saguenay River in Québec, turning the spruce forests of the north into pulp and paper for urban consumers (and especially to feed the voracious demand for newsprint through these years). Requiring huge capital investment and a steady supply of timber, these mills marked and produced a significant set of changes in the forest industries of northern North America. Their workforce included skilled managers and trained professionals, and the companies that owned them invariably acquired rights to enormous tracts of forest to ensure a sufficient supply of wood. (National Archives of Canada)*

Brunswick, the ten largest claimants (all of them sawmilling interests) held 40 percent of the province's leased land in 1896; thirty years later pulp and paper firms dominated the top ten, which accounted for almost two-thirds of the province's forest lease holdings.

Whether pulpwood was taken from Crown leases or from private land by contractors who took low prices per cord for their labor because they were dependent upon the returns it brought and had little leverage against company policies, short-term goals shaped exploitation (Roach 1987). Although professional foresters began to voice concerns about the consequences, the "logging outlook" almost always trumped a "forestry point of view." Woods operations, said John Wilson, an Oxford-trained forester employed by a major pulp and paper company, were typically left in control of "competent loggers who know little and care less about the future of the forest" (quoted in Swift 1983, 73). In a similar vein, the Chief Forester of the Canadian Commission of Conservation insisted in 1920 that there had been too little regulation to ensure that cutover lands were left in a productive condition: "The whole effort thus far has been centred upon attempts by woods superintendents to keep down logging costs and make a satisfactory showing in comparison with the costs of other companies and the costs of previous years" (quoted in Swift 1983, 72). Few worried about the long-term implications, although they were not entirely unappreciated. As early as 1905, Ellwood Wilson, a forester working for the Laurentide Company in Québec, noted that balsam fir was replacing spruce on the company's timber licenses on the St. Maurice River and concluded that prevailing systems of exploitation "were slowly depreciating the value of the limits" (quoted in Swift 1983, 70–71).

Conservation, in its progressive, utilitarian form, made some gains in early twentieth-century Canada, not least in blunting some of the exploitative attitudes toward Canadian forests, in encouraging the employment of professional foresters, and in fostering fire protection to limit the waste of timber. But the early twentieth century was no "golden age" of judicious forest management. Many apparent improvements in forest practice extended little beyond the paper upon which they were defined. "For thirty, forty, or even fifty years," lamented the Director of the Canadian Forest Service in 1933, "timber licences and leases have contained clauses which imply at least the elements of forest conservation, but all too rarely have they been enforced" (quoted in Gillis and Roach 1986, 231). The assault on the forest had left "deserted villages, impoverished settlements and devastated lands" in its wake (quoted in ibid., 230). Writing from Québec late in the 1920s, forester Alan Joly de Lotbinière, the scion of one of the great patrons of early Canadian forestry, insisted that "unless immediate steps are taken to effect a change in our forest policy there will be a timber shortage in a few years." Smaller and smaller trees were being exploited in areas that were "being cut more intensively each year." In this view there was no doubt that "the smaller trees which really belong to the future are now being taken" (quoted in ibid., 212). The situation was not improved by the catastrophic economic depression of 1929. In efforts to sustain, or revive, slumping economies, governments reduced the levies they charged companies for use of the forest, relaxed provisions against cutting undersized trees, and removed financial support for forestry programs. From the Maritime Provinces to the Pacific coast, the consequences of years of "destructive logging" were all too

The pulpwood cut in the Saguenay River, Québec. *The "wooden world" of early nineteenth-century settlers lived on in the "wooden rivers" of springtime in the pulp era of the early twentieth century. Here 90 million feet of logs bound for downstream mills fill the Saguenay River. Similar scenes occurred on several rivers that were important to the early nineteenth-century timber trade, because the annual drive on the spring freshet was the easiest and most economical method of moving logs and timber from interior forests to mills and harbors. The "drive," filled with excitement and danger as men risked lives and limbs to keep logs moving downstream, became a storied part of lumbering, celebrated in songs and stories across the northeast. The sheer magnitude of the annual onslaught on the forest represented by pictures such as this forms a less-remarked part of the story of logging. (National Archives of Canada)*

evident. In Ontario the politics of development held sway, and forestry became equated, in effect, with firefighting. Pulp and paper interests resisted the short-term costs they associated with conservation, and argued for essentially uncontrolled exploitation to generate economic growth. The government acquiesced. Between 1905 and 1935, not one company was forced to alter harvesting techniques to achieve silvicultural goals. The province rested the future of the forest on a ringing declaration by Thomas Southworth, clerk of forestry for the province in 1896: "practically all that needs to be done in order to maintain our timber supply in perpetuity . . . is to retain in the possession of the Crown all such timbered land as is not adapted to agriculture, and to protect it from fire" (quoted in ibid., 89).

By contrast, the Québec Forest Service had imposed serious silvicultural requirements upon the larger pulp and paper operations in its jurisdiction. Yet these arrangements quickly fell away when cutthroat competition and economic hardship seized the industry late in the decade. Nurseries were sold, silviculture was abandoned. As in Ontario, "natural regeneration" became the hope for the future. A relatively sophisticated, scientific, and effective system of forest management—arguably the country's most advanced—was quickly rendered subservient to the immediate dictates of politics and the market. By the 1940s forester L. Z. Rousseau was comparing forest management on the northern fringe of Québec's pulpwood zone to the agricultural practices of "the African peoples who, having used up the fertility of one tropical area, move their primitive exploitative agriculture to another corner of the forest" (quoted in ibid., 128).

Across the continent in British Columbia, provincial forests had been established in relatively remote areas of good timber, as reserves for the future, and the Forest Branch (established in 1912) had a staff of almost 300 (plus half as many summer fire rangers) by 1930. New technologies, from internal combustion engines to aircraft and radio communication, enhanced the Branch's effectiveness in fighting fires. Silvicultural research was initiated and a forest inventory begun. But here as elsewhere the Depression undercut commitments to conservation. Neither business interests nor finance ministers were inclined to look kindly on remedies that cost money, or required expenditures on behalf of voters yet unborn.

From the dark depths of the 1930s, Chief Forester E. C. Manning reflected that if forestry was defined simply as the replacement of old crops of trees, forestry in British Columbia had been "more of a fancy than a fact." The Forest Branch had been "so engaged in collecting revenues and protecting present values in mature timber that a new crop has received insufficient attention through lack of funds and staff" (quoted in ibid., 155). A few months later his claims were borne out by investigations into the regeneration of logged-over lands on Vancouver Island. Studies of almost 20,000 acres (8,000 hectares) on which logging had occurred between 1921 and 1938 showed that barely 5 percent (less that 1,100 acres, or 440 hectares) carried trees that would one day produce marketable timber. Even on tracts that had been cleared for twelve to seventeen years, fully 90 percent of the land remained below the threshold of satisfactory restocking.

17

INDUSTRIAL FISHING

On both Atlantic and Pacific coasts, the application of new technologies and the centralizing tendencies of capital investment transformed the fisheries. The changes ran deeper in Nova Scotia than they did in Newfoundland in the half-century after 1880 but both areas were marked, in Harold Innis's words, by a transition from commercialism to capitalism as an increasingly industrial and capitalist fishery impinged upon and to some extent replaced the older mercantile-family fishery that had held sway on these coasts since the seventeenth century. In British Columbia, by contrast, the commercial, nonindigenous fishery developed later and was shaped, almost from the first, by capital, industrial technology, and science.

Writing of the economy of the Maritime Provinces in 1932, S. A. Saunders found it difficult to gain a measure of the fish business. After fifty years of instability and change precipitated by declining salt fish markets, efforts to concentrate and industrialize the province's fisheries seemed only to have compounded uncertainties. The dried fish industry struggled with obsolete production methods, suffered from defective organization, and produced barely half of what it had fifty years before. In Guysboro County dried cod production fell from long-standing levels of 30,000 cwt a year to barely 1,000 cwt in 1927. But the "inertia . . . prominent in human society" made it difficult to envisage a solution to these problems. A new fresh fish trade held promise but in this business it seemed that the region was at a "distinct disadvantage." Trawlers were a mixed blessing. Lobsters held potential but they could not carry the future of the fishery. Industrialism had failed to create a healthy, modern industry; its impacts were "neither widespread nor growth-sustaining" and the entire business seemed on the verge of collapse (Apostle and Barrett 1992, 47–50; Saunders 1932; Innis 1940, 425–442).

Markets for dried fish fell away in the late nineteenth century, with the abolition of slavery in the Spanish colonies, an economic recession in the sugar-producing Caribbean islands, and shifting consumer demands in the United States. Returns to the family fishermen who were the main producers of salt fish declined precipitously. Thousands left Nova Scotia for higher wages in the United States. At the same time, the merchants who orchestrated the dried fish trade consolidated and reorganized their businesses. On the Gulf of St. Lawrence, firms from the Channel Islands, able to exercise close control over fishermen and their production, enjoyed relative success, but even here there were amalgamations and consolidations and a gradual reduction in the number of fishing stations.

On the Atlantic coast, the production of heavy-cure dried fish produced with mechanical dryers was steadily consolidated in Halifax and Lunenburg. These developing focal points of the trade depended, to some extent, upon fishermen who worked the inshore and outport merchants who assembled and shipped cargoes of green fish in the schooners and steamers that tied dispersed producers and centralized processors together. But from Lunenburg, in particular, a growing fleet of schooners carried men and dories to fish the offshore Banks. Working on shares, and subject to high outfitting costs and low prices for their catch, these dorymen, like their inshore counterparts, bore much of the risk involved in harvesting the seas, while the companies at the heart of the business capitalized on successful voyages and seasons. In the long run these arrangements, and the consolidation of influence and profit that they facilitated, reduced many fishermen, their families, and their communities to penury (see also Innis 1940).

Outside the Gulf of St. Lawrence and beyond Halifax and Lunenburg and their tributary settlements, the production of salt fish became a by-product of new fresh and frozen fish trades. Urbanization associated with the development of coalfields, steel mills, and other industries increased local markets, and improved rail transportation allowed fresh fish packed in ice to reach wider markets. Early in the twentieth century, improvements in cold storage and refrigeration did likewise. All of this required considerable investment in storage and packing facilities, processing plants, and transportation infrastructure.

Government subsidies—for the construction of rail lines and, in 1913, for the operation of a refrigerated express train to Montréal—helped expansion, but the concentration and consolidation of control and activity were inescapable. Processing facilities were expensive. Plants needed access to the ocean and the railroad. Larger plants were more efficient. Technological improvements, such as the filleting machine developed in 1922, mechanized production and increased standardization, routinization, and processing capacity. With the expansion of the fillet trade, fish meal and fertilizer plants were developed to utilize the half to two-thirds of each fish that would otherwise have gone to waste. Three companies quickly came to dominate the fresh and frozen fish business, each with operations centered in a couple of ports, Halifax and Port Hawkesbury, Saint John (New Brunswick) and North Sydney, and Canso and Digby. Further consolidations concentrated two-thirds of Nova Scotia's fish-freezing capacity in nine plants located between Halifax and Shelburne by 1939.

To supply the large quantities of fish that their plants required, firms began to use steam vessels using side-haul otter trawls to increase the catch. Inshore fishermen protested immediately. In their view, trawlers ruined the fishing grounds and damaged the reproductive capacity of the stock, while their use inshore interfered with the gear set by inshore fishermen. Companies were accused of using cheaply acquired, second-hand, foreign-built trawlers to depress prices to the detriment of local fishermen. When the Royal Commission of inquiry into the fisheries investigated these issues in 1927, it divided, and threw little light on the crucial issues, its Minority Report being "hampered," in the words of S. A. Saunders, "by the lack of available data," and its Majority Report by "no appreciation of Economics" (Saunders 1932, 56).

The depletion of the New England lobster fishery drew a substantial infusion of capital into the development of that industry in the Canadian Maritime Provinces in the late nineteenth century. By the 1880s the six leading American lobster companies operated seventy-one canneries in the region and the live lobster trade that began in the 1870s continued to grow as improvements in transportation provided better access to American markets. By the turn of the century, the lobster fishery may have surpassed the cod fishery in value, even in Nova Scotia. Relatively high prices, competition among buyers, and cash payments to fishermen attracted labor into the lobster fishery, helped to break the long-established hold of the truck system in fishing communities, and contributed to the decline of the Nova Scotia codfish trade in the final decades of the nineteenth century.

Yet the lessons of New England were ignored. Despite reports of an "alarming decrease" in the lobster fishery, and the claim in 1876 by Canada's Commissioner of Fisheries that this was due to "over-production and [the] wasteful capture of spawners and undersized lobsters," little was done to arrest the plunder (Canada, Sessional Papers 1876, quoted in Apostle and Barrett 1992, 51). Fishermen protested efforts to regulate or restrict the catch and canneries threatened to close if their supplies were limited. Between 1900 and 1924 Nova Scotia's production of one-pound cans of lobster declined from 5.25 million to less than 2 million. More and more traps were put in the water and the reduction of space between the slats was "nearly universal . . . making the trap more of a jail than it was originally" (quoted in Innis 1940, 437).

Still, one fisherman reported in 1910 that he required a thousand traps to capture the same poundage as he would have taken in 1885 with a hundred. The Royal Commission appointed to investigate the region's fisheries in 1927 heard of "surreptitious canning and packing in barns and kitchens and woods, of illegally caught lobsters," and of "an utter lack of observance of the existing lobster regulations and little individual or community sentiment in support of their enforcement" (quoted in Apostle and Barrett 1992, 52). They also reported significant catch declines. By 1927 the live trade to the United States was worth a good deal more than canned lobster production. Half the canned output went to the United Kingdom and almost a third to the United States, but as Saunders noticed in 1932, many of the near-300 canneries in the region were small and probably produced an inferior commodity (ibid., 53).

Late in the nineteenth century the government of Newfoundland sought to spur industrial development and to address increasingly evident problems in the cod fishery. Development efforts focused on the construction of a railway across the island, the stimulation of manufacturing, and the development of land-based settlement, as well as the exploitation of Bell Island iron-ore for export and the forests of central Newfoundland for pulp and paper. In the fishery, government policies were aimed at increasing exploitation of the offshore banks, opening up new northern fishing grounds off Labrador, and improving the efficiency of the industry by using telegraph lines to inform fishermen, quickly, of the most productive areas in which to fish. The use of new gear, especially the cod trap (invented in 1866), the seine, and the bultow, was also broadly encouraged by declining yields and the general reluctance by administrators to legislate against its use. By and large, then, official policies favored the expansion of fishing effort.

Declining export prices for salt cod fish—they fell by almost a third in 1880–1884 and 1895–1899—and a corresponding drop of salt fish shipments (which declined 20 percent by volume) hit Newfoundland hard. Shrinking returns and a rapidly increasing population meant that the fishery was unable to provide acceptable incomes to the substantial proportion of the population who remained dependent upon it. A rising clamor to assist the fisheries by "legislation, experience, science and every other means at their command" finally led the government to establish a Fisheries Commission in 1888 (Cadigan 1999a, 164). Under the leadership of Norwegian Adolph Nielsen, the commission attempted to increase inshore cod stocks by establishing a hatchery at Dildo in Trinity Bay. By their own count, workers hatched and released almost 833,000 billion ova in the vicinity of the hatchery before 1895, when it was reported that an abundance of cod covered the bottom of Trinity Bay "in a thick mass for long distances" although there were few cod in neighboring Conception and Bonavista bays (Ryan 1986, 73). By 1895, twenty-three lobster hatcheries likewise claimed to have hatched and planted 2.6 billion lobster ova. Yet neither of these efforts had much impact on catch rates and production in the long term, and they were soon ended by the election of a new government in the colony.

In the meantime, the commission had done little to restrain exploitation of marine resources. Although use of the bultow was proscribed briefly in 1888, this ban was lifted in 1890. By Sean Cadigan's account, all of this amounted to the squandering of an important opportunity. By failing to act decisively against more intensive harvesting of the seas at a time when many expressed reservations about the new fishing methods in the last third of the nineteenth century, Newfoundlanders missed the chance to protect and nurture the resource. Caught up in a highly commodified staple trade, indebted to merchants, and thus compelled to fish, even those who deplored the effects of new gear were forced, by its greater productivity, to adopt it once a few among them chose to do so. Thus the ineffectuality of the state "contributed to the growing economic culture of the open-access resource in the cod fishery" (Cadigan 1999a, 167).

In the decade around the turn of the century, the cod and seal fisheries accounted for approximately 85 percent of Newfoundland's exports and employed over half of the colony's workforce. Yet an approximate doubling of the number of fishermen since 1860 meant that yields per fisherman declined by 60 percent in fifty years (Alexander 1983). Early in the twentieth century the Fishermen's Protective Union emerged to demand fair prices and improved conditions for its members. But the war stayed any progress toward reform. Total output increased only slowly in the first decade of the twentieth century, and in the thirty years thereafter real output from the fishery declined by some 2.3 percent per year. This trend was evident across all sectors. Between 1900 and 1935 the Newfoundland Banks fishery declined. In 1889 it accounted for over a fifth of the colony's total catch; by 1912 it contributed little more than a tenth of the total; and in 1923, at 70,000 quintals, it amounted to barely a twentieth. Other countries also saw markedly decreased returns. Between 1942 and 1958, Portuguese side trawlers working the Grand Banks caught 50 percent less fish despite a tenfold increase in fishing effort.

Catches in the Labrador fishery increased from 150,000 to 200,000 quintals in the 1850s to as much as 800,000 quintals a year early in the twentieth century, then fell off dramatically. By the 1920s only half as many schooners voyaged to the Labrador as in 1910. In the 1930s and 1940s, yields declined precipitously, from 215 to 57 metric tons per trap. Inshore catches also trended downward between 1910 and 1940, although they exceeded the combined catch of all countries working the offshore banks until the 1950s, and were relatively stable through the first half of the 1940s, when inshore catches were at the lowest levels recorded in the previous hundred years. Withdrawal from the Labrador likely meant a redirection of fishing effort to Newfoundland waters, so that drop in catch per unit effort was likely even greater than decline in landings, from 335,000 metric tons in 1917 to 180,000 metric tons early in the 1950s. In sum, neither northern cod nor the families whose livelihoods depended upon their exploitation were thriving.

On the Pacific Coast, remote from major centers of population, a small-scale commercial fishery in which native peoples supplied Maritime fur traders and the Hudson's Bay Company was established in the early nineteenth century; by the 1840s the Hudson's Bay Company was shipping small quantities of salted salmon to the Sandwich Islands (Hawaii) and Asia. But the development of a significant commercial fishery for nonlocal markets depended upon the preservation of fish in airtight containers, a process known as canning. Employed in Scotland and the northeastern parts of North America in the early nineteenth century, canning techniques were brought to California in the 1860s and quickly spread northward to Alaska.

They imposed an industrial dimension on the fishery from the first. Canneries were land-based operations that produced millions of standardized cans "using those methods of mechanization—minute division of labour, repetitive operation, and line assembly—that are . . . the hallmark of . . . industry" (Ralston 1968/1969, 38). Canning operations began on the Fraser and Columbia rivers in 1867, but dependent on merchants in Victoria (and a long sea voyage) for access to markets in industrial Britain, the early Fraser canneries lagged far behind the Columbia in production. In the 1880s the Fraser River pack averaged about 200,000 cases (of 48 pounds [22 kilograms] each). New plants on the Skeena and Nass rivers of northern British Columbia and in Alaska glutted British markets.

By 1889 a surfeit of small, locally financed, independent canning plants on the coast was ripe for consolidation. Within a couple of years, two limited liability companies underwritten by British capital had acquired several of them. Many others consolidated under Victoria-based leadership. By 1891 there were only two independent canneries on the Fraser River. Little over a decade later a new consortium financed in eastern Canada and the United States and known as the British Columbia Packers Association took over twenty-two existing operations and immediately controlled over half of Fraser River sockeye production. Through these years at the turn of the century, the industry expanded dramatically. On average, the five big years of the Fraser River sockeye run (which follows a four-year cycle) between 1897 and 1913 yielded a pack of almost 800,000 cases, and the 1897 pack from British Columbia alone exceeded a million cases.

Heavily focused (particularly before 1905) on the annual run of sockeye salmon returning to their natal rivers to spawn, the canning industry was highly seasonal. It was also—despite the growing monopsonistic concentration of cannery ownership—a highly competitive, uncertain, and even wasteful business (Meggs 1995). Most fish were caught near the mouths of their spawning rivers by hundreds of individuals in small boats. Harvests were correspondingly unpredictable, and heavy runs or exceptional fishing effort could produce gluts of fish that local canneries were unable to handle. Moreover, only the thickest part of each fish was canned. Vast, albeit unknown quantities of useless dead and discarded fish were dumped back into the waters from which they had come each summer. Because canneries lay idle most of the year—the sockeye run lasted about six weeks on many rivers—they depended upon the availability of short-term labor.

Up and down the coast, native peoples were crucial to the industry. Indian agents recorded the annual migrations of hundreds of families, and in some cases almost entire bands, to work in the fishery. By Dianne Newell's account (1993, 54) at least 1,700 native peoples were drawn to work for the Fraser River canneries in the 1880s, some of them from as far afield as the Skeena River. For women and children, this meant factory-style work in the fish plants (scraping and cleaning fish, putting salmon into cans, attaching labels, and so on) or making and mending nets. Men and boys more typically worked on the water catching fish, but far from home and using new gear, many of them found their traditional knowledge of the fishery substantially devalued. On both land and water, native peoples also met competition. Large numbers of Chinese workers were brought into the canneries under contract, and increasing numbers of Japanese immigrants and Euro-Canadians entered the fishery itself.

Early efforts to regulate the fishery were desultory. The Federal Fisheries Act did not become operative in British Columbia until 1877, and then its provisions were minimal: Nets were banned from fresh water; set nets could not extend more than a third of the width of the river; and fishing for salmon was prohibited for a set period each week. Yet even these scant regulations meant that the highly productive (and destructive) fish wheels and traps used by those fishing American waters were precluded from British Columbia. Given the spawn-and-die cycle of the anadromous Pacific salmon, it was generally understood that some fish should be allowed to pass to the spawning grounds. But with limited understanding of fish biology and little knowledge of stock sizes, no one knew how many spawners were needed to maintain stocks (Evenden 2004b). Many were unconcerned by the question, believing that "artificial propagation," the development of fish hatcheries, would both ensure continuing abundance and allow fisheries officials to eliminate the enormous differences between "big-run" and other years in the migration cycle. Not until 1888 were licenses required to fish for salmon in British Columbia.

In recognition of their traditional dependence on the resource, native peoples were allowed to fish without licenses for subsistence purposes; thus the state conceived the concept of the "food fishery"—the definition of which turned on the prohibition of sales and became the focus of bitter contestation late in the twentieth century (D. Harris 2001). As numbers of fishers, and competition, increased, the government limited access to the Fraser fishery; in 1890 only 500 licenses were issued, and most of these were granted to canneries. Protests led the government to abolish the limit on licenses in 1892, but they were made available only to British subjects (thus excluding independent American fishers and Japanese sojourners from the business). Still, canneries retained substantial control of the industry; in 1893 they owned over three-quarters of the boats on the Fraser and paid fishermen to work them.

Through the next two decades, efforts to manage the fishery saw the introduction of restrictions on gear, the requirement that all canneries and salmon-curing operations be licensed, and the decision that all fishing licenses in northern districts be attached to canneries. There was also a good deal of scientific research initiated in efforts to better manage the stock as expansion of the industry continued, and it became the most lucrative fishery in Canada.

Then disaster struck the Fraser fishery. In 1913 railway construction through the narrow canyon above Hope precipitated a rock slide at Hells Gate. This blocked the sockeye migration in one of the "big-run" years that occur on a four year cycle and disrupted it for several years thereafter (Evenden 2000, 2004a). The spawning run of interior pink salmon was also interrupted, and the fishery on which hundreds of upriver native peoples depended was devastated. Many faced famine and misery as a result, but before the end of the First World War the scarcity of fish in the Fraser and mounting pressure from the commercial sector induced the government to place further restrictions on the native food fishery. Still the Fraser fishery was slow to recover. The annual catch (which had approached 21 million in 1909 and exceeded 31 million in 1913) was under 7 million in 1917 and did not exceed 4 million before the Second World War. Indeed it remained below 3 million into the early 1930s (ibid. 2004a, 46–48).

Scientists have suggested that the disruptions of 1913 broke the four-year synchronicity of runs to many different spawning streams and that year-to-year differences between returning fish populations have been smoothed out as a consequence. Be that as it may, recovery was slow. The removal of 45,000 cubic meters of rock from the river channel between 1913 and 1915 reduced the obstacle created by the slide, but passage upriver was still very difficult for fish. The construction and improvement of fishways through Hells Gate between 1946 and 1951, as well as early and mid-season fishing closures during this period, certainly helped rebuild fish populations. Further investment in fishways coupled with increasingly sophisticated fisheries management during the 1960s did likewise. But Fraser River salmon catches have never equaled those of the early twentieth century. After 1913, fishing effort shifted elsewhere, to the Skeena and other northern rivers.

Commercial fishing, Prince Rupert, BC. *The domination of the Pacific coast fishery by large canneries was well under way by the time this picture was made in Prince Rupert in 1913. With the decline of the Fraser River fishery as a result of a rockslide at Hells Gate in the Fraser Canyon, more and more of the catch came from northern waters. (M. M. Stephens/Library and Archives Canada)*

Despite the near-catastrophe on the Fraser, however, the 1920s were productive years for the industry as a whole. In three of the five years between 1926 and 1930, the province's fifty or so canneries sent approximately 100 million pounds of salmon to market. But these returns came at a cost. They were built on more intensive fishing with more efficient gear and reflected a marked shift toward the exploitation of pink and chum salmon. In the helter-skelter clamor for profit in the boom years, concludes Dianne Newell, "everyone ignored the weekly closed time, closed seasons, and fishing boundaries" (1993, 100). Time would show that the Nass River runs peaked in the 1920s and those on the Skeena in the 1930s.

Completion of the transcontinental railroad late in the 1880s opened new markets for fish packed in ice. For Pacific coast halibut fishermen this was a boon. Annual landings increased forty-five fold in little more than a quarter century, to reach 69 million pounds by 1915. Much of this expansion was generated by the collapse of east coast stocks and the transfer of labor and capital from the Atlantic to the Pacific in the 1890s, and the industry grew increasingly centralized as corporate control was extended thereafter. Independent fishermen using sailing (and after 1903 gasoline-powered) schooners moved into the halibut fishery in considerable numbers as the pelagic seal hunt was restricted, but most cold-storage plants were owned by large and increasingly integrated concerns. By 1914 the New England Fish Company operated several storage facilities and had eighteen steamers (each with as many as fourteen dories and forty fishermen) seeking halibut along the length of the coast from southern British Columbia to the Gulf of Alaska. The largest of these steamers could bring in 300,000 pounds of fish from a single trip.

Even before the end of the nineteenth century, some were concerned about the depletion of specific halibut stocks. A few years later, in 1906, *The Pacific Fisherman,* a trade journal published in Seattle, averred that "Banks which half a dozen years ago were bountiful in their yield of halibut were . . . as free of fish as a billiard ball is of hair" (Thistle 2004). But larger vessels ranging more widely, coupled with more intensive fishing effort, meant that the total catch continued to increase. By 1914 the British Columbia Department of Fisheries had hired an ichthyologist to study the fishery, and his first report suggested that the slow-maturing halibut were being exhausted by exploitation. These conclusions were cautious, however, hedged about for want of better data. Cautious science was buffeted by competing interpretations and corporate self-interest that ultimately steered legislation toward closed-season conservation measures that actually "favoured expansion and exploitation of the stock over restraint and restoration" (ibid., 109, 124).

18

INTO THE HARD ROCK

Northern Mining

North of the southern Ontario peninsula and west of the St. Lawrence lowland, forest, rock, and water are the defining characteristics of a vast and difficult expanse of territory, widely but imprecisely known as the Canadian sub-Arctic. Stretching from the coast of Labrador to the Rocky Mountains, rimming the grassland/parkland of the prairies, and extending northward to the tree line, this heartland of the early fur trade was a thinly peopled realm in 1870. A developing society intent on agricultural settlement had found little attractive about this region. When settlers were encouraged into the southern fringes of the area (sometimes called the "Near North"), by the development of Colonization Roads into the Shield after 1850, the results were dispiriting (Wynn 1979b). Both soils and climate were marginal for agriculture, and sooner or later most homesteads were abandoned.

Still, the agricultural dream died hard. Veterans of the Boer and Great Wars as well as those in need of relief during the Great Depression were encouraged to settle in the so-called Clay Belt of the north. Yet Thomas Gibson of the Federal Bureau of Mines had it more or less right in 1894 when he pointed out that such farmland as there was in the north was interspersed with "countless rocky ridges and great stretches of peat morass" (quoted in Nelles 1974, 54). After vigorous debate and some skirmishes, the Ontario section of the region was effectively left to the lumbermen, who regarded its extensive forests as their entitlement, and who extracted great wealth from them in the late nineteenth century.

By and large, the Ontario government conceded the wisdom of excluding agriculture from areas better suited for forestry even if they never quite agreed with Gibson that the very idea of agricultural settlement in these areas was "preposterous." In Québec, by contrast, there was no such resolution and forestry and agriculture often became "hopelessly and fatally entangled as competitors for the same [northern] land" (Zaslow 1971, 160). Rarely, however, did the forest industry secure final and exclusive rights. Railways had to be constructed through the northern fastness to link distant parts of the country together, and the railway, the federal government, and provincial authorities quarreled over resource entitlements along the way.

The promise of mineral wealth in the north soon became apparent. Geologists surveying the territory regularly reported the presence of greenstone ("Huronian rocks") that were thought to indicate mineral deposits. Tantalizing discoveries were made around the fringes of the Shield in the 1860s, including copper at Bruce Mines and gold at Madoc. Ontario quickly enacted legislation to encourage mineral prospecting and to facilitate the staking of claims and the mining of properties.

It was the remarkable Silver Islet Mine near Thunder Bay that fueled mining fever, however. The mine was based on a vein discovered and claimed by a prospector working for the Montréal Mining Company in 1868, and then sold on to a group of Americans for the princely sum of $125,000. Facing tremendous challenges, the Americans established their mine on a bald rock a mere 80 feet in diameter that rose only a few feet above the surface of stormy Lake Superior. By 1884 the tiny rock was covered by tightly packed buildings, mining apparatus, a dock, and a lighthouse. The shaft descended 1,230 feet below the lake. In a decade and a half it yielded silver worth $3 million. Then it was closed.

This was an iconic beginning. From the harvest of windfall riches and the sale of the initial claim to American interests, through the investment of massive capital for development, to the eventual abandonment of the site, Silver Islet epitomized many of the essential characteristics of northern mining. The utilitarian nature of the settlement, its location in the shadow of the mine, and its hastily erected dwellings were also broadly typical of ventures that followed. So, too, were the profits garnered and the fact that they mostly went to persons other than those who toiled in these places. Indeed, one might extend the iconic reach of Silver Islet by recognizing its location, as Jane Urquhart does in her novel *The Underpainter*, as the toe of The Sleeping Giant, a rocky human-shaped peninsula extending into Lake Superior that, she writes, the local Ojibway believe to be the form of a man "turned to stone as punishment for revealing the secret location of silver to white men greedy enough to demand the information" (Urquhart 1997, 2).

For most contemporaries, indigenous stories were of little concern. Riches beckoned, and many people were ready (and greedy) enough to find the country's destiny in the subterranean treasures of its subarctic reaches. While nationalists celebrated the societal and personal benefits of Canada's "invigorating" northern climate—as "Norsemen of the new world," Canadians were expected to be vigorous and robust, to "toss pine trees about in their glee" and to be spared the debilitating effects of time spent in enervating southern climes

Silver Islet Mine, Thunder Bay, Ontario, in the 1870s. *This illustration serves in many ways as a metaphor for mining development in the north. The improbability of this scene is striking: The buildings shown occupy almost all of the island, which rises only a few feet above the not-always-so-calm waters of Lake Superior. Yet rich returns of silver were extracted from this unlikely spot for two decades at the end of the nineteenth century. Ingenuity, determination, dogged persistence, and luck: These were all part of the story behind hard-rock mining in the Shield (as well as the Cordillera). (National Archives of Canada)*

(Berger 1966)—promoters added hard-headed commercial boosterism to such rhetorical flights of fancy. Government publications, such as the annual reports of the Bureau of Mines, and financial journals such as the *Monetary Times*, began to represent the north as an industrial heartland. A decade or so later, the Reverend F. A. Wightman announced in a volume titled *Our Canadian Heritage* (Toronto 1905) that the country encompassed "the best half of the best continent God ever made" and looked forward to the day when Moose Factory on James Bay would take its inevitable place as "the new Chicago of the North" (quoted in Nelles 1974, 55).

Developments in the Sudbury Basin helped underpin such visions. Following reports of unusual rocks discovered by surveyors of the Canadian Pacific Railway working west of Lake Nipissing in 1883, prospectors from the nearest towns quickly staked claims. But they lacked the capital to develop them. By 1885 the Canadian Copper Company (CCC) had acquired rights to most of the ore bodies around Sudbury. There were difficulties to overcome. Sudbury ore was rich in sulfur and in little-known nickel, which had to be separated from the copper, and markets had to be found. The former challenge was addressed by copying the refining processes developed by an English firm and the latter by convincing the U.S. Navy to use nickel for armor plating. In 1890 the CCC mine brought 130,000 tons of ore to the surface. Late in the 1890s the leaders of the Orford Copper Company, who had partnered the CCC in the initial development of the Sudbury area, developed their own mines and furnace at Copper Cliff. By 1902, however, the two operations were formally joined as the International Nickel Company of New Jersey (INCO). Production of the nickel matte that was shipped from Sudbury to a refinery in New Jersey skyrocketed from 3.4 million pounds in 1896 to almost 50 million pounds by 1913.

Few worried about the environmental impacts of this expansion. When Francis Clergue, the American mastermind behind the industrial development of Sault Ste. Marie, spoke (as he often did) of the remarkable growth he had fostered there, he tarried not a moment on such matters. By his account, his search for raw materials for the development of a sulfite pulp mill left him so askance at the cost of importing sulfur from Sicily that he took himself off to Sudbury where "they were racing sulphurous acid gas off into the air at a value of $2,000 a day" (Nelles 1974, 58). Rather than attempting to use the CCC waste, he acquired a nickel-mining concession, had his scientists develop means to produce the (highly toxic, water-polluting) sulfur-liquor he required for his pulp mill, and planned to sell the ferro-nickel alloy that remained to the Krupp steel works in Germany.

But his alloy proved too rich in nickel. Needing iron to mix with his nickel, he was led to discover the Helen hematite mine. This was quickly developed as the enormous Algoma iron mine and led, almost naturally, to the development of mills for making steel at the Sault. As many have noted, Clergue was the quintessential promoter. His capacity to link "familiar commonplaces of abundance with the popular belief in progress through social and industrial evolution," and to couple this with abundant and evident enthusiasm for modern science, technical efficiency, and individual capacity, produced, as historian Viv Nelles has observed, a persuasive conviction "that Ontario possessed all the essential requirements of a self-sufficient industrial state" (Nelles 1974, 56).

For the north, however, the future was as resource hinterland rather than industrial heartland. Railroad promoters cutting ambitious lines across northern rock, forest, and muskeg found their bonanza if they happened upon rich mines, forests, or waterpower sites. So the Temiskaming and Northern Ontario Railway achieved instant and otherwise unlikely viability when workers laying track east of Lake Temiskaming discovered rich silver deposits, and thousands of prospectors arrived almost overnight in the new town of Cobalt. The railway

The Cobalt townsite mine in the late nineteenth century. *This remarkable illustration of the early days of Cobalt captures many of the essential characteristics of hard-rock mining settlements across the north. Notice the difficult terrain of thin soils and scrawny forest, the dependence on the railroad (here the Temiskaming and Northern Ontario line), and the central importance of the mining company buildings—the power house, the assay office, the bunkhouse, and the dining hall. This was truly a work camp in the wilderness. Within a few years it would be one of the world's major silver producers. (National Archives of Canada)*

company leased a section of its right-of-way to those who rushed into the territory and, as new discoveries were made nearby, it constructed new branch lines to service them.

Early returns for the railroad and the miners were lavish. Initially, large rich silver nuggets could be taken at or near the surface. Little investment was required and profits were huge. One mine produced better than a tenfold return on a capital outlay of $225,000 in two years. Within a decade, however, costly underground operations and expensive concentrating mills were essential to success. In 1913 Cobalt concentrators handled 2,720 tons of ore a day, and extracted only 20 to 25 ounces of silver from each. Still, the early riches and the magnitude of the reserves—Cobalt yielded more than a million dollars of silver a year after 1905 and over $17 million in 1912—made the Precambrian Shield synonymous with potential mineral wealth (see Nelles 1974, 156–179; and http://cobalt.ca/history.htm).

The consequences were many. Stock exchanges made venture capital available for mining development, Toronto received a financial fillip from its involvement with northern mining, and the impetus of the Cobalt rush was converted into experience and support for educational institutions that facilitated later mining developments. Within a few years, the excitement generated at Cobalt had spawned new discoveries of profitable gold deposits in Porcupine, Timmins, and Kirkland Lake. According to the Toronto Board of Trade, it was "the prospector and the miner, rather than the home seeker," who were vital to increasing the "population and wealth" of "New" Ontario (Nelles 1974, 121), and contemporaries waxed eloquent about the richness of the province's mining potential.

Only a minority asked the question posed of the Sudbury nickel deposits by the director of the Bureau of Mines in 1910: "of what particular and special benefit are these deposits to our country?" (ibid., 332). Refining of the nickel matte was done in New Jersey until a new electrolytic treatment process prompted the development in 1918 of a refinery in Port Colborne that used Niagara power (ibid., 332). In the meantime, as the Toronto *Telegram* observed, "A few boarding houses around two or three holes in the ground, plus Sudbury, represent[ed] all that Ontario ha[d] to show for a monopoly of 90 percent of the world's nickel supply" (*Telegram* quoted in *Canadian Mining Journal*, 1 September 1916, and in Nelles 1974, 328). That and the environmental damage associated with mineral extraction. Contemporary photographs of the Sudbury operations show extensive areas of disturbed and despoiled ground with the machinery of mining (from headworks to locomotives) in the foreground and the slender bare limbs of trees on the far horizon. Initial treatment of Sudbury ore required that it be roasted to reduce the sulfur content and yield a copper-nickel amalgam for refining elsewhere. The roasting process was conducted in the open, using vast quantities of wood for fuel. The forest was clearcut for miles around, and the release of sulfur and arsenic fumes during the roasting process poisoned the landscape and killed the vegetation over an even more extensive area.

Elsewhere, mining had a range of equally deleterious impacts. Enormous quantities of crushed rock, left as waste after valuable minerals were extracted from them, had to be disposed of. The quantities were staggering. In 1911 Cobalt produced over 850 metric tons of silver. At a yield of 25 ounces of silver per ton of rock, this implied over a million metric tons of crushed waste rock. In some places this material was piled into enormous dumps or tailings piles on the surface. In others it was dumped into conveniently close lakes. This put the debris out of sight, and even perhaps out of mind, initially, but the quantities were such that even large water bodies soon began to show the effects of infilling. Lake depths were reduced dramatically over the years and some lakes were reduced in circumference.

Such changes had important, albeit largely invisible, consequences. They affected water temperatures, oxygen, and nutrient levels, and fine tailings particles in suspension clouded the waters, thus shrinking the photic zone and reducing aquatic life in the lakes. Less evident but even more serious were the chemical and ecological consequences of such waste disposal practices. The silver ore of the Cobalt region contains arsenic and mercury, and both were dumped into lakes with mine tailings, polluting the water and being taken up in the food chain. The Timmins and Kirkland Lake mines used the capacity of dilute cyanide solutions to selectively dissolve gold from the waste rock or gangue material in a capital-intensive process that was common through much of the twentieth century. Cyanide is highly toxic but does not bioaccumulate, so that prolonged exposure to subtoxic levels does not generally pose health risks. However, cyanide leached into soil or groundwater or spilled catastrophically can do great damage. Tailings ponds in which cyanide-rich ore wastes were typically stored to allow the cyanide waste to degrade by photodegradation, hydrolysis, and oxidation were both unsightly and possibly imperfect at preventing leaching.

Before Ontario claimed preeminence in mining on the basis of the Cobalt discoveries, British Columbia called itself "The Mineral Province of Canada" and accounted for a quarter of the country's production by value. Silver, lead, and zinc were discovered in the rugged mountainous terrain of southeastern British Columbia in the 1890s. Within months prospectors, miners, and capitalists had flocked into the region. The pace of growth—and in many instances, decline—was breathtaking.

In the Slocan Valley, mines developed on the flanks of Idaho Peak led to the construction, within five years, of spur lines linking the boomtown of Sandon on narrow Carpenter Creek with two transcontinental railways, one in Canada and the other in the United States. In the same span of time, over 3,000 mining claims were staked in the Slocan Valley. From sites high on the mountain where exposed ores were worked with picks and shovels by men who carried their rich nuggets out in backpacks, mining evolved rapidly. Adits were driven into hillsides and piles of waste rock accumulated below their mouths. Mule and horse trails were built to haul out the ore, then replaced in bigger operations by aerial tramways. Miners' cabins in the hills were replaced by company bunkhouses. Concentrating mills were built just above the railway tracks.

Sandon, the focus of mining activity around Carpenter Creek, sprouted into a town of 2,000 people, with seventeen hotels and fifty stores by 1897. Fires, some of them set to aid prospecting, burned forests from slopes. "The summer air," notes Cole Harris in his wonderfully evocative account of "Industry and the Good Life around Idaho Peak," "was hazy with smoke. Every winter and spring, avalanches pouring down newly bare slopes took some men and buildings with them" (Harris 1985, 199). New Denver developed on the delta of Carpenter Creek as a small town to provide a range of services to the local region. Laid out on a grid plan, with sidewalk- and tree-lined streets, neatly kept houses and gardens behind picket fences, an elementary school, post office, hospital, and bank, it was a small place with some ambition, a village that denied the recency of its establishment. Nearby, a few individuals cleared farms on benchlands above the lake that provided the only small patches of land with agricultural potential for miles around to supply the local demand for produce. The mining boom substantially remade this once-isolated area, and did so almost overnight.

Decline was less dramatic, but hardly less arresting. The peak of silver and lead production came in 1897. Ten years later output was down by approximately three-quarters, and the heyday of the "Silvery Slocan" had passed. In 1910 a forest fire destroyed railway trestles, mine buildings, and other property. Sandon counted only 400 residents, and by the 1930s it was a ghost town. For smaller, more sedate New Denver, the decline was not as sharp, but it, too, felt the loss of mining activity that took 28 million ounces (875 metric tons) of silver and 200 million pounds (6,250 metric tons) of lead from the rugged and increasingly despoiled terrain at its back in twenty years after 1891.

In the Boundary country of southern British Columbia, rich copper deposits (and lesser amounts of gold and silver) brought rapid development to a cluster of towns between the Kootenays and the Okanagan (see Fig. 15). Prospectors staked a number of claims in 1891. On steep valley sides high in the hills, high hopes rested on little patches of rock named Old Ironsides, Knob Hill, Stemwinder, Silver King, Snowshoe, Rawhide, Monarch, Gold Drop, Fourth of July, War Eagle, and so on. Initially, access was from the south, from the state of Washington, along the Kettle River valley and the Dewdney Trail. But many of the ores were low grade, and their development required significant investment. For five years excitement was suspended by assessment work. Then new syndicates moved in to begin extracting the ore and building the smelters necessary to process it. Individual claims were bought up and consolidations were agreed upon. Within a few years, the Miner-Graves syndicate formed to work the original claims at Knob Hill/Old Ironsides had become the Granby Consolidated Mining, Smelting and Power Company. The first shipments from workings between the 4,700 and 5,000 feet contours were made in 1900 to the newly built smelter in Grand Forks.

The base camp for mining operations, reached by a branch of the Canadian Pacific Railway in 1898, was incorporated as the city of Phoenix in 1900.

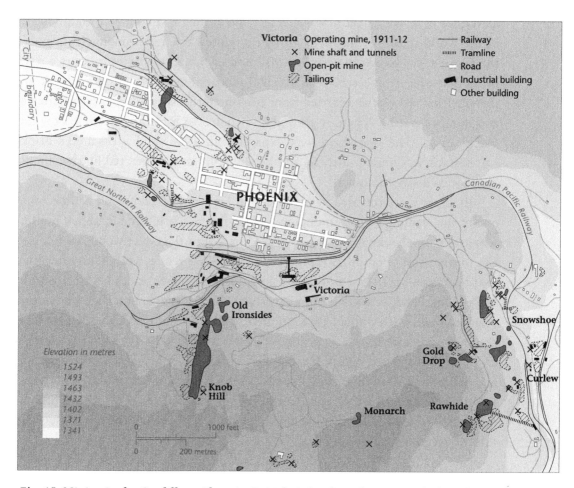

Fig. 15. *Mining in the Cordillera: Phoenix, British Columbia, about 1911. (Adapted from* Historical Atlas of Canada *3: pl. 22)*

Squeezed into a narrow valley at an elevation of some 4,500 feet, the settlement had a population of 4,000 (including miners in mountainside bunkhouses), and after the arrival of the Great Northern in 1904 was served by two railroads. Maps of the region in 1911 show the steep valley sides traversed by tracks leading to the mines. Mine shafts and tunnels dotted the slopes south of Phoenix. At many locations miners had stripped the overburden rather than construct shafts and tunnels; on the hillsides and in the town itself small open-pit mines extended their circumferences, and nearby tailings piles accumulated. Photographs reveal a ravaged landscape of clearings, burned forest, excavations, waste dumps, tramways, industrial buildings, and debris slides.

A few miles from Phoenix, the British Columbia Copper Company, a New York–based operation, built a smelter at the mouth of Copper Creek near Greenwood to process ore from its recently acquired Mother Lode mine.

Described by a Vancouver newspaper as "one of the most complete and modern in the world today" (*Vancouver Province*, quoted at Greenwoodmuseum n.d.), the furnace was blown in early in 1901, and operated twenty-four hours a day. In its first year it processed over 100,000 metric tons of ore. Molten slag from the smelting process was carried from the smelter in specially designed railcars and dumped nearby. There it glowed and cooled to form a vast block of solid black waste. The operation was expanded in 1904, and a 36-meter brick smokestack was built to lift the smoke and pollutants generated by the smelting process above the town.

In a few years the area between Grand Forks and Greenwood had become one of the leading copper-producing regions in the world. The Granby Smelter was said to be the largest in the British Empire. But the Greenwood smelter began to run short of ore in 1912; the First World War reduced others to intermittent operation; in 1918 a worldwide downturn in copper prices led to closures. Before long Phoenix was all but abandoned, Greenwood approached the status of ghost town with a mere two hundred residents, and the scarred hills were gradually reclaimed by vegetation, at least until the 1950s, when resurgent copper prices spurred highly mechanized, extensive, open-cast mining that all but obliterated the remains of Phoenix and its surrounding mines. In Greenwood, however, the BC Copper Company smokestack remains, "a prominent landmark [notes the local museum's website] . . . overlooking the huge Slag Pile" (Greenwood Museum n.d.).

19

TRANSFORMING THE INTERIOR

Canadian politicians and "western expansionists" made much of the need to "open" the prairie west—the province of Manitoba and the North-West Territories established when the Dominion acquired Rupert's Land—to commercial agriculture and to "fill" it with white settlers, in the years after 1870. These eastern ambitions have been characterized by scholars as integral elements of a broader national policy, as reflections of a sense of imperial destiny, as essential to the extension of Canadian dominion from the Atlantic to the Pacific oceans, and as a form of defensive expansionism necessary to prevent American encroachment into, or annexation of, "the last best west." Whatever motives were in play (and all of these and more fueled the drive to develop the territory) the rhetorical commitment to "opening" and "filling" the prairies also carried the unfortunate implication, as historian John Herd Thompson noted in 1998, that this was "closed" and "empty" space.

This was far from the truth. The western interior was home to considerable numbers of native peoples (30,000 to 40,000) and Metis (12,000), and the fur trade had long since drawn the region into the orbit of global commerce. But the 1870s were difficult times for the indigenous inhabitants of the plains. Bison herds, upon which both natives and Metis depended heavily, declined precipitously as commercial exploitation quickened. Local economies were severely disrupted. Hunger, and the specter of Indian-white relations to the south, helped bring native peoples to the treaty tables set up by the Canadian government. In seeking treaty settlements Dominion officials followed the British principle of recognizing aboriginal title and negotiating its transfer before making land available for settlement. Between 1871 and 1877 formal treaties, numbered 1 to 7 for the order in which they were signed, covered most of the territory between the Precambrian Shield and the Rocky Mountains, and south of the 55th parallel. In all cases, indigenous people conceded large tracts of land in return for a number of relatively small "reserves" set aside for their use.

Native peoples were not simply pawns in this process. They insisted on their claims to the land, and protested its transfer. Some refused to sign treaty documents, at least until impending starvation broke their will. But many also recognized changing circumstances—that "our country is no longer able to support us"—requested "cattle, tools, agricultural implements, and assistance in everything when we come to settle," and negotiated the terms of successive agreements so that later treaties included larger allocations of land and greater annual per capita payments than earlier ones (Tobias 1983). For all that, doubts and apprehension remained, and disputes about the "meaning" of the treaties (were they onetime transactions or contractual symbols of a continuing, and evolving, relationship?) have continued into the present. In immediate and practical terms, however, the results of the treaty process were simple: Native rights to land were limited to the reserves; beyond them, the prairies were "open" for settlement (Carter 1990).

Quickly, the government moved to impose order on "the one great country [laid] before it" (Wynn 1987a, 377). Following the American lead, the Canadian prairie was divided into a grid of farm holdings, defined by 640-acre sections and 36-section townships. Exceptions were few. The long, thin, river lots granted Metis settlers by the Hudson's Bay Company were accommodated, and the Hudson's Bay Company was allowed 50,000-acre blocks of land in the vicinity of its trading posts. Special surveys were implemented on certain particularly dry lands in Alberta where, it was recognized, irrigation would be necessary. In southeastern Manitoba, several thousand Mennonites were given exclusive use of sizable blocks of land in a special group settlement scheme. For the rest, however, straight lines and rigid uniformity prevailed as "one vast system of survey" was marched off across the land. In 1872 the Dominion Land Act set the terms of access to this inland empire: Homestead entries could be registered on quarter sections (160 acres) for a $10 fee; title would be granted after three years' residence and the completion of specified improvements. Ultimately, however, only about half of all prairie farmland was available as free homesteads; some 60 million acres allocated to railroads, schools, swamp-drainage projects, and the Hudson's Bay Company were marked for sale.

All told, the basic infrastructure necessary for the orderly occupation of the territory was built remarkably quickly. Yet settlers failed to come, at least in the numbers anticipated and desired. In 1886 there were barely 160,000 people in the western interior. Well over a third of these were concentrated in the Manitoba lowlands, within 75 miles (120 kilometers) of Winnipeg, a city of 20,000 (Warkentin 1967). Most of the rest were in southwestern Manitoba and along the valley of the Qu'Appelle River. Beyond this, settlement was sparse and change was slow. There were, on average, fewer than 3,000 homestead entries a year across the entire expanse of the Canadian prairies through the two decades straddling 1886.

Here, as in so many other facets of Canadian development during the late nineteenth century, the railroad was crucial. Until tracks were laid across the forest, rock, and muskeg of the Shield, the western interior was extremely difficult to reach from the St. Lawrence Valley. A line linking Winnipeg to Port Arthur, on Lake Superior, was opened in 1882, and it ran some distance west of the prairie city along the Qu'Appelle valley, but an all-Canadian rail route between Montréal and Vancouver was not opened until 1886. For strategic, economic, and scientific reasons—including reinforcement of Canadian interests in the border country along the 49th parallel, competition with American lines, ease of construction, concern that the growing season to the north was too short to ripen available strains of wheat, and the presence of coal on the chosen route—this Canadian Pacific Railway followed a southerly course through the grasslands, rather than one through the aspen parkland region that in the 1850s Henry Youle Hind had called the "Fertile Belt" in contrast to the unpromising region traversed by the Canadian Pacific, regarded as a northward extension of the "Great American Desert." Moreover, it soon became evident that prairie homesteaders were reluctant to take up land distant from the railroad because grain could not be moved, profitably, more than a few miles by wagon. Dozens of north-south branch lines were built in response. By the 1920s dozens of railways ran within twenty miles of one another, crisscrossing the 800- by 400-mile expanse of prairie.

The pace of prairie settlement quickened markedly after 1896, when a new federal government streamlined immigration procedures and mounted a campaign to recruit immigrants from new areas of Europe, particularly the Slavic countries. Large quantities of prairie land granted to railroad companies as incentives to construction were also forced onto the market. These efforts helped to remake the prairie region. The number of people living in Manitoba and the territories that became the provinces of Alberta and Saskatchewan in 1905 quadrupled in fifteen years after 1901. More homestead entries were filed between 1900 and 1904 than in the previous quarter-century. Almost two-thirds of all homestead entries recorded between 1872 and 1930 were made between 1900 and 1914. The number of prairie farms climbed from just over 55,000 to almost 200,000 in the decade after 1901, and more than 40,000 homesteads were established each year between 1909 and 1912. Between 1901 and 1921, the population of the prairie provinces climbed from less than 8 percent to more than 22 percent of the Canadian total, which itself increased by almost two-thirds (from 5.3 to 8.7 million) during the same period.

Prairie settlement followed a complex spatial and temporal choreology (Friesen 1984). Although it is widely regarded as a distinctive region, the western interior is far from homogeneous. Behind and beyond Henry Youle Hind's bold distinction between the Fertile Belt and the Great America Desert (or Palliser's Triangle as it became known in Canada) lies a more intricate pattern of variation in relief, precipitation, temperature, vegetation, and soil characteristics. Many of these characteristics are highly interdependent. Small differences in aspect (the orientation of slopes with respect to the sun) can significantly affect microclimates and thus vegetation and soil types, and thus differentiate opportunities for, and the implications of, settlement. For all that, the division between the interior plain and the "stony country" of the Shield, the western edge of which runs from east of Lake Winnipeg to west of Lake Athabaska, is the most elemental ecological line in the western interior. On the far western edge of the interior plain the land rises into the foothills of the Continental Divide.

The interior plain itself is conventionally divided in three: (1) a wide flat plain, about 250 meters above sea level, traversed by the Red River, the former bed of glacial Lake Agassiz; (2) the "second prairie level," or Saskatchewan Plain, extending between the Manitoba Escarpment and the Missouri Coteau at an average elevation of 500 meters; and (3) the area between western Saskatchewan and the Alberta foothills (at an elevation above 1,000 meters). Most of the area south of the 55th parallel is drained to Hudson Bay by rivers flowing west to east in deeply incised valleys. Everywhere, the climate is cold and dry: The average growing season ranges south to north from about 120 to less than 100 days.

The frost-free period is highly variable year on year; so, too, is precipitation, which averages 250 millimeters a year in the southwestern prairies and little more than double this in moister parts of the region. In these semiarid and sub-humid conditions, a shortfall of precipitation, or rain in the wrong season, will stress most crops. Climatologists have identified thirteen serious droughts (at least five years with less than 60 percent of the average annual precipitation) in the area since the fourteenth century. But the variability of climatic circumstances (and the impact of this upon human endeavors) is better suggested by records from nearby (and slightly moister) Bismarck, North Dakota, where there were twenty-three dry years, twenty-four wet years, and eight years with average rainfall between 1851 and 1905 (Kraenzel 1955, 19).

Native vegetation patterns reflected these influences. In the driest parts of southeastern Alberta and southwestern Saskatchewan, where brown chernozemic soils are less productive than their darker counterparts, hardy, drought-resistant perennial grasses form a "short-grass prairie" that also includes various forbs and small shrubs. West, north, and east of this zone, slightly cooler summer temperatures and higher levels of precipitation allow taller grasses to grow alongside the hardy short-grass species to produce "mixed prairie" on dark brown chernozems. In Canada the "classic prairie" of the more southerly plains, with black chernozem and gleysol soils and some grasses reaching 1.5 meters in height, occurs only in a small area of south-central Manitoba, along the Red River. Here and there more precipitation allowed broadleaf trees to grow. They might have been more common, had not buffalo herds and large numbers of other ungulates (among them elk, mule deer, and pronghorn antelope) as well as prairie fires (both natural and anthropogenic) favored the dominance of grasses and forbs. On the northern and eastern margins of the mixed grassland, prairie gives way to parkland, where aspen groves are interspersed with patches of grassland. To the north, where the growing season generally shrinks to less than 100 days (at approximately 55 degrees of latitude in central Alberta, 53 degrees in central Saskatchewan, and 51 degrees in western Manitoba), the parkland crescent gave way, in turn, to boreal forest.

Writing of the American West, Walter Prescott Webb memorably observed that the onward march of American settlement had been halted, a last temporarily, at the 98th meridian, because two of the three legs (wood, water, and land) that supported farming in the humid east were lacking on the plains. Early in the nineteenth century, explorers of the Canadian prairies recognized the same shift in available resources and doubted the prospects of building a yeoman empire in the grasslands. David Thompson, one of the most able geographers of his era, concluded that the prairies had been "given by Providence to the Red Men for ever, as the wilds and sands of Africa are given to the Arabians" (Thompson 1798, quoted in Warkentin 1964, 103). By the end of the century, after several decades of frontier settlement in the trans-Mississippi west, however, the challenges presented by subhumid conditions seemed less insuperable.

Promotional literature intended to attract immigrants to the Canadian northwest walked a fine line between enthusiasm and misrepresentation in efforts to overcome lingering doubts and prejudices. Manitoba was "Sunny and Fruitful"; the western interior was "the largest flower garden on the continent." In one clever enticement, the region was made to stand for all that newcomers might desire: **C**ontentment, **A**bundance, **N**ationalism, **A**ffluence, **D**emocracy, **A**mbition, **W**elcome, **E**ncouragement, **S**ociability, **T**hrift. It was a friendly egalitarian paradise, in which "every settler is given, practically free, a large farm in a country which produces the best wheat in the world" (Francis 1989, 138).

Turning the concerns later identified by Webb to the advantage of settlers, western booster John Macoun (who insisted that Palliser's Triangle was more suitable for agriculture than most of his contemporaries allowed) prepared two sketches. One offered a view of a pioneer farm in the eastern forest—all looming trees, tangled brush, and debris-snarled stream. The other provided a picture of a new prairie homestead with nary a tree in sight, other than a few planted "ornamentals" and the dense thicket of a thriving orchard; neat and highly productive fenced fields all but assured the settler's happiness and fulfillment beneath bright, wide skies. Echoing the point, an advertisement for properties placed on the market by the Hudson's Bay Company simply stated: "the Land is

Prairie, not Bush Land." For all that, trees and their clear association, by Hind and others, with the more productive soils of the region shaped the emerging geography of prairie settlement. Could anyone, especially those familiar with English landscape tastes, resist the lure of "Park Lands in the Fertile Belt," offered by countless posters and pamphlets?

Although the Canadian Pacific Railway ran through many miles of short-grass prairie, large numbers of pioneers moved north, by branch lines and cart and wagon from the stations at which they disembarked. Before 1901 most settlement concentrated in the parkland belt and the slightly moister northern margins of the prairies. Maps showing townships with at least two persons per square mile before this date mark a clear, albeit somewhat broken and untidy, swath of occupation extending northwest from Winnipeg toward Edmonton and from thence south to Calgary. During the next five years, this pattern was reinforced as the rising tide of newcomers took up land. Only relatively late, after 1906, did significant number of homesteaders begin to move into the drier lands on the outer fringes of Palliser's Triangle.

Many factors shaped this emerging pattern. Western Canadian settlement was part of a larger North American process, and so long as subhumid land was available, as it was in the United States through the 1880s, settlement gravitated to these areas rather than confront the risks inherent in northern climatic marginality. Except in the southeast, the prairie growing season was too short for reliable ripening of those varieties of wheat most commonly available before the final decades of the nineteenth century. Precipitation levels (and their annual variation) also made prairie wheat cultivation a risky business. By contrast, the Manitoba parkland received over 15 inches (400 millimeters) of rainfall during the growing season, had good brown soils, and offered conditions that were broadly familiar to settlers from more humid climates.

Moreover, as John Lehr has shown, most of the 170,000 immigrants from Ukraine who entered the west before 1914 were drawn north by early decisions to locate Ukrainian block settlements in the woodlands (for which the newcomers expressed a preference), and the desire of later arrivals to "go where the others have gone before." Indeed, Ukrainian immigrants often frustrated immigration agents for "their refusal of prime wheat-growing prairie land in favour of the less-fertile woodlands on the northern fringes of the parkland belt" (Lehr 1990, 141). Ultimately, however, rising wheat prices in industrializing Europe; falling transportation costs; declining interest rates; technological improvements in milling equipment, farm machinery, grain varieties, and cropping techniques; and the closing of the American frontier combined to shift the equations of profit, loss, and risk and to draw cultivation and settlement into and across the grasslands.

Settler's Home, *painting by Edward Roper.* *Boosters of prairie settlement produced many enticing images of comfortable homesteads surrounded by wheatfields and wrote much about the inherent fertility and ease of cultivation of the treeless plains. The reality was usually a good deal more forbidding—as this image and the recollections of many of those who came to the western interior reveal. Open skies were little consolation for isolation, the difficulty of breaking the prairie sod, and (in many areas) the scarcity of wood for fuel and building. Enervating toil, hardship, and uncertainty were the common lot of pioneer settlers both east and west. (National Archives of Canada)*

Still, as thousands discovered, settlement on the open prairie was forbiddingly difficult. First dwellings were often simple huts made of prairie sod. The wind and drifting snow made winters hard. In summer driving winds could desiccate crops and blowing sand might flatten them completely. Fires swept across the land and burned out more than a few families. Newcomers often found it difficult to adapt to this environment. There was, reflected one eventually successful English settler,

> never enough time on the prairie. Ploughing cannot start until the frost is out of the ground . . . . Once the land is ploughed, seed must be thrown in as fast as possible at what the farmer guesses or senses is the optimum soil condition. Planted too early the seedlings can be damaged by late frost, frequent in May. Planted too late they risk frost damage before the grain has matured in August. Then the crop must be cut and stooked, again in a wild rush to secure optimum conditions, and threshed as soon as possible to get a good price. (Quoted in Friesen 1984, 306)

A Danish immigrant who joined her husband in the upper reaches of the Qu'Appelle Valley in 1911 wrote home: "Here it's still so desolate and frightening on the wild prairie. It is like the ocean. We are a tiny midpoint in a circle. . . . understand that it looks terrifying, more than you can imagine." Later came the realization that the family was locked in "a struggle for existence, a struggle against the loss of culture's benefits"; it had embarked on a crusade "against nature's fury, the unyielding soil, the harsh climate" (quoted in Friesen 1984, 255–256; see also Nelson 1976).

Joined by tens of thousands, this crusade had significant environmental consequences. In the broadest terms favored by contemporaries who approved the changes, a "sea of grass" was replaced by "waving fields of golden grain." More specifically, an empire of grass gave way to an empire of farms as natives and Metis yielded space and influence to ranchers and cultivators. A geometric grid of fences and the linear tracks of roads and rail lines marched across the open range, dividing "the commons" and sprinkling it with towns. Buffalo gave way to cattle, to horses, and to sheep, and native grasses including blue grama, speargrass, wheatgrass, and Junegrass succumbed to countless ploughs and the cereals sown in their wake. But this was only a part of it. Each of these changes implied others, whether deliberate or incidental, perceived or (at least temporarily) invisible, all of which cascaded through the complex ecosystems of the prairies to make the environment anew. Here, the face of the country was changed so rapidly, so extensively, and so dramatically as to render its mid-nineteenth-century form scarcely recognizable in the landscape of the 1920s.

In Palliser's Triangle the challenges and consequences of grassland utilization by nonindigenous peoples were intense and starkly evident. Early in the nineteenth century this most arid section of the western interior was Assiniboine and Blackfoot territory. Metis traveled its fringes in pursuit of buffalo as the century wore on. By 1879, however, native peoples had surrendered their claims to the area and all but a few of the Metis had withdrawn as the buffalo population plummeted. The coming of the Canadian Pacific Railway in the early 1880s spurred a short-lived economy centered on the collection of buffalo bones for sale to manufacturers of fertilizers and refiners of sugar. In quest of bones, collectors set fires to expose them where they lay amid the blackened grasses. The returns were surprisingly large. One trader at Wood Mountain purchased 33,000 tons of bones in six years after 1882. But by 1890 the grasslands had been picked clean of skeletal remains. Without enormous herds of buffalo grazing and trampling the vegetation and with fewer anthropogenic fires, the short-grass prairie flourished. Soon, reported the Surveyor General of Canada, the cattle-ranching industry was "making wonderfully rapid strides" into and across the area (quoted in Potyondi 1995, 50).

The expansion of the cattle kingdom into the Canadian prairies was part of a globe-encircling incorporation of mid-latitude grasslands into the production of meat for the markets of industrializing, urbanizing Britain and Europe. The entire trade was heavily reliant upon technological developments ranging from canning processes to refrigerated shipping, but the Canadian industry, which included live stock shipments across the ocean, was crucially dependent upon the Canadian Pacific Railway, which linked western cattle producers with transatlantic shipping routes. In anticipation of the railroad, the Dominion Lands Act was amended in 1881 to allow enormous leases on very generous terms: Lessees could acquire up to 100,000 acres, at one cent per ace per year, for twenty-one years provided they stocked their land with cattle at the rate of one for every ten acres within three years.

Within a year, the government had received applications for 154 leaseholds covering four million acres, most of them in the foothills of the Rocky Mountains southwest of Calgary. Four years later there were 12,000 people, 100,000 cattle, and 25,000 sheep between the Cypress Hills and the Rockies. The ranching economy was very much confined to the semiarid margins of the prairies. It occupied the northern margins of the rangeland over which it had spread and prospered to the south.

In technique (horse-riding cowboys, stock marked with distinctive brands, round-ups, and so on) the early Canadian industry bore many resemblances to its American counterpart, but it differed in capital structure and in the forms of administrative control to which it was subject. Especially in the far southwest of the prairies, the early industry comprised a small handful of very large operations and many smaller enterprises; most of the leading cattlemen were drawn from the middle class of central Canada and the lesser landed gentry of Britain. Implementation of the relatively long lease system gave cattlemen on the Canadian ranch lands a greater sense of tenure security than prevailed in the "free grass" regime to the south, even as the almost nominal rents sustained the basically profitable economics of a system in which relatively cheap calves became valuable steers after several years of grazing on nature's capital. The Canadian government also sought to protect ranchers' interests against the claims of squatters and small farmers, at least into the 1890s. In 1886 it allowed riverbank reserves to give stock access to water and to "prevent Ranchmen being harrassed [sic] by a very objectionable class of squatters whose aim is largely to levy black-mail" (Breen 1983, 57; see also Evans 2000, 2004). They also prohibited sheep grazing and reduced cattle-stocking rates (to one beast per twenty acres) when ranchers worried that squatters and the ready availability of American cattle were threatening the Canadian range with serious deterioration. All of this, as well as the relative unattractiveness of the rangeland for cultivation and the presence of the North-West Mounted Police, helped spare the Canadian cattle kingdom the "range wars" and vigilante violence that flared up between "sodbusters" and "cowpokes" to the south.

Some believed that it also mitigated the environmental effects of ranching. As pressures to allow settlement of the rangeland intensified after 1896 and the leasing system was changed, many former leaseholders reverted to free grazing. By 1899 reports from the Pincher Creek district suggested that "the range has become so eaten out owing to want of any regulations . . . that some . . . have had to move their cattle away to other and less crowded portions" of the district (Breen 1983, 120). Similar patterns were apparent near High River and elsewhere. A report prepared in the wake of these developments by an author convinced that large parts of the southwestern prairie were unsuitable for agriculture without irrigation, harbored no doubts about the merits of the lease system. Seeking "the most equitable and efficient use of grazing land," its author was quick to quote an American agricultural scientist who argued that the "once beautiful" San Simon district had been "despoiled and hopelessly ruined in . . . fifteen years," due to the "ruinous methods which seem inevitable upon a public range, which, being everybody's property is nobody's care" (quoted in Breen 1983, 125). Yet others concluded that the late nineteenth-century destruction of the Pincher Creek and High River range was a deliberate and strategic ploy by ranchers who herded "their cattle in the vicinity of the settlers' farms for the express purpose of injuring those settlers by eating out the grass" (see Breen 1983, 119–120).

That Canada avoided the tragedy of the commons, at least on the scale that it was experienced to the south, probably owed as much to the relative underdevelopment of the Canadian ranching economy as it did to the leasing system. In southwestern Saskatchewan thousands of American cattle ranged illegally across the 49th parallel in the 1890s, and in the next decade several American ranchers treated the area south of the Canadian Pacific Railway's tracks as "a land of free grass similar to the one they had enjoyed for many years south of the line" (Potyondi 1995, 56). Still, a careful (if necessarily somewhat imprecise) estimate of peak cattle numbers in the heart of this area finds it "impossible to conclude that overgrazing was a concern" between 1880 and 1907. Extensive open grasslands (under lease or not) and relatively low stocking rates per acre allowed cattle to graze at will, drifting with "the weather, the quality of the grass and the availability of water" (ibid., 65). Relatively light grazing may even have improved the quality of the forage by improving the vigor of the grasses and by enriching soil nitrogen. Certainly one American rancher entering the Cypress Hills region from the overgrazed pastures of Montana in 1904 remarked that "several years growth of grass rippled in the wind, knee deep to a horse as far as the eye could see" (T. B. Long quoted in Potyondi 1995, 66).

Within months, however, the ranching industry was locked in a struggle for survival. The quickening flood of immigrants into the west, coupled with several years of above-average precipitation and the appointment of a new Minister of the Interior eager to see the spread of small, mixed farms through the southwestern grasslands, changed the terms under which ranching operated. Shelter and water reserves were opened up for sale as regular Crown Land; squatters and homesteaders on the rangeland were given various entitlements; and new grazing leases (revocable on two years' notice) were granted only if the land was deemed unfit for agriculture. The minister acknowledged that "it is a most difficult matter to decide where possible cultivation ends and where permanent grazing rights should begin." But his inspectors were firmly instructed that climatic conditions were not to bear on assessments of the land's suitability for farming; only areas that were "too gravelly, stony, sandy, or of too rough a surface" for agriculture were to be leased for grazing (Breen 1983, 143, 141).

Ranchers responded by purchasing freehold to secure their access to water, but they were virtually powerless against the new Dominion Land Act of 1908, which opened for preemption and sale all odd-numbered sections across 28 million acres of ranching country between Moose Jaw and Calgary. These were dark years for ranchers. As early as 1906 the *Medicine Hat News* reported that ranchers south of the Canadian Pacific Railway tracks were being driven out by settlers (ibd., 148). But even this story, published in the heart of ranching country, conceded that the ranchers' "Loss Is the Country's Gain and They Must Retreat." Two years later, a diorama displayed at the Dominion Exhibition in Calgary and the Canadian National Exhibition in Toronto was titled "Another Trail Cut Off." Intended to reveal Alberta's rapid transformation from ranching to grain growing, it showed "a field of standing grain with a cowboy in the distance." The cowboy was understood to be "following a familiar trail" which reappeared in the foreground of the diorama, "on the other side of the wheat field but . . . [was] suddenly stopped by a wire fence and a field of grain" (ibid., 150). Between 1906 and 1910 the area planted to spring and winter wheat in the area south of Calgary and west of Medicine Hat all but tripled, from 96,000 to almost 280,000 acres.

A devastating winter in 1906–1907, "the worst ever experienced in the ranching country," added to the cattlemen's woes. Bitter temperatures, a heavy snowfall, the failure of the chinook, and a shortage of feed decimated herds. Overall, official sources put losses at 50 percent; some large concerns lost 80 percent of their cattle. Several of the most prominent ranchers in the region

Rocky Ranch, *by **Edward Roper.** Ranching in Alberta occurred in semiarid grassland country as well as (as here) in the foothills of the Rocky Mountains where the country was less open and the problems of watering stock were generally less severe. The mountains arrest the eyes of those who look at this picture, but the scenic delights of such settings were surely only partial compensation for the remoteness and the hard work entailed in running cattle on such country. (National Archives of Canada)*

quit the business. Those who remained were in a declining industry. Although smaller herds after 1907 helped offset the loss of rangeland to farmers, markets were poor (at least until the temporary rise caused by the First World War), and adjustments were required in almost every facet of the business. In Saskatchewan, ranchers were required to fence their range to prevent damage to farm crops, but farmers also strung barbed wire with remarkable energy.

Within a few years, reports Barry Potyondi, "the open range had become a web of enclosed pastures" (1995, 68). This increased grazing pressure on the grassland, as confined herds tended to crop the grass more closely than those free to move at will. In response, the proportion of blue grama grass increased, other species declined, and weeds such as cheatgrass, sagebrush, and thistles took firmer hold. Less palatable to and less preferred by cattle, these weeds flourished and seeded in abundance. Persistent heavy grazing exposed the soil surface to the sun, leading to an increase in evaporation. Baked soil surfaces also reduced infiltration and increased run off. Desiccation and a less vigorous vegetative cover were the results. Fires, long a scourge of ranchers, continued to burn over the shrinking range, and low prices tempted many ranchers to hold rather than sell their stock, thus further increasing pressure on the grassland.

By 1920 many ranchers were in serious difficulties; they had to import winter feed and ship their cattle to pastures elsewhere. At the same time, drought conditions sorely affected farming in the marginal, former rangeland areas. Ranchers made some gains in the 1920s. The lease rate was cut in half, and twenty-one-year closed leases were reintroduced in 1925, but market conditions remained difficult, and ranching held a much-diminished place in the western economy throughout the decade.

Important though government policies and the sheer press of immigrant numbers were to the spread of farming into the rangelands of southern Saskatchewan and Alberta, the advance of small holdings into the semiarid grasslands was critically dependent upon science and technology. The chilled steel plough, barbed wire, the steel windmill for raising water, and the self-binding reaper were adopted or adapted from the United States. Red Fife wheat, hybridized near Peterborough, Ontario, in 1860, was the main choice of prairie farmers through the late nineteenth century, but even the early-maturing variety introduced in 1885 remained susceptible to early frosts.

By 1905, however, decades of trial and error attempts to cope with climatic marginality had been coupled with almost twenty years of scientific inquiry into the challenges of Canadian agriculture conducted in federal research stations and experimental farms across the country. Researchers at the station in Indian Head, Saskatchewan, were especially enthusiastic advocates of "naked summer fallowing" and "backsetting." Developed by Angus MacKay in the 1880s, "summer fallowing" meant plowing the land in late May/early June and keeping it free of vegetation throughout the rest of the year. Intended to keep down weeds and enrich the soil, the technique also served to retain soil moisture, allowing (farmers said) the precipitation of two years to grow one crop and making "possible the growing of grain in areas in which it is doubtful whether any other system . . . would have produced equally good results" (Potyondi 1995, 89–93).

Backsetting was a method of breaking new land. It required shallow ploughing of the prairie sod before the summer rains and a second ploughing in late summer, when the grass and roots had begun to rot. This second ploughing produced about three inches of soil atop the rotting sod. It was then harrowed for planting in the spring. Farmers with an eye to short-term progress were often impatient with backsetting, as the extra ploughing meant less land was broken, but it also helped retain moisture and generally produced better yields in the second year. When the Marquis wheat cultivar, developed at the Dominion Agricultural Station in Ottawa from Red Fife and Hard Red Calcutta, was stabilized and distributed, early in the twentieth century, its early-ripening attribute reduced the risks associated with early frosts, and both yields and the quality of wheat improved. By one estimate, the cultivation of Marquis wheat may have increased annual farm incomes by as much as $100 million by 1918 (Brown and Cook 1974, 53).

Neither backsetting nor summer fallowing solved the problems of agriculture in the driest parts of the southwest. There, irrigation was essential. In the 1890s Mormon immigrants from the south began a series of projects in the area around Cardston in conjunction with the Alberta Coal and Railway Company and several Utah sugar producers. Sugar beets were their main crop. A few years later the Canadian Pacific Railway embarked on an ambitious venture to irrigate and develop land along its main line near Calgary. Two blocks, each over a million acres in extent, were the core of the project, and plans called for the irrigation of over 400,000 acres in each block. Water from the Bow River was diverted into a canal to serve the western block and a large dam and reservoir were created below Calgary to water the dry eastern plains. The investment was considerable. Dams, canals, and aqueducts had to be built; the scheme had to be "sold"—promotional materials advertised "the finest winter wheat lands in America," the best pasture "in the world" and, more ambiguously, soils "almost beyond belief"; settlers had to be placed on the land.

Clustered near the railroad and the canals, those who paid the premium for irrigated acreage helped transform the landscape. Dwellings, equipment sheds, fences, grain elevators, and shelter belts of exotic trees replaced short-grass prairie. Fields of wheat, and fodder crops, vegetable gardens, and numbers of cattle, hogs, horses, sheep, and chickens marked the emerging mixed farming area. But returns were disappointing. At most, in the early 1920s, there were just over 1,000 homesteads on the eastern block; at an average of about 100 acres each in size, and recognizing that not all homesteads "took water," this meant that less than a quarter of the land originally considered irrigable was being utilized. In 1931 the railroad calculated that it had incurred a $15 million loss on its $41 million investment in the two projects (Evenden 2006).

Yet the balance sheet of Euro-Canadian settlement in the dry lands of the southwest, and in the prairies more broadly, was not to be measured in dollars alone. Early in the 1920s, settlers, politicians, scientists, and others began to realize that the drive to colonize the prairie west had instigated ecological changes that threatened the very prospect of people's persistence on the plains. Indeed, the great Canadian historian W. L. Morton, born and raised in Manitoba, was brought to wonder whether "traditional European culture" could adapt and survive on the prairie (Morton 1946, 26–31). The drought of 1918 to 1922 revealed the dark side of settlement. Summer fallowing pulverized the soil and

Marquis wheat. *Commemorated on a postage stamp, Marquis wheat helped to move the mail. It was far more important in the early years of the twentieth century when it moved the frontier of settlement northward. Developed by scientists at the Dominion Agricultural Research station, this cultivar ripened earlier than other available varieties and reduced the wheat crop's (and thus the prairie farm economy's) vulnerability to early frosts. By reducing risk it enabled farmers to succeed where otherwise they might have failed, and the improved productivity it brought added millions of dollars to Canada's gross domestic product by 1920. (Canada Post Corporation, 1988. Reproduced with permission.)*

exacerbated the effects of changes in soil structure induced by cultivation. Generally, the soils of the dry grassland have less carbon and nitrogen than those in moister areas, but under grass they have high organic content and good crumb structure.

In virgin soils, over 80 percent of minerals are held near the surface in crumb aggregates. But under continuous cultivation this proportion falls below 5 percent. The organic content and nutrient levels of soils drop markedly. This can lead to compaction, increased surface runoff, and a reduction in the soil's capacity to hold moisture. Discing and harrowing (as summer fallowing required) reduced compaction, but only at the expense of soil structure. It allowed more moisture to penetrate the surface—but only when moisture was available. In drought years, when the wind blew, soil "drifted" away.

With it went the bases of the farm family's livelihood, and many decided, after investing years of their lives in the effort to convert the land to wheat production, to drift away themselves. Almost two-thirds of the original homesteaders in one township in the Maple Creek area of Saskatchewan abandoned their land between 1910 and 1929. Across the central prairies more generally, dozens of townships suffered population declines of 20 percent or more between 1921 and 1926. But the abandonment of farmland did not mean restoration of the prairie. As one surveyor working in the vicinity of the international boundary reflected, perhaps a tad too pessimistically, "After the land is broken up it is fit for nothing, the native nutritious grasses being exterminated and a rank growth of weed taking their place" (C. F. Miles quoted in Potyondi 1995, 94).

Irrigation had very different, unanticipated, but no less consequential, environmental effects. Putting water on the prairie created unforeseen problems for humans and new habitat for a broad range of species. Muskrats and beavers found good haunts in irrigation canals, although their tunneling and damming were constant threats to irrigation systems. Fish and birds and weeds were afforded new niches in or alongside the waterways. Water seeping from canals or released too liberally onto fields was lost through inefficiency, but it also produced waterlogged ground and/or led to the buildup of salt compounds (alkali) on the surface when it evaporated. This reduced the value of the land, killed crops, and encouraged the spread of unpalatable salt grass (compare with Fiege 1999).

Other human-induced changes also helped to transform the ecology of the prairies during these years. In the parkland, settlers felled trees for fuel for building fences and for construction, as well as to open the land for cultivation. On the grassland, many of them met their initial fuel needs by collecting and burning "buffalo chips." Later they substituted the dung of domestic stock. This was not the cleanest of fuels. Using the material thus also deprived the land of the nutrients that would be returned to the soil through in situ decay. Where lignite cropped out in the grasslands, farmers opened small strip mines to supply themselves and their neighbors with the soft coal that could be burned alongside harder, more expensive varieties brought in by rail.

More significantly, farmers and ranchers both mounted serious campaigns of destruction against various forms of prairie wildlife. Wolves and coyotes were regarded as particular scourges by stockowners. Bounties were paid on wolves from the late nineteenth century and there were wolf bounty inspectors throughout Saskatchewan after 1907. But the annual wolf kill may have peaked before this. In 1911 bounties were claimed on 270 wolves, small business by comparison with the number of coyote claims, which topped 7,500 in that year and reached almost 36,000 by 1918–1919.

Gophers were the bête noir of farmers. The numbers were staggering, as were the lengths to which people went to exterminate them. Agricultural officials stressed the importance of eliminating gophers and suggested that half a bushel of grain was "saved" with every kill. Traps were set, and poison bait was widely used to kill both gophers and coyotes. Strychnine was mixed with wheat and molasses and placed—"a tablespoon . . . will be enough"—at the mouths of gopher holes in the spring. In 1920 the Saskatchewan Department of Agriculture distributed 125 kilograms of strychnine for this purpose. Failing to win the war, the province declared May 1 "Gopher Day" and gave every schoolchild in the province the day off to roam the fields in search of the pests; each tail brought a 1-cent reward, the most successful individual hunters were given special prizes, and schools competed for the "Gopher Shield" awarded to the institution with the greatest harvest of tails. In 1920, two million of the creatures had their reckoning on Gopher Day (Potyondi 1995, 83, 115; Stegner 1962).

Like most furious campaigns, this one produced collateral damage. Magpies were almost wiped out by poisoned bait left for other creatures. The long-billed curlew and the swift fox were also brought close to local extinction. In addition, prairie chickens and grouse were decimated by hunting and by the destruction of their nesting grounds in the grassland. And weeds, insect pests, and diseases raced across the prairie. Stem rust and stinking smut spread through the plains from Texas northward, their march facilitated by the near monoculture of wheat. Wireworms, cut worms, and sawflies, all indigenous to the region, proliferated in the twentieth century. By one estimate, the impact of these insects, mainly on the spring wheat crop, cost Saskatchewan $54 million between 1926 and 1931.

As Barry Potyondi has noted, these interventions and developments, and the larger transformation of the prairie environment of which they were only a part, were driven by strong desires for productivity and control. Economic growth was the leitmotif of this story, which took its shape from the conviction that the elimination or nullification of "any natural force or creature deemed inimical" to development—what contemporaries might have thought of as "environmental management" and what we might term the subjugation of nature—was essential to success.

20

RUSHING NORTH

Gold, the Yukon, and Alaska

In the 1850s the northern Pacific Coast was native territory into which Europeans had entered in small numbers in pursuit of furs and trade and geographical knowledge. The Hudson's Bay Company was established on Vancouver Island, which had become a Crown Colony in 1849 and operated trading posts that were no more than tiny enclaves in native space across the broad mainland territory known to fur traders as New Caledonia. Then news of gold along the Fraser River reached San Francisco in the early spring of 1858. Gold fever had been reverberating around the Pacific since the fabled California rush of 1849, and within months some 25,000 to 30,000 hopeful miners and camp followers poured into the Fraser drainage basin. Victoria, the gateway for many who took ship from the south, was transformed almost overnight from an outpost of the fur trade to a city of 5,000 people.

Anxious about the incursion of so many newcomers into uncontrolled territory, the British government declared the mainland north of the 49th parallel a Crown Colony. Members of the Royal Engineers regiment were dispatched to establish the rudiments of connection and control: roads, townsites, and settlement surveys. New Westminster was established on the north bank of the Fraser River, just above its delta, as the capital of British Columbia. Beyond, land in the Lower Fraser Valley was made available for settlement, as farm lots were laid out and preemption claims were recognized. In the vicinity of New Westminster and upriver around Chilliwack, homesteads were built, clearings extended, and livings made (at least temporarily) in the provisioning trade. But only a few came to farm. Most of the newcomers pushed upriver, beyond (Fort) Hope, where riches were believed to lie. Their quest carried them into the Fraser canyon, then quickly on to the Cariboo and the celebrated gold rush town of Barkerville (Harris 1997).

In the canyon miners found gold nuggets in river gravels, but they also encountered forbidding terrain and met staunch resistance from the Nlaka'pamux people (formerly known as "the Thompson" or "Thompson Salish") whose territory this was (Harris 2002; D. Harris 2001). At Yale, a few miles upriver from Hope, the large, fast-flowing Fraser is deeply incised and confined by steep, rocky walls. Small tributaries enter the main river from east and west, and gravel bars and outwash fans—dynamic elements of the fluvial system—provide some of the few points of effective access to the river. Here native people had lived and fished for centuries. Although they apparently exchanged goods and labor early in the summer of 1858, natives and miners were soon brought into conflict by their competition for space, for it was on these very bars and fans which native people had used for centuries that miners hoped to find their fortunes. Most miners were well armed; many were aggressive; significant numbers among them simply dismissed the Nlaka'pamux as "savages"; and they wanted the gold before others got to it. The native peoples, for their part, faced displacement from their ancestral territories and the disruption of their lives. By August, avers one authority, "miners and natives were at war in the canyon" (Harris 1997, 110–111). Most diggings were abandoned, miners congregated and seethed in Yale, there were raids on native settlements, and killings on both sides. But the interventions of Governor Douglas and Chief Cexpe'ntlEm (Spi'ntlam) forged an uneasy peace.

Miners were soon back in the canyon. There they exacted unthinking vengeance on the environment. Hidden in riverine deposits, placer gold had to be retrieved from vast quantities of rock and gravel. The most rudimentary means of achieving this—the classic technique of the legendary solitary miner—involved swirling a small quantity of gravel and sand with water in a pan until gold nuggets were separated from the detritus. But this was slow and hopelessly inefficient, except in the richest of deposits. It soon gave way to more complicated technologies. All of them involved water, and the passage of a slurry of sediment through "rockers," cradles, or wooden sluice boxes. In these devices, transverse barriers or "riffles" helped separate the gold (which sank quickly because of its high specific gravity) from waste materials. In some more complicated versions that combined riffles with slots, mercury was used to trap gold flakes.

Simple as they sound, the use of these techniques wrought environmental havoc wherever they were used. In the canyon, in the Cariboo, and in the Klondike, "the work of gold mining" was a "work of disassembly" (Morse 2003, 91). Entire ecosystems were taken apart. With thousands of miners fervently pursuing their dreams of wealth, rivers, creeks, river bars, and terraces were ripped asunder. The landscape was almost literally demolished. Gravel bars and old river terraces were stripped of vegetation. Streams were rerouted. Flumes (for the construction of which trees had to be felled) were built to carry water over gullies and from lakes and rivers to particular sites on the "diggings." Vast quantities of dirt were set in motion, sometimes with picks and shovels as sediment was lifted and dumped into sifting devices, sometimes with high-pressure hoses. In some parts of the middle Fraser, twelve, fifteen, perhaps even twenty feet of extensive terraces were "hydrauliced" away. Most of this dirt ended up fairly quickly in the Fraser River, which whisked it away to be deposited (along with persistent traces of mercury) on downstream bars and in the islands of the delta at the river's mouth. To this day, the areas from which it came are barren boulder fields and often ill-drained flats from which small buttes rise, seemingly inexplicably, where the prior and sacred claims of Christian churches and administrative decrees (but not the claims of long-standing indigenous use) deflected the destructive work of flumes and hoses.

Within a few years, however, most of the miners had moved on. Some, mostly Chinese, remained working around the edges of the washed-out boulder beds, abandoned ditches, and derelict flumes of the initial rush. By 1860 miners were pressing northward following gold discoveries east of Lac la Hache and on the Quesnel and Keithley rivers; the next year they were drawn onward into the Cariboo country by the rich gravels of Williams Creek and neighboring streams. Barkerville (and its satellite settlements of Richfield and Camerontown) quickly emerged as the focus of mining activity and the focus of dreams (Galois 1970). Thousands made the difficult journey northward through the Fraser canyon or across Harrison Lake and then overland via Lillooet.

By 1863 stagecoaches ran along the rough and narrow, but still impressive, wagon road through the middle Fraser. Some described the booming settlement as the largest community west of Chicago, which was almost certainly an exaggeration, though ten thousand people, most of them men, likely lived in and around the three towns when they were at their peak. Photographers were among them. "[P]arties writing home," said an advertisement in the *Cariboo Sentinel* of August 13, 1870, for L. A. Blanc's photographic gallery in Barkerville, "would do well to . . . select from his varied supply a souvenir for their friends." Panoramas of the gold fields taken by Frederick Dally and others provide a remarkable record of the environmental impacts of the Cariboo rush. As students of the photographic record have noted, pictures of "gold rush communities drew attention to tightly grouped wooden buildings flanked by denuded hillsides. Shops, offices saloons hotels and log cabins shared the creek floor with flumes, mineheads, Cornish wheels [which pumped water from the mines] and rock tailings" (Schwartz 1977–1978, 41). Other pictures pose miners armed with picks and shovels at the entrance to a mine or alongside a flume.

None of these is more revealing than Dally's "Hydraulic Gold Mining, Williams Creek," taken in 1867–1868. The top quarter of the frame is a wasteland of tree stumps, debris, and faintly evident artifacts of mining. The bottom quarter shows half a dozen miners leaning on shovels, a flume off to their right, amid gravel, boulders, large balks of timber, and other detritus. Between foreground and background, occupying the center of the photograph, is a cliff. A rough and rickety ladder ascends its steep slope, and water spurts from various outlets in its face. Huge rocks, sizable boulders, great volumes of unconsolidated sediment, and cluttered channels running at various angles further complicated the scene. This is a washout. The men stand on a horizon created by their hydraulic operations some forty feet below the original, stump-strewn surface of the earth. This was one of those views that the *Cariboo Sentinel* touted as a valuable representation of "the scene of labor in which the wanderer from home is engaged." It was also a scene of the utmost havoc.

Perhaps the most storied of the great nineteenth-century gold rushes was the last, to the Klondike in 1897–1898. Remote, exceedingly difficult to access, and possessed of a climate hostile to outdoor activity through much of the year, the Klondike presented unique challenges to miners—and amplified their impacts upon the environment. Reflecting the mythic dimensions of the rush, Robert Service's popular ballads peopled the Klondike with larger-than-life characters, such as dangerous Dan McGrew and Tennessee Sam McGee, gave readers a sense of its allure, and made of it a Darwinian battleground, where only the strong could thrive and only the fit would survive. Likewise, Buck, the "unduly civilized dog"—"suddenly jerked from civilization and flung into the heart of things primordial"—in Jack London's beloved *The Call of the Wild*, stood metaphorically for all who went to the Yukon.

Getting there was a large part of the ordeal. Crossing the Coast Range from the Alaskan port towns of Skagway and Dyea, which were the main gateways to Yukon Territory before construction of the White Pass and Yukon Railway, required an ascent of the precipitous Chilcoot or White Pass trails. This entailed miles of arduous climbing, much of it through mud in spring and summer and on ice and snow during the winter. Here, said James Hamil, who went this way in 1897, men made horses and mules of themselves. Every pound of provisions—and the Canadian authorities required everyone entering their territory to have a ton of supplies—had to be carried in "over jagged rocks, down through deep canyons and over mountains of ice which never disappear." The last mile of the journey was "almost perpendicular," and part of it had to be crawled on hands and knees; it was "an utter impossibility . . . to attempt to describe the hardships" to be endured (quoted in Morse 2003, 4, 49). And that was only the start of the journey.

Once in Dawson, or more accurately up along one of the many tributary creeks of the Yukon River prospected and worked for gold, miners faced the

Yukon gold rush. *Klondike miners, circa 1897. These miners working on W. M. Cowley's claim, above Bonanza in the Klondike, were among the thousands who made the arduous journey north in search of gold. Seeking "Eldorado," they found themselves battling harsh conditions, facing the challenge of mining in permanently frozen ground, and contributing their piece to the work of tearing up the earth in the quest for riches. (Hulton Archive/Getty Images)*

challenge of adapting familiar mining techniques to northern circumstances. This meant, as Sam McGee might have said, that strange things were done in the land of the midnight sun. These are the latitudes of continuous permafrost, where the ground below the relatively thin active layer (subject to summer thaw) remains at subzero temperatures. Because miners sought gold in the lower depths of riverine deposits as well as in the subterranean beds of former streams, they endeavored to extract as much of this ground as possible during the winter in anticipation of the spring thaw that liberated the water they needed to run their sluice boxes to separate gold from gravel. Thus, they lit fires to thaw the earth. Working in poor light and very low temperatures, they melted and excavated, melted and excavated, creating ever larger piles of soon-refrozen earth beside their deepening holes in the ground. If and when they struck an old river course, miners would attempt to melt their way laterally along what they took to be promising shows of gravel. This took considerable skill. Depths of thirty feet were not unknown, and drainage was a problem, especially as water melted in one place would freeze away from the fire.

Smoke hung in the air, and in the shafts and tunnels. Noxious gases were sometimes released by burning. Thawing of the permafrost could produce subsidence, or the hummocky terrain known as thermokarst, at the surface. And enormous quantities of wood were consumed. According to Tappan Adney, whose book *The Klondike Stampede,* published in 1900, is one of the most luminous accounts of the Yukon rush, half a cord was required to thaw five cubic yards of gravel. At this rate a single small operation might use between thirty and sixty cords of wood a season. Add to this the fuel used to power paddlesteamers on the Yukon River, to heat dwellings, and to construct buildings and flumes, and it is clear that the miners' assault on the slow-growing boreal forest of the region was immense.

This had consequences. When snowmelt and rain came, runoff was quickened, erosion increased, flooding became more frequent, and streams and miners' flumes sometimes ran short of water. To overcome these problems, dams and more extensive flumes were built, some of which served to divert water from streambeds entirely, so that they could be worked for gold. Quickly, the northern landscape was turned into a wasteland—and when its immediate riches were extracted, it was abandoned. "It was quite a sight," wrote one diarist in 1900, to "see how people are tearing up the earth" (quoted in Morse 2003, 110). Two years before, Tappan Adney, journeying up Bonanza Creek, near Dawson City where the first gold finds were made in 1896, had reflected that:

> Dams of crib-work filled with stones, flumes, and sluice boxes lay across our path; heaps of "tailings" glistened in the sunlight beside yawning holes with windlasses tumbled in; cabins were deserted—the whole creek, wherever work had been done, was ripped and gutted. Nothing but flood and fire is so ruthless as the miner. (Adney 1900, 404)

Wildlife—the moose and bear and elk and caribou, as well as the wolves—that had once used these streams and their banks for food and water and shelter were displaced. Birds no longer found nesting areas or cover amid the devastation. Fish could not survive the thick muddy flows that streams of once-clear water became. Native peoples were sorely afflicted by diseases (particularly influenza and measles) introduced to the area by the sudden incursion of white southerners. In western and northern Alaska, reports medical historian Robert Fortuine, a "great sickness" killed between a quarter and a third of the native population in 1900. As Kathryn Morse has noted (2003, 112), however, "when gold miners looked at the torn-up, muddied, lifeless landscape they left behind, . . . they looked through the lens of their culture; instead of degradation, they saw riches."

According to Tappan Adney, the typical Klondike miner was "wrapped in a blanket of his own thoughts. Sometimes such men are cranks. Every man becomes a crank who stays long in this country" (quoted in Fetherling 2004). Here, Adney, a journalist who knew how to turn a phrase, was thinking less of the miners' blindness to despoliation than of the particular individualism of many of those he encountered in the north. Perhaps in fear of the environment's transformative powers, he chose not to linger there. After following the rush from Dawson City to Nome, Alaska, he returned to New Brunswick and New York. At various sites along the Yukon River in Alaska there were active, if ultimately ephemeral, camps that yielded small amounts of gold and left permanent marks upon the landscape. In Nome, however, Adney encountered a place almost as remarkable as Dawson City. Far easier of access than the Klondike, the site attracted 20,000 people within two years of the first strike on Anvil River. Here even the beaches were full of gold—almost literally; in 1899, 2,000 prospectors took $2 million from the sands of the shore. A few years later there was another major gold find on the Chena River near its confluence with the Tanana. More capital was required to extract the buried gold from these deposits, but the results were much the same. People rushed into the area; Fairbanks sprang up as a service center for the diggings; and in 1906 gold production from Alaska peaked at 6.5 million ounces, with almost 60 percent of this coming from the Fairbanks district and most of the rest from the Seward peninsula.

Adney to the contrary, those whom gold fever brought to the Klondike and Alaska were less cranks than adventurers and not so much misfits as dissenters. Some, certainly, went north in desperation after the financial slump of the mid-1890s. Others were part of that footloose tide that flowed through mining camps around the Pacific; one, wrote a compatriot, "had mined within a few miles of Cape Horn and now he is here on the Yukon" (quoted in Morse 2003, 129). Probably most believed that they would strike it rich. Some hoped to do so by providing services, from hotels to blacksmithing to sex. Whatever their motives for heading north, thousands left disappointed and disillusioned. After struggling across the Chilcoot Pass some learned that all of the famed creeks were claimed; others realized that neither the work nor the place was for them. By some estimates, between 100,000 and 200,000 people set out for the Klondike; slightly more than 40,000 reached Dawson City; and of these only about 4,000 actually "struck gold."

All the same, many of those who came discovered something more important than gold. They found themselves. In their minds at least, isolation was a fair price for the independence they gained. The rough-and-ready north provided an attractive alternative to the constraints of the south, and the hard, diverse, and challenging work of mining was a satisfying alternative to "sitting in a cage in a bank and thinking . . . [of oneself as] an important citizen instead of a slave" (quoted in Morse 2003, 123). In *Klondike Fever* (1972, 429), the Canadian historian Pierre Berton, who was born in Dawson City, quotes one participant in the rush reflecting: "I made exactly nothing, but if I could turn time back I would do it over again for less than that."

Seen in this light, the north offered people an opportunity to pursue a particular variant of the "strenuous life" in nature that many of those who remained in more southerly, urbanizing, industrializing areas of North America sought to enjoy through summer vacations spent hunting, fishing, or canoeing in nearby woods and mountains. Part of a widespread, albeit far from universal, reaction against what many contemporaries perceived as the enervating, debilitating effects of urban existence, this embrace of nature and aversion to the city took many forms. In the broadest of terms, it is fair to say that cities lost their allure as they grew larger and the implications of life within them became clearer. The Methodist preacher and leader of the Social Gospel movement in Winnipeg J. S. Woodsworth articulated one version of this story when he wrote:

> The higher the buildings, the less sunshine; the bigger the crowds, the less fresh air . . . [.] We become weary in the unceasing rush, and feel utterly lonely in the crowded streets. There comes a wistful longing for the happy life of "God's out-of-doors" with the perfume of the flowers and the singing of the birds. (Woodsworth 1911, 11)

Others suggested the taking of "rest cures" in canoes or the creation of urban parks as antidotes to the developing problem of urbanites "paralyzed by introspection and self-doubt, [and] obsessed with easing [their] own psychic tensions" (Lears 1981, 56). In Canada, the "call of the wild" (and Jack London's title assumes new resonance in this context) assumed many forms. Readers of *Rod and Gun* were assured that the activities alluded to in the title of that magazine would reconnect those suffering from urban malaise with their "Stone Age inheritance"; advertisements for lakeside resorts promised that those who patronized them would return with "manlier heart[s]" and "tougher muscles" to cope with the "rush and bustle of the city"; and writers and artists produced animal stories and "wilderness paintings" in such numbers that people might "be forgiven for thinking that Canada was essentially an undeveloped" land (Jasen 1995, passim; Bumsted 1992, 131).

Too much can be made of this. And too much has been. Alaska and the Yukon Territory have often been seen as last frontiers, as places and societies defined by the pioneering spirit of early arrivals (and all who followed them).

Thus the popularity, in both areas, of the image of the lonely prospector, the grizzled sourdough who stands, gold pan in hand, as the epitome of self-reliant individualism and the embodiment of the personal freedoms that recent generations of northerners often claim as their birthright. But this misses an important dimension of northern development. Both Alaska and Yukon were colonial territories. They owed allegiance to different states, but both depended on governments and investors elsewhere for their economic growth and development. Certainly, the influx of individual gold seekers was crucial to the early growth of the non-native population in these territories. There were only about 2,000 non-natives in the Yukon when William Carmack's discovery triggered the Klondike rush. In 1880 the non-native population of Alaska was probably less than 500; ten years later it was little over 4,000. By 1900, Yukon had over 40,000 non-native residents. Alaska counted over 30,000 newcomers in 1910.

Yet hardly a soul among them entered the north without assistance. Most arrived in the Gulf of Alaska on steamships out of Vancouver, Victoria, Seattle, or San Francisco. Those who crossed the Coast Mountains and continued to Whitehorse on the 110 miles of the White Pass and Yukon Railway after 1900 rode on rails financed by English investors. As Kathryn Morse has shown, northern lives depended upon the capacity of processors in Chicago, Omaha, Toronto, New York, or California to render perishable foods "portable through time" (in Daniel Boorstin's phrase) by canning, dehydrating, and condensing them. Little wonder that one contemporary characterized "the five gallon evaporated potato can" as "the national washtub of Alaska" (quoted in Morse 2003, 143).

By the same token, merchants in southern port cities organized and profited from the trade with northern Pacific ports, and much of the wealth generated by fervent activity in the goldfields was accumulated or spent in Vancouver, Seattle, or other places far removed from the sub-Arctic. Once the initial bonanza yields of placer gold gave way to the exploitation of subterranean reserves, mining quickly assumed an industrial cast and required sizable investment. Indeed, the development of the Treadwell gold deposits, discovered in 1880 in the vicinity of what became Juneau, established this pattern of frontier settlement long before the more storied rushes of the late 1890s. As hopeful gold seekers made their ways to the Yukon, over 800 crushing stamps pounded the ore from Treadwell mines. Two thousand men worked there. The capital as well as the labor for such enterprises came from the south.

Moreover, neither miners nor investors entered terra incognita when they went north. Surveyors had defined territories (if not always precisely), and officials of the state (members of the Mounted Police or the military, customs agents, assay officers, Gold Commissioners, and land agents) helped to keep miners and their possessions safe. In Alaska the American government established a number of agricultural experiment stations to investigate the region's potential for settlement. And both Canadian and American governments underwrote the construction of telegraph lines that linked remote settlements to the "outside," with immense consequences. As Stephen Haycox summarizes the situation in Alaska, the "telegraph facilitated economic development by linking corporate agents to their directors in the world's capitalist centers. It put the government's field agents and administrators . . . in touch with the government in Washington D.C. . . . [and it spawned] an instant newspaper industry" capable of reporting the same events as filled papers "in Emporia Kansas and Presque Isle, Maine" (Haycox 2002a, 214).

Alaska's status as a dependent territory sustained by external interest in returns from its natural resources, and subject to administrative control from the center, became fully evident early in the twentieth century. Interest in the wealth of Alaska quickened as progressive reform ideas gained popularity late in the nineteenth century and encouraged both the reassessment of American public lands policies and vigorous efforts to conserve natural resources. New agencies, including the U.S. Forest Service, the Bureau of Reclamation, and the National Park Service, were established, and the Antiquities Act and the Mineral Leasing Act gave Washington new authority over the public domain. In short order, these developments—which understood conservation as the wise and efficient use of resources rather than as their preservation—impinged upon many facets of Alaskan development.

As interest in the wildlife and natural resources of Alaska increased—not least through the publication of John Muir's *Travels in Alaska*, along with fourteen volumes of photographs and information about the region gathered by participants in a remarkable expedition financed by railroad magnate Edward H. Harriman, and organized in part at least to allow him to kill an Alaska brown bear on Kodiak Island—Muir's followers urged preservation of this remote wilderness, influential members of the Boone and Crockett Club lobbied for stricter game laws, and leaders of the Progressive Conservation movement saw opportunities to implement their conviction that the resources of the public domain belonged to all Americans (see also Cruikshank 2005). In Alaska, noted one of the prominent figures of this faction, "we have our last chance to preserve and protect rather than to restore" fish and game populations (quoted in Haycox 2002b, 44).

Although such economic development as there had been in the late nineteenth century had rarely considered the environmental consequences of resource exploitation—bowhead whale and walrus populations were exploited almost to extinction in the western Arctic, and the chlorination process used to extract gold at the Treadwell Mines killed the vegetation along the shores of

Gastineau Channel for a mile above and below the works, for example—such impacts affected only a tiny proportion of the extensive northern territory. Yet the path ahead was neither clear nor straight. If the beginnings of the commercial salmon fishery in Alaska are dated to 1878, when the first large cannery was built on Prince of Wales Island, expansion was rapid through the next decade. By 1890 there were more than twenty canneries operating in the territory and the pack was worth almost $3 million. Five years later competition among the thirty-seven or so canneries (many headquartered in Seattle) had led to the organization of the Alaska Packer's Association, with control over two-thirds of the industry.

Its grip slipped as the industry continued to expand. In 1900 the pack was over two million cases and was valued at nearly $10 million. By the turn of the century, fish traps had begun to replace small boat gill netters. Depending upon the use of large surface-to-sea-floor nets that diverted salmon runs from the mouths of streams into large closed net traps, they were extremely efficient at catching fish, but threatened to wipe out entire runs if used without significant interruption. As a countermeasure, the federal government required every company fishing in Alaska to establish hatcheries and to return to the rivers four times as many red (sockeye) salmon as they had taken the previous season. But enforcement was virtually nonexistent. In 1903 the newly created Bureau of Fisheries assumed responsibility for Alaska salmon and obtained funding for two hatcheries. But it was less successful in regulating and limiting the fishery. By 1911 there was growing concern about the falling size of salmon runs. Faith in the restorative capacity of hatcheries and the high demand for canned salmon through World War I stayed significant action, however. By 1920 the crisis was clear. Hatcheries had not sustained stocks, which were dangerously low. In 1922 the president took the unprecedented step of creating two ocean reserves in which fishing was prohibited. Two years later, the White Act mandated that half of the salmon run on every stream had to be permitted to reach the spawning grounds.

When remarkably rich copper deposits were discovered high in the Wrangell Mountains, they quickly attracted the attention of the Guggenheim Corporation, the largest private mining company in the United States. Conscious of the extraordinary difficulties of bringing the copper to market, but also aware of coal and oil deposits in the Bering River area, they hatched large plans for the development of these resources. A railroad would be built, where some engineers thought no railroad could go, to bring copper ore to the coast. A smelter would be built on Prince William Sound. Local coal would power the smelter, the railroad, and the mine and be sold abroad. Eventually the railroad would be extended clear through to the Yukon River and the mining potential of that region would be brought within the company's orbit. "This integrated scheme," judged historian Stephen Haycox, "represented colonial capitalism at its height" (Haycox 2002, 223; see also Cronon 1992).

Anxious about the widespread engrossment of oil and coal lands by mining companies across the United States, President Roosevelt withdrew all coal deposits on public lands from development in 1906. This dramatically undercut the Guggenheim Company's plans. But the quality of the copper deposit (it assayed as high as 85 percent, when mines in Utah were exploiting concentrations of 2 percent) led the corporation to persist. In partnership with J. P. Morgan, it created a confederation of companies called the Alaska Syndicate to build and operate a railroad from Cordova to Kennecott, to open the Kennecott Mine, and to run the Alaska Steamship Company, which would carry the copper ore from Cordova to the Guggenheim smelter in Tacoma. The Syndicate also acquired the Northwest Fisheries Company, which owned a dozen salmon canneries, for good measure.

Local politicians protested, in the vein of Progressives across the country, at the Guggenheim-Morgan monopoly and denounced the Alaska Syndicate for treating the territory as "its own plaything." But the railroad was completed at astronomical cost ($23 million for fewer than 200 miles), and the mine operated for twenty-five years. Five hundred men were employed there, and Kennecott developed as a classic company town. By one reckoning, the 4.6 million tons of ore that were taken from the mine yielded an average of 13 percent copper, and nine million ounces of silver were also extracted. Profit levels were a handsome 35 to 50 percent on ore sales. For all that, both the mine and the railroad were closed when the price of copper fell during the 1930s.

Much to their general chagrin, residents of Alaska also saw large parts of the public domain being set aside for "conservation purposes" early in the twentieth century, and they railed at the ways in which southern interests hamstrung the development of their territory. The nationwide coal reservations of 1906 were bad enough. However much they might have been justified in the contiguous states, in Alaska they seemed perverse. The coal-leasing system that Roosevelt expected Congress to establish did not materialize until 1914, and the lingering restrictions on coal exploitation not only undercut the Guggenheim Company's grand plans; they quadrupled the price of coal for all Alaskan consumers and were readily seen as arresting progress on an undeveloped frontier. In 1911 disgruntled residents of Cordova shoveled tons of imported British Columbia coal into Orca Bay in protest, an action that the *Philadelphia Bulletin* described as a measure of the people's "not unreasonable impatience with the dilatory federal policy relating to the development of Alaskan resources" (quoted in Borneman 2003, 241).

Earlier, some 4.5 million acres of land in southeastern Alaska had been set aside as the Alexander Archipelago Forest Reserve. In 1907 another 2.25 million acres on the mainland, between Portland Canal and the Unuk River, were designated the Tongass National Forest. Two years later these two areas were combined into a single administrative unit under the Tongass name. In south-central Alaska, the Chugach National Forest was also established in 1907, to encompass the Kenai peninsula and lands bordering Prince William Sound. Reflecting Gifford Pinchot's conservationist agenda intended to enable the prudent management of forest resources, these initiatives were widely criticized for locking up valuable resources, preventing agricultural settlement, and shackling the growth of the territory. A few years later, a large area in the Alaska Range was set aside as Mount McKinley National Park.

In some broad sense a northern expression of the movement to establish national parks that had begun with Yellowstone in 1872 and continued up and down the western Cordillera with the establishment of Yosemite, Glacier, Mount Rainier, and Sequoia National Parks in the United States and Banff, Jasper, Glacier, and Yoho National Parks in Canada, Mount McKinley National Park was also very directly the result of one man's concerted commitment to its creation. Hunter and naturalist Charles Sheldon was fascinated by mountain sheep. He first came to Alaska in search of trophy heads in 1906 and then spent 1907–1908 living in a remote cabin, carefully observing the Dall sheep of Denali. For almost a decade thereafter he lobbied for a park to preserve the ecosystem and protect the sheep of this area. Enlisting the Boone and Crockett Club and Theodore Roosevelt in the cause, Sheldon finally saw success in 1917, although four more years went by before funds were appropriated for management of the park.

In sum, most Alaskans took a dim view of these developments. Many were quick to point out that the area designated Chugach National Forest had few trees. Countless evils—"empty houses, deserted villages, dying towns, arrested development, bankrupt pioneers, and the blasted hopes of sturdy, self-reliant American citizens" among them—were attributed to meddling and misguided figures such as Pinchot and John Muir (quoted in Borneman 2003, 241). One article in the *Alaska-Yukon Magazine* of February 1912 neatly summarized widespread resentments in its title: "Conservative Faddists Arrest Progress and Seek to Supplant Self-Government with Bureaucracy."

• • •

In three-quarters of a century, before the stock market crash of 1929 and the deep and debilitating economic depression of the 1930s, the overwhelmingly "wooden world" in which most British North Americans lived largely vanished. By the second quarter of the twentieth century, the days of "Buck" and "Bright"—as three-quarters of the working oxen in Upper Canada were said to be called—of "a Robinson Crusoe sort of existence" on the family farm, and of the neighbor who was forced by necessity to make an ivory tooth for his wife and an iron one for his harrow in the same day were but memories, or stories of other times, for most Canadians. Railroads, the telegraph, and electricity (which powered streetcars, drove machinery, banished the dark, and cooked the dinner) were the everyday realities of most lives. Armed with increasingly powerful technologies, Canadians had taken enormous strides in the exploitation and subjugation of nature. Old and new resources were extracted in quantities and with a speed and efficiency probably unimaginable a century before. Although many Canadians, and Alaskans, still worked in and with nature on a daily basis, fishing, logging, ploughing, mining, and so on, and depended to a considerable degree upon the strength of their muscles and the stamina of their bodies to earn their livings, science, engineering, and plain old grassroots ingenuity had done much to ease the physical burdens of existence.

They had also transformed the landscape in ways both astonishing and subtle. The ravages wrought by the combination of growing technological might with spendthrift attitudes toward the environment were fully evident in many places. Great piles of rock, excavated from the earth and discarded as waste, enormous dams and extensive lakes where rivers once flowed unfettered, large tracts of ground disturbed by loggers and gold miners, wide swaths of vegetation killed by noxious emissions, all of these were clear evidence of humans' growing dominance over, and exploitative attitudes toward, nature. So, too, expanding cities, railroads carved through mountain passes and carried over steep valleys on intricate trestles, swamps drained, and grasslands turned to the plough spoke of people's immense capacity to change the face of the earth. But there were other, less obvious changes in human-environment relations hidden behind these highly visible transformations. Improved communications and a pervasive quest for market efficiencies contributed to the economic rationalization of production, to the concentration of particular activities in certain areas. In manufacturing, this led to agglomeration, the emergence of specialized manufacturing regions or districts. In farming it meant a narrowing of the range of commodities produced in any one area and the development of specific concentrations of agricultural output, emphasizing grain in one area, cattle in another, dairying in a third, fodder crop production in a fourth, and so on. In human terms it also meant a growing separation of urban from rural, and of the consumption from the production of foodstuffs. These were emergent trends rather than confirmed and immutable patterns by 1930. But they marked a major re-working of the fundamental bases upon which lives rested and of people's relations with nature in the northern part of the North American continent—reworkings that would be confirmed and extended in hitherto unimagined ways in the years after World War II.

PART 5

NATURE TRANSFORMED

A week after Britain declared war on Germany in 1939, Canada did likewise. By 1945 over a million Canadians had served overseas (and more than 40,000 had been killed). Patriotism accounted for many enlistments. But as Gabrielle Roy's magnificent evocation (in *The Tin Flute*) of life on the poor streets of Montréal during these years suggests, signing up for military service also promised an escape from the dull and gnawing oppression of poverty. Canada had been particularly hard hit by the Great Depression that followed the stock market crash of 1929.

National prosperity rested upon commodity exports. Wheat exports that exceeded a million bushels a day, on average, in 1929 were decimated by the sudden contraction of the international grain market in 1930. Protective tariffs, erected by one country after another in efforts to protect home production in a time of economic crisis, also cut down Canadian exports of pulp and paper, minerals, and manufactures. Both agriculture and industry were very hard hit as commodity prices plummeted, and in the prairie west economic problems were exacerbated by drought and dust storms. By 1933 over a quarter of the nonagricultural labor force was unemployed; nationally, the average per capita income was barely half of what it had been in 1928. By 1937 two-thirds of the rural population of Saskatchewan depended upon "relief" (or public assistance).

The war changed all of this. As enlistees left for "the front," production in support of the war effort increased. The federal government assumed new leadership in domestic affairs. The Department of Munitions and Supply took on a highly active role in the economy, establishing Crown corporations and operating factories for the production of wartime requirements, from cartridges and shells to trucks and aircraft, and using the War Measures Act to regulate and direct the use of various materials. Federal government expenditures increased 7.5-fold in six years after 1939. By war's end, Canada had contributed "16,000 aircraft, 6,500 tanks, almost a million rifles, a quarter of a million machine guns, thousands of cargo and escort vessels and almost a million motor vehicles" to the Allied cause (*Historical Atlas of Canada* 3: 117).

At the same time, skyrocketing demand for agricultural products (such as wheat, flour, and bacon) produced a three- to fourfold increase in Canadian exports of those commodities between 1939 and 1945. In the remote northwest of the continent, the Alcan [Alaska-Canada] Military Highway was driven through the mountain and forest wilds of northern British Columbia, Yukon Territory, and eastern Alaska in 1942–1943. All of this spelled rapid economic expansion. By 1941 unemployment levels approached historic lows, and a year later labor was in such short supply that tens of thousands of women were recruited to industrial jobs and people had to seek government permission to change employment. By 1944 the number of women in paid employment in Canada had all but doubled, and exceeded one million.

Although the incentives offered to encourage married women into the workforce (such as income tax concessions and the provision of day nurseries for children) were withdrawn in 1945, the effects of wartime policies and of the economic expansion of the early 1940s reverberated in countless ways through the postwar years and did much to shape patterns of Canadian development through the last half of the twentieth century. On the policy level, agreements for the defense of North America and arrangements under the U.S.-U.K. lease-lend accord, intended to ease the balance of payments burden generated by Canada's wartime production, increased the integration of American and Canadian economies and provided a platform for the extension of this pattern in the 1950s.

As the Cold War escalated and American fears of nuclear attack by the Soviet Union increased, Canada became important strategic space. The Americans built

Distant Early Warning (DEW) radar stations in a line across the north, and Canada signed on to Defense Production Sharing Arrangements and the North American Air Defense Command (NORAD) with its southern neighbor. American fears of resource depletion generated heavy American investment in Canadian resources, and American companies also assumed major roles in some sectors of Canadian manufacturing (including the production of chemicals, electrical equipment, and rubber).

Domestically, the Canadian Government adopted a qualified free-market stance toward the economy, but sought to sustain "a high and stable level of employment and income" for Canadians by following a Keynesian strategy of countercyclical expenditure (or pump priming in periods of economic downturn) largely effected by public spending on social welfare policies such as unemployment insurance and family allowances. The federal civil service more than doubled in size during the war years, and the effectiveness of the Canadian response in Britain's "hour of need" did much to seed the conviction that planning—economic, social, and cultural—was both possible and important. Federal initiatives (sometimes in conjunction with provincial governments) sought to improve the infrastructure of national development.

Regional inequalities were also a concern. The roots of this issue were deep and structural, but they had been exacerbated by the pattern of wartime disbursement. Expenditures on war production facilities had been very heavily concentrated in Ontario and Québec. In the six years of the war, almost $369 million was invested in Ontario and over $253 million in Québec. Of the other provinces, only British Columbia ($25.6 million) received more than $19 million. Employment followed investments. At the peak in 1943, Montréal, Toronto, and Québec City had over 155,000 jobs in Crown plants; no other city had more than 10,000. After the war, plant conversions entrenched the essence of these disparities. Federal grants intended to boost the economies of poorer regions were continued in the immediate postwar years, but it took a decade and more before these arrangements were systematized as "equalization payments," by which monies from the richer provinces were redistributed to those that fared less well economically in efforts to establish broadly comparable minimum standards of health, education, and social welfare provision across the country.

Fueled by the "baby boom" and the general consumer demand (for everything from housing through deferred purchases to education) produced by the return to peacetime conditions, sustained by government policies, and fed by the benefit of a more robust economic infrastructure than that of any of the war-torn European countries, Canada's wartime prosperity continued after 1945. On the international stage, Canada emerged as a "middle power" but, more importantly, it also became a favored destination of emigrants, particularly from European countries in which the war and its aftermath (including the growing tensions of the developing "Cold War") induced people to contemplate futures elsewhere. Two million immigrants entered Canada in the fifteen years after 1946. Together, the influx of newcomers and high rates of natural increase drove Canadian population totals upward. Between 1946 and 1961 they increased by 50 percent, from twelve to eighteen million.

Some previously sparsely settled areas, such as the northern interior of British Columbia, experienced what one authority has characterized as "supergrowth" (an increase of more than 75 percent) in the decade after 1951 (*Historical Atlas of Canada* 3: plate 59). For all that, more and more Canadians, an ever-growing proportion of the population, lived in cities. In 1961 fourteen million were enumerated as urban dwellers and fully 46 percent of all Canadians lived in places with more than 100,000 residents. Although the Canadian economy performed relatively sluggishly for a brief period late in the 1950s, the two decades after midcentury were marked by widespread prosperity across the country.

Still, the good times were not universal. In general, urban dwellers did better than their country cousins. All sectors of the resource economy faced wrenching adjustments driven in large part by technological innovations and market changes. Some older agricultural areas, in which the traditional economy faltered, did less well than more fertile regions near large, fast-growing cities. And changes in the fishery and the forest industries left many workers in straitened circumstances even as exploitation of the resources quickened.

In broad terms, the quarter century or so after the Second World War marked the heyday of high modernism in Canada. Following James C. Scott, this high-modernist moment might be seen as the expression of an ideology "best conceived [of] as a strong, one might even say muscle-bound, version of the self-confidence about scientific and technical progress, the expansion of production, the growing satisfaction of human needs, the mastery of nature (including human nature), and, above all, the rational design of social order commensurate with the scientific understanding of natural laws" (Scott 1998, 4). Rooted in the remarkable scientific and technological advances of the late nineteenth and early twentieth centuries (the cornerstones of Mumford's Neotechnic Revolution coupled with developments in the petrochemical realm), high modernism evinced extraordinary faith in the human capacity to alter the world, embodied an almost unshakeable conviction that such alterations were good, and demonstrated remarkably uncritical optimism about the benefits that would flow from the implementation of its basic tenet, that human existence could be improved by the rational planning and orderly organization of both patterns of settlement and processes of production and consumption. Typically, high-modernist initiatives depended upon the close collaboration of capitalist entrepreneurs and the state. They involved what historian Paul Josephson called geo-engineering through the use of brute-force technologies and they ran roughshod over local practices, social differences, ecological diversity, mutuality, and informality as they turned upon the implementation of simplified, schematic, formal designs in the cause of social and environmental renovation.

The confidence, even hubris, of the high-modernist ethos was put to severe test in the last twenty-five years or so of the century. The certainties upon which it had rested began to unravel in the 1960s, with price inflation, growing aversion to the entanglement of the United States in Vietnam, and the rise of a "counterculture" that assumed an anti-establishment stance and preached the virtues of individual liberation. Although Prime Minister Trudeau made "The Land Is Strong" his election slogan in 1972, the Canadian economy (with those of the United States and several European countries) faced the vexing problem of "stagflation." This seemingly intractable combination of high levels of inflation and unemployment threw Keynesian principles of economic management into question.

At much the same time, the destabilization of world currency markets (produced by the devaluation of the American dollar in 1970), the rise of protectionism in the United States and Europe, the growth of industrial economies in the so-called Asian-tiger countries, and the sudden dramatic escalation in the price of oil engineered by Arab domination of the Organization of Petroleum Exporting Countries (OPEC) in 1973 produced economic and political turbulence and cast doubt on the capacity of governments (and their servants) to manage economies and plan development. In old industrial communities in eastern and central Canada (as in the northeastern United States and the coalfields of Britain) factories were closed and thousands of well-paying jobs were lost. For Canada, industrial decline in the American "rust belt" also meant dramatically reduced demand for minerals such as nickel and iron ore. The multiplier effects of reduced consumer spending reverberated through the economy.

Following the dogma of neoclassical economists, much of the developed world swung to the right politically. In Canada, as historian Desmond Morton observed, pollsters reported that "Canadians considered 'Big Government' their worst enemy, with 'Big Labour' close behind" (Morton 1987, 535). In retrospect, economists would recognize that the so-called Fordist coalition in which labor, government, and corporate interests had come together to further their mutual interests had shattered, and that the stability that it had engendered was being replaced by a new, more volatile arrangement of just-in-time production and "flexible accumulation." Now the virtues of free enterprise were celebrated, and business, including "Big Business," was admired. For a while the federal government fought against the tide in efforts to deal with the oil crisis. But by and large, governments withdrew, more or less rapidly and more or less completely, from many of the activities in which they had taken leadership roles in the 1950s and 1960s. In the last two decades of the century, both federal and provincial "states" were "hollowed out" as government expenditures on regulation and enforcement were slashed.

None of this brought "big development" or mega-projects of the sort that had dominated the high-modernist era to an end. The brute-force technologies brought to new levels of sophistication and efficacy during the third quarter of the century lost none of their power in the fourth, but their deployment was tempered to some degree by economic constriction, uncertainty, and the loss of confidence this produced. Still, grand ambitions, and the designs they spawn, are notoriously difficult to slay. Governments found themselves wedded to commitments from which it was virtually impossible to withdraw, most spectacularly perhaps in the case of the massive James Bay hydroelectric scheme announced in 1971.

In the oft-proclaimed "globally competitive" environment unlocked by deregulation and free-trade agreements, companies were subject to amalgamation and takeover, but they responded to development opportunities whatever their ownership, for this was their business, and they did so, for the most part, in fields opened for the operation of private enterprise, and hedged about with far fewer restrictions on the environmental impacts of development, as a result of the rightward drift of public policy.

21

CORRIDORS OF
MODERNIZATION

Under the perceived threat of Japanese invasion, the government of the United States established the Alaskan Defense Force in 1940 and built large military bases at Fairbanks, Anchorage, and on the Aleutian Islands (the westernmost sections of which were closer to Tokyo than Seattle). Several thousand personnel were stationed there before the attack on Pearl Harbor (Hawaii) in December 1941. But they had no land connection with the rest of the continent. Alaska was supplied by sea (and to lesser extent by air) from the south. Early in February 1942 the Committee of Public Roads in the U.S. House of Representatives called for the construction of a highway.

By mid-March Canada had agreed to arrangements under which Americans could construct a defense project on Canadian soil. Suddenly the northwest frontier was abuzz. American troops and civilians arrived with enormous quantities of heavy equipment—tractors, blade graders, steam shovels, rock crushers, bulldozers, and over 5,000 dump- and other trucks "in a region where there were hardly that many people" (Coates and Morrison 1992a, 41). In Dawson Creek, Fort St. John, Whitehorse, Big Delta, and elsewhere, an "army of occupation" began to build a road and engineer the transformation of the north.

The pace of construction was breathtaking. Some 1,500 miles of rough pioneer road (suitable for heavy trucks moving at low speeds) were pronounced complete in less than eight months. The road passed through a good deal of difficult terrain, where climate and surface conditions posed severe challenges. Typically a route was blazed by a locating party before a large bulldozer cleared a narrow path along this line. Then other large bulldozers widened the road to 60 or 90 feet, working laterally to push back the trees; ten or twelve machines engaged in this task would move through two or three miles of forest a day. They were followed by smaller bulldozers that scraped away moss and other debris before another contingent of men and equipment built culverts and ditches, laid down corduroy (logs set transversely in wet ground), and rough-graded the surface. A final pass, by yet another crew, graveled and leveled and smoothed the road. All of this produced a right-of-way some 18 to 24 feet wide.

The Alcan (Alaska) Highway. *This highway was built with astonishing speed through forbidding country (much of it in Canada) for reasons of American security during World War II. Here a bridge across a tributary of the Peace River suggests the magnitude of the project and the nature of the terrain through which it passed. (Library of Congress)*

But this reckoned without northern conditions. Large parts of the road broke up and washed out in the spring thaw of 1943. The U.S. Public Roads Administration (PRA) and thousands of civilians, both American and Canadian, working for civilian contractors, then rebuilt, rerouted, and upgraded the pioneer road. Late in 1944 the PRA had spent over $130 million on the Alaska Highway, which still fell short of the planned gravel-surfaced and well-graded road with two lanes, timber bridges, and proper culverts along its entire length.

Still, the new road rolled back the frontier as it hauled once-tiny fur-trading posts "into the maelstrom of the mid twentieth century." The telegraph line that paralleled the road worked similar magic. Formerly isolated settlements suddenly had access to larger centers and news of the world at their fingertips. Others places, bypassed by construction, languished. Dawson City, the bustling center of the Klondike gold rush and long the most important Canadian town northwest of Edmonton, was 200 miles from the Alaska Highway.

When the American army made Whitehorse the center of its Yukon operations, Dawson City's fate was sealed. People, then businesses, then (early in the 1950s) the territorial administration moved to Whitehorse. The former capital languished in decline until millions of federal dollars were spent on its reincarnation as a tourist destination. In Dawson Creek, the British Columbia town at which the Alaska Highway began, by contrast, the tax base more than doubled as the population skyrocketed from about 700 in 1939 to over 5,000 in 1944. Although the highway was described, disparagingly in 1947, as "the most publicized and the least used road in the entire world," it served, in time, as the backbone for the extension of a road network that enhanced movement through and access to many parts of this far corner of the continent and played a major role in the transformation of this wild northern fastness (Coates and Morrison 1992a, 167, 223).

The social and environmental effects of highway construction were enormous. The sudden influx of soldiers and civilians from the south brought new diseases among the indigenous people of the region. Influenza, measles, dysentery, whooping cough, and mumps ran through native communities and took their toll of lives as they had in earlier "virgin soil" epidemics elsewhere. The Teslin and Lower Post bands were said to be "distressingly affected by the new contacts" (Coates and Morrison 1992a, 78); at one point 128 of 135 Teslin residents suffered from the measles. In the scramble to build the road, heavy equipment tore the overburden off the permafrost, causing it to melt. Fires that had started accidentally by carelessly discarded cigarettes, by the escape of smoke-fires intended to keep off mosquitoes, and by other means sped through slow-growing forests, destroying thousands of acres of vegetation. More trees fell for firewood, building construction, and telegraph poles. Waste was dumped in streams, and fuel oil spills were not uncommon. Fish and wildlife also suffered. Construction activity disturbed stream beds and banks and drove animals away from the road; work crews with few recreational opportunities hunted and fished with abandon.

Still, the British Columbia Game Commission did not consider the "killing to be excessive" (Coates and Morrison 1992a, 94, 70–101). Nonetheless, the designation of a large game preserve in southeastern Alaska led the Canadian Government to set aside the Kluane area, between the international boundary and the Alsek River, in 1942 to protect bighorn sheep, mountain goats, and other large animals. Officially, this meant that native people were excluded from a traditional hunting territory, but here as with other strictures intended to limit the aesthetic effects of forest exploitation, regulations were difficult to enforce and largely disregarded.

By and large, the campaign to build a highway through this northern realm began "in a general state of ecological unconsciousness, in which planners knew little about the area they were 'developing,' and gave scant thought to the long-term environmental impact of their activities" (Coates and Morrison 1992a, 85). By 1943, however, scientists impelled by the challenges and concerned about the consequences of northern development were conducting surveys of soil conditions, vegetation, and animal, fish, and bird populations. Systematic inquiry into the particularities of permafrost did not begin until after the war was over. By then, biological studies had revealed the fragility of northern ecosystems and given the lie to perceptions that the region was a wildlife cornucopia. New restrictions on hunting and fishing were implemented in Yukon and Alaska. Indigenous peoples lost access to traditional sources of subsistence, and in the Yukon as elsewhere across the Canadian north, they were drawn, increasingly, into the orbit of national (and provincial) social, educational, and resettlement programs, the reach of which was greatly facilitated by the new highways and airfields that served as conduits of modernization through the Arctic and sub-Arctic.

At the other end of the country, a very different, but equally consequential, transportation artery was developed in the 1950s. A conjunction of economic and strategic needs with the powerful promise of brute-force technologies and the full bloom of high-modernist confidence in the human capacity to transform nature brought Canada and the United States together in a project of hitherto unrivaled magnitude—the building of the St. Lawrence Seaway. Despite the arrowlike appearance of the St. Lawrence River on maps of the continental northeast, the natural channel of the river is interrupted by falls and rapids and elsewhere follows a sinuous line between islands, banks, and shallows. Large vessels moved slowly along this course and required highly skilled pilots. Canals, most notably the Lachine and Welland canals, had long carried shipping around otherwise impassable sections of the river, but despite the widening, deepening, and lengthening of the Welland Canal locks (to 80 feet, 25 feet, and 859 feet) in 1932, these works were as small beans to the Seaway project, which entailed the opening of a deep waterway into the heart of the continent. An American publication of 1957 rather excitedly described the magnitude of the endeavor in midflight (construction began in 1954 and the Seaway was opened in 1959):

> It has taken 15,000 men to perform the labor, the basic digging, dredging, hauling and building that ended with the taming of the St. Lawrence River. They have used hundreds of machines worth sixty million dollars, including, in round numbers, 500 heavy trucks, 250 bulldozers, 150 of the biggest shovels and draglines, and 15 dredges. They have excavated 200,000,000 cubic yards of earth and rock, and impacted 10,000,000 cubic yards in the form of dikes,

around 20 miles of them, some more than 50 feet high. They have built dozens
of cofferdams . . . . They have cut channels, removed islands, filled in points of
access, laid down roads, set up bridges, [and] relocated everything from telegraph poles to towns . . . . (Thomas 1957, 13)

Other accounts give different figures—some claim fifty miles of dike, for
example—but all leave no doubt about the sheer magnitude of the enterprise.

Several islands were flooded. Others were sliced apart. Near Cornwall, in
the International Rapids section of the Seaway where the river falls 92 feet in
less than fifty miles and engineers confronted a difficult challenge of damming
and canalizing, they simply "ordered everything removed that stood in the way
of a channel with a minimum width of 600 feet," and drove a straight ten-mile
path overland. This required excavation of about 36 million cubic yards of material. Long Sault Island, a "landmark ever since man first came upon the
river," simply "had to go," and was broken up to the point that "no-one would
ever recognize the few fragments" that remained after "one of the great excavation projects of the scheme" (Thomas 1957, 21). Construction of the Snell and
Eisenhower locks along the ten-mile canal known as the Wiley-Dondero Ship
Channel entailed pouring over a million cubic yards of concrete. Together they
dealt with most of the elevation change between Prescott and Cornwall.

The associated dams backed up the river for almost thirty miles, creating a
hundred-square-mile lake that submerged the Long Sault Rapids. The Long
Sault Dam that helps control this section of the St. Lawrence is a massive con-
vex arc, 2,960 feet long and 114 feet high, built with 12 million pounds of steel
and 650,000 cubic yards of concrete. Upriver, the Iroquois Dam was even larger.
These works, claimed the President of the St. Lawrence Seaway, marked "the
beginning of a series of man-made alterations to nature . . . which were to become famous throughout the world" (Chevrier 1959, 94–95).

In the end, some 28,000 acres of land (20,000 of them in Canada) were inundated by the reservoir above the Long Sault Dam. Recounting this, author and
broadcaster Lowell Thomas voiced the confident mantra of high modernism as
he simultaneously minimized and validated the consequences of human enterprise. Only some 8,500 acres were needed for the "St. Lawrence flowage basin."
To be sure, all of this had to be cleared of brush and trees. This might sound
rather drastic if one thought "of the changes wrought in the primeval natural
beauty of the St. Lawrence, and of the dispossession of the creatures of the wild
native to the area." But the proper focus was different. The "authorities were
careful to salvage as much as they could in the way of timber." They were
"careful to maintain the scenery as far as possible." And besides, these changes
were a "matter of modern progress and international prosperity" and none who

believed in these things could deny that "the correct decision was made" (Thomas 1957, n.p.). In its entirety, after all, the Seaway allowed vessels up to 750 feet in length, with beam and draft of 75 feet and 27 feet respectively, to travel from Montréal clear through to Duluth, at an elevation almost 400 feet above sea level.

For those who shared this perspective, the Seaway warranted its costs. In 1959, the first season of its operation, traffic on the St. Lawrence increased by two-thirds; over 6,500 ships used the Seaway; Toronto freight volumes were up 150 percent; in Hamilton overseas shipping increased 700 percent; and Kenosha and Duluth–Superior shipping increased twenty- and seventy-five-fold. The multiplier effects were enormous. By one account, from May 1958, American and Canadian ports were investing $100 million in harbor improvements in anticipation of the Seaway opening. The cost of shipping an enormous range of goods—from China clay to marble, fruit to wine, and pulpwood to sugar on the import side of the ledger, and from synthetic rubber to soybeans, and from automobiles to frozen meat in the export columns—fell considerably. This provided an enormous fillip to agriculture, manufacturing, and consumption in the Great Lakes Basin, and changed both environmental circumstances and patterns of land use as the region's producers expanded and specialized production to supply worldwide markets, and its consumers drew many of their needs from ever further afield. Among the major industrial developments located alongside the new transportation artery were the Reynolds Metals (aluminum) and General Motors fabricating plants near Massena in the United States, a steel mill between Montréal and Sorel, a sugar refinery in Toronto, and a widespread expansion of steel manufacturing and shipbuilding capacity.

For all that, trade on the Seaway was dominated by two commodities. Grain from the American Midwest and the Canadian prairies accounted for over half of the traffic moving downstream in the years after 1959. The St. Lawrence drew trade away from the Hudson and Mississippi corridors and would have provided a sterner challenge to rail shipments of grain westward to Vancouver and Prince Rupert were it not for winter ice on the St. Lawrence and long-standing rail-freight arrangements that kept the costs of overland movement to Pacific ports artificially low until the so-called Crow Rate was replaced in 1984. Iron ore dominated upriver traffic. Declining reserves of iron ore in the Mesabi Ranges of Minnesota had provided a major impetus to Seaway construction. When rich deposits of iron ore in the remote Ungava District of the Labrador-Québéc border region proved up, the economic and strategic synergies between the ore fields and the Seaway proposal became clear.

A conglomerate of American mining and steel interests joined with the Canadian Hollinger Company to form the Iron Ore Company of Canada in exploiting the new reserves. They built a railroad from Sept-Iles on the St. Lawrence to Knob Lake, 575 kilometers inland, and there erected the town of Schefferville in 1953 as the center of mining operations. A few years later Québec Cartier Mining opened a new mine on Lac Jeannine, a new seaport at Port-Cartier, and a new 310-kilometer railroad between them. By 1959 iron ore shipments from this part of Québec were worth $92 million. Most of this moved along the Seaway to supply new and established demand in the Great Lakes region. With reserves estimated at 420 million tons, the Iron Ore Company of Canada was soon extracting and shipping about 8 million tons a year. In 1979 the output was valued at more than $280 million. In 1983 over 45 million tons of cargo moved through the lower reaches of the Seaway. This was four times the annual average for the St. Lawrence in the 1950s.

These are compelling numbers upon which to build an argument in favor of the Seaway. But they tell only a part of the story. Not all communities benefited equally. As ever, reductions in the tyranny of distance shifted the horizons of effective interaction, broke down local monopolies, changed the parameters of economic competition, and initiated a long chain of environmental and societal consequences. In Ontario, long-established grain-handling ports on Georgian Bay and Lake Huron (Midland, Collingwood, Owen Sound, and Goderich) declined as they were bypassed by vessels destined for new elevators and the expanded port facilities of Montréal. As jobs disappeared and local economies contracted, local producers struggled with shrinking markets for their produce. Farmers variously turned to new forms of production, abandoned fields to forest, and disposed of their properties to urbanites in search of country properties.

And many of the gains that flowed from the Seaway did so only temporarily. In 1983 a worldwide surplus of iron ore and the decline of heavy manufacturing in the Great Lakes "rust belt" led to the cessation of mining in Schefferville and the near-complete shutdown of the town (High 2003). Railway interests and those in charge of east coast ocean ports resented what they considered the unfair subsidy offered users of the Seaway by massive public investment in the project. Bottlenecks and changing shipping requirements also required continuing investments in the project, and in 1977 crippling interest and operating costs necessitated a restructuring of the Seaway Authority's debt into equity held by Canada.

Moreover, the inundation of several thousand acres is not justly counted as merely "the price of progress." Many of those acres were farms. They contained fields on which generations had bestowed labor, homes to which people had grown attached, trees that they had planted and nurtured. Other acres included villages and churches, and graveyards in which people had laid parents, grandparents, spouses, partners, and children to rest. They were not, in other words, just simple acres, any one of which might be valued according to the metric of the marketplace or substituted for another elsewhere. They were spaces in which people had invested themselves to create places full of associations and memories. These could not be relocated or transplanted easily—if at all. Up and down the St. Lawrence, planners, engineers, "authorities," faced the challenge of "moving people out of the way" of their grand development scheme (Mabee 1961, 204). In all 1,100 permanent residents faced relocation in New York, about 1,500 in Québec, and some 6,500 in Ontario. In addition, more than a thousand families faced the loss of summer cottages. Many were deeply unhappy at the prospects. In Lisbon, New York, "farmers barred surveyors from their fields . . . in Iroquois, Ontario, farmers ordered bulldozers off their land. One old lady kept Seaway agents out of her lamp-lit stone house with a gun" (Mabee 1961, 204). But most accepted the inevitable—and such compensation as they were offered.

Three entirely new settlements (Iroquois, Ingleside, and Long Sault) were developed to replace towns and villages flooded by the Seaway, and in Morrisburg new development replaced submerged sections of the town. In each of these locales, standard modern dwellings were built in greenfield sites, and in Morrisburg virtually the entire business district was relocated. When some homeowners expressed a strong preference for their old houses over the new structures, arrangements were made to relocate some 525 threatened dwellings. Even this action turned into an advertisement for the capability of engineering technology. With TV cameras rolling, old houses were lifted by giant machines, transported a mile or so, set down on new foundations and connected to utility networks within an hour. Astonishingly, families who had just finished breakfast when the house movers arrived took morning tea in their relocated kitchens as though nothing had changed. Could there be a more telling demonstration of modernity's capacity to obliterate space and time?

On the ground, rather than the TV screen, it was much harder to believe that the balms of planning and technology relieved all the strains and bruises of change. Abandoned dwellings were torn apart, pulled down, and set to the torch. Smoke swelled across the sky. Fences were flattened and telephone poles were toppled. Mills, barns, and churches were laid low. Archeological sites, occupied thousands of years earlier, were flooded. Roads and rail lines were rerouted through fields and farms, and new bridges, underpasses, side roads, and access lanes were constructed to serve them. Where those who lived along the river had once heard the sounds of water tumbling through rapids, chainsaws and bulldozers screamed and growled, and when the work was done and the water rose behind the dams, only the eerie silence of slack water remained. The past was, almost literally, "shaved off the landscape." Watching the waters rise across familiar fields and paths between Dominion Day and Independence Day 1958, one longtime resident of the Cornwall area encapsulated the sentiments of dozens when he turned and said to his tearful wife: "There goes our youth" (Mabee 1961, 220, 222; see also Parr 2001, 2004).

Other environmental impacts of the Seaway development were less immediately apparent. Thirty years after rising waters altered the visual and sensory environment near Cornwall, zebra mussels were discovered in Lake St. Clair (between Lakes Huron and Erie). This temperate freshwater species (*Dreissena polymorpha*) is indigenous to eastern Europe. In 1988 it was new to North American waters. By all accounts it likely reached the Great Lakes in larval form, carried across the Atlantic in 1985 or 1986 by a vessel that emptied its ballast tanks of water taken aboard in Europe. Since then, the mussels have spread to many other lakes and rivers in northeastern America. The larvae (known as veligers) are nearly invisible but they grow quickly and reproduce aggressively (a single adult female can produce as many as 100,000 eggs a year). Attaching themselves securely to hard surfaces (such as rock, glass, fibreglass, metal, and native mussels) they accumulate rapidly and often clog "water-intake pipes, and screens of drinking water facilities, industrial facilities, power generating plants, golf course irrigation pipes, cooling systems of boat engines and boat hulls," producing inconvenience and incurring costs as plants and facilities have to be shut down (Zebra Mussels 2006).

In addition, zebra mussels filter large quantities of phytoplankton and small zooplankton from the water. This has produced remarkable improvement in water quality, especially in Lake Erie, but it has also radically reduced the food supply for juvenile fish and other native species. Small, shrimplike invertebrates, known as Diporeia, present in the Great Lakes since the Ice Age, have disappeared, native clam and mussel populations have declined dramatically, and the quality of lake whitefish, an important commercial species for which Diporiea was a primary food, has fallen. Moreover, zebra mussels are startlingly efficient accumulators of such organic pollutants as polychlorinated biphenyls (PCBs) and polycyclic aromatic hydrocarbons (PAHs). Concentrations 300,000 times greater than those in the ambient environment have been found in some zebra mussels. These concentrated organic compounds are discharged as psuedofeces, "loose pellets of mucous mixed with particulate matter" that the mussels filter from the water. Eaten by other species, these psuedofeces may hasten the passage of pollutants up the food chain.

The problem of invasive species in the Great Lakes is neither entirely new nor entirely attributable to the opening of the Seaway. The earliest recorded aquatic species invasion of the lakes was in the 1820s, by sea lamprey that entered via the Erie Canal and which were largely responsible for the loss of native lake trout early in the twentieth century. Since the 1820s, over 160 aquatic invaders have established themselves in the Great Lakes. They include fish, invertebrates, plants, algae, and pathogens. Among them are the spiny water flea, the fish hook flea, a bottom dwelling fish known as the goby, Eurasian ruffe, Eurasian watermilfoil, hydrilla, and water hyacinth. By current estimates over 70 percent of the nonindigenous species introductions to the Great Lakes since 1959, including several of the aforementioned, have come from ballast tank discharges (most of the remainder are accounted for by unintentional escapes from aquaculture operations, aquaria, and so on). Of eleven introductions in ballast water since 1986, eight are indigenous to the Black, Caspian, and Azov seas. These include zebra mussels, quagga mussels, round gobies, fish hook waterfleas, and echinogammarus amphipods, which together "constitute a very significant component of the biomass and productivity of the Great Lakes food webs" (AIS 2006).

Without question, the Alcan Highway and the St. Lawrence Seaway were the largest and most spectacular of the transportation projects that reshaped economies, landscapes, livelihoods, and environments (sometimes inadvertently) in the northern part of the American continent in the postwar years. But they were not alone. There were important extensions and improvements to the highway network after 1945. The construction of new road segments between North Bay and Port Arthur–Fort William in 1946 made it possible to drive through Canada from the St. Lawrence to the Pacific by car—although large sections of the route were unpaved and the going was especially difficult through parts of the western Cordillera.

The Trans-Canada Highway Act of 1949 sought to improve the route through a federal-provincial cost-sharing agreement, and much work was done in the 1950s, but when the "highway" was opened in July 1962, 3,000 kilometers remained unpaved. Generally, the best roads were those between major cities. Toronto, Hamilton, and Buffalo had been linked by Canada's first "superhighway," the Queen Elizabeth Way, on the eve of the war, but similar improvements came later elsewhere. In 1946 there were fewer than 20,000 miles of paved road in the entire country. By 1951 only some 5,000 miles had been added. But the total length of paved highway more than doubled in the next decade, to exceed 55,000 miles. The number of motor vehicles increased even more dramatically, from approximately 1.5 million at war's end to 5.5 million by 1961. Over 2 million of these vehicles were registered in Ontario, and over a million were in Québec, but Ontario, British Columbia, Alberta, Saskatchewan, Manitoba, and Prince Edward Island each registered more than 300 vehicles per thousand persons in 1961 (*Historical Atlas of Canada* 3: plate 53).

All of this meant improved mobility, connectivity, and communication. It opened markets, lubricated economic expansion, integrated the urban system, hastened the emergence of dominant metropolitan centers, widened horizons, remade landscapes, and (with improvements in air travel and other forms of communication) transformed sociospatial patterns across the country. It also initiated a steady increase in the amount of "machine space" in Canada—space devoted in this instance to producing, sustaining, using, and storing ever more numerous automobiles and trucks.

22

POWER LINES

T he challenge of providing power to drive economic expansion and extend the modernist project led postwar Canadians to the exploitation of new energy sources and the development of the infrastructure to move energy to its markets. Oil and gas pipelines and the high-voltage transmission lines that carried electricity from generating stations to consumers marched across the country, as the proportion of Canadian energy derived from oil (and gas) and hydroelectricity increased from one-third to over 80 percent of consumption between 1946 and 1961. Wells, refineries, dams, and transformer stations were the nodes of this increasingly extensive network, and its extension and elaboration brought new—and for many years markedly cheap—energy to consumers whose needs varied from the voracious demands of smelters and pulp mills to those of urban dwellers using small quantities of electricity and oil to light and heat a dwelling and power a private automobile.

Taken as a whole, this ramifying grid of power lines was an immensely potent instrument of socioeconomic and environmental transformation. By making power available at the flick of a switch or the surge of a pump, they separated most consumers of power from the circumstances of its production. In essence, cash exchanges (in the form of bills paid to the utility company or dollars handed to gas-station attendants who filled the tanks of thirsty automobiles) largely replaced the physical labor of chopping and splitting wood, hauling coal, stoking the stove, cleaning the fireplace—and the contact with nature that these tasks provided. Electric appliances, from kettles and irons to vacuum cleaners and washing machines, reduced the physical demands of everyday domestic existence. Raw energy was transformed, by the largely mysterious and substantially invisible labor of a few anonymous individuals, into a convenient and usable commodity readily available to most people. It was, in a classic manifestation of the high-modernist manner, abstracted and generalized.

In an effort to address the "fuel problem in Alaska," the American government decided in the fervent disquiet of wartime to bring oil by pipeline across the Richardson Mountains from the Norman Wells field in Canada's Northwest Territories to a refinery in Whitehorse, from whence it would be distributed through other pipelines to Skagway, Watson Lake, and Fairbanks. The difficulties were immense and the project was enormously risky. Production from the Norman Wells field was small, and over eighty new wells were drilled—increasing output far beyond demand. Little was known about building oil pipelines through such terrain as these had to cross. Trees were bulldozed and six-inch steel pipeline was laid on bare ground. Leaks and fractures were not unusual. The remote refinery had to be built from scratch. In the end, the project was an expensive failure. One commentator described it as a "junkyard of military stupidity." Approximately $140 million was invested in the project, but by March 1945 the refinery was closed and the pumping stations along the supply pipeline had been turned off. Production resumed briefly under military auspices, but the Canol venture was a white elephant and shut down for good in 1946 (Coates and Morrison 1992b, 34–37, 61–67).

Petrochemical industries expanded rapidly during the war in Sarnia (Ontario), manufacturing synthetic rubber, and near Calgary (Alberta), using natural gas to produce ammonia. But the discovery of a major oil field at Leduc, south of Edmonton (Alberta), in 1947 was a major stimulus to expansion in this sector. Quickly, the Imperial Oil Company bought the Whitehorse refinery and relocated it to Leduc. An American firm bought the Canol pipeline, dismantled it, and moved it south. Through the next two decades, frenetic development marked the Alberta "oil patch." Initially, exploitation focused on southern parts of the Western Canada Sedimentary Basin, which encompasses the Mackenzie Valley, Alberta, southern Saskatchewan, and southwestern Manitoba. Alberta was by far the most prolific oil producer. In southern Saskatchewan natural gas became the fuel of choice during the 1950s; by 1960 almost three-quarters of nonfarm households there were served by gas. Foreign companies quickly took over most Canadian companies operating in the oil business, and by the early 1970s over half of both the assets and the sales of the industry were accounted for by the seven largest multinational oil corporations. Almost 90 percent of Canadian petroleum revenues accrued to foreign-controlled companies.

With the bulk of Canada's population and industry in the St. Lawrence–Great Lakes lowland, much Alberta oil went initially to U.S. markets in the Midwest and Pacific Northwest, while eastern Canada depended upon oil imports. In the 1950s heated debates over the construction of pipelines to link western oil and gas producers to eastern markets created political turmoil. The Liberal Government decided that a gas pipeline linking western producers with central Canada and running across Canadian territory was a national necessity. In 1956 they forced through Parliament a bill authorizing construction of the Trans-Canada Pipeline, and two years later it operated over 3,700 kilometers between Burstall, Saskatchewan, and Montréal. In the interim, the government fell, at least in part as a consequence of the way in which it handled the pipeline issue.

Taking no chances, the newly elected Conservative Government estab-lished a Royal Commission on Energy to deal, among other things, with the de-mand from Alberta's independent oil producers for an oil pipeline between Ed-monton and Montréal. Large multinational oil corporations were generally opposed, as they profited from supplying eastern refineries with imported oil. Following the Commission's recommendations, a National Energy Board was established to regulate interprovincial pipelines and energy sales, and in 1961 the National Oil Policy gave domestic oil producers a protected market west of the Ottawa River, while allowing imported oil into the five eastern provinces.

This was a considerable boon to an already expanding industry. Early in the 1960s Canadian oil production was four times greater than it had been a decade earlier and gas production was up eightfold. The Inter-Provincial Oil pipeline carried Alberta oil into southern Ontario through the American Midwest and the Trans-Mountain pipeline served the Pacific coast. Refineries served a range of local demands for heating oils, fuels, and asphalt, and the hugely capital-in-tensive petrochemical industries expanded rapidly. In addition, considerable quantities of Canadian oil and gas were sold into American markets; late in the 1950s about one-quarter of the total output went south; and in 1971 Canada ex-ported 57 percent of its oil and almost half of its gas production to the United States.

When the OPEC crisis quadrupled international oil prices in 1973, the fed-eral government froze Canadian oil prices and imposed a large export tax on sales to the United States to secure Canadian supplies for Canadians and pro-vide revenues to lessen the price shock faced by residents of the import-depen-dent eastern provinces. Albertans were outraged. Although prices were allowed to climb through the next several years, they remained some $3 per barrel be-low international levels in 1978. Then world prices spiked again as a result of the revolution in Iran. Troubled by the wealth transfers (from consumers to pro-ducers, from eastern to western Canada, and from Canada to the United States because of the high level of American investment in the oil patch) and threat-ened by this price rise, the government introduced the National Energy Pro-gram (NEP) in 1980. It aimed to increase Canadian ownership in the oil indus-try, to ensure Canada's self-sufficiency in oil, and to secure a larger proportion of energy revenues for federal coffers. Canadian prices were held below world levels, and both oil exploration and reductions in oil consumption were encour-aged. Again, western politicians, American oil executives—at least one of whom attributed his annoyance with the NEP to the fact that "we have never treated Canada as a foreign country"—and many residents of the producing provinces protested federal interventions in the marketplace (Page 1986, 306).

Although the history of these tumultuous decades is generally written around the theme of federal-provincial confrontation, it also has a significant environmental dimension. Through the late 1960s untrammeled growth was the order of the day in the oil and gas sector. Led by private enterprise and encouraged by government, the industry expanded until it accounted for almost three-quarters of Canadian energy consumption by 1970 (a proportion accounted for by coal and coke only fifty years before). Oil-fired stations were built to generate electricity, and many consumers in coal-rich Maritime Canada were encouraged to switch their homes to clean and cheap electrical heating. Pipelines were extended to gather and distribute oil and gas, and early in the 1980s Canada had the world's second-longest pipeline network (almost 200,000 kilometers), second only to the United States. Trucks and automobiles offered convenience and scheduling flexibility, and powered by economical and seemingly abundant gasoline or diesel fuels, they assumed an ever-larger share of the transportation burden across the country. Plastics and other synthetic materials, produced by the country's growing number of petrochemical processing plants, found a thousand uses in everyday life—and in time became a major component of the growing mounds of garbage discarded by a society increasingly given to the use of elaborate packaging and the disposal of old and broken goods.

Through most of the 1950s and 1960s, Canadian energy consumption increased at a rate of approximately 5 percent per annum. Early in the 1970s Canada was among the most highly oil- and gas-dependent countries in the world. And artificially suppressed domestic oil prices did little to change this through the next decade. As elsewhere, the environmental costs accumulated. Although the shift away from leaded gasoline was driven by health concerns, atmospheric pollution, including periodic accumulations of dangerous levels of ground-level ozone in and around major cities, continued with little abatement (despite growing attention from scientists and environmentalists) through the 1980s. Tires and "wrecked" automobiles proved difficult to dispose of and accumulated in vast quantities as the years went by.

Petroleum's share of Canadian energy consumption fell slightly during the 1970s, to just over 50 percent, but overall energy use climbed—albeit at a lower annual rate than in the 1960s—and real consumption increased by well over 20 percent. Oil consumption did fall during the 1980s, as the NEP encouraged a shift to more efficient vehicles, improvements in home heating systems, and a move to other sources of fuel. Still, demand remained voracious, and known conventional reserves were projected to depletion within a few decades. Efforts to find new sources of oil and gas carried the industry into new and difficult environments as energy prices rose and new incentives were offered for exploration. In Alberta attention turned to huge deposits of bitumen found in "oil sands," which can be processed to yield so-called synthetic oil. In eastern Canada, exploration of the continental shelf offshore from Newfoundland and Nova Scotia led to the opening of successful wells, although the expense of developing and utilizing ocean-drilling platforms is high and the dangers of working in these waters were tragically realized with the collapse of the Ocean Ranger oil rig and the loss of eighty-four lives during a storm on February 14–15 1982.

The Arctic slope of Alaska and the Beaufort Sea north of Yukon and the Mackenzie River delta were also important foci of oil and gas exploration and development. In 1968 the Atlantic Richfield Company discovered enormous oil and gas reserves in Prudhoe Bay, on the north slope of Alaska. Estimates of the amount of oil available ranged from nine to fifteen billion barrels, far, far in excess of the billion barrels generally regarded as necessary for exploitation in these northern realms. This was North America's largest known oil field. How were these reserves to be secured and how were they to be brought to market in the conterminous United States? These were difficult questions. By remarkable fortune, the Prudhoe Bay discoveries lay between an extensive government reserve to the west of Colville River, established in 1923 as a Naval Petroleum Reserve, and some nineteen million acres in the northeastern corner of Alaska, set aside in 1960 (after several years of lobbying by wilderness and wildlife groups) as the 8.9-million-acre Arctic National Wildlife Range (later doubled in size and reclassified as a Refuge [ANWR]). This seemed to open the way to the development that most Alaskans craved. But native land claims remained unsettled, even after Alaska's admission to statehood in 1959, and environmental interests argued for protection of wilderness and northern ecosystems beyond the ANWR. Both groups seemed, to resident newcomers, to be strangling the potential of the new frontier.

To settle questions of title to the oil and gas reserves, to clear legal obstacles to pipeline construction, and to address the appalling economic and social disadvantages endured by indigenous Alaskans, the U.S. government transferred (by the Alaska Native Claims Settlement Act of 1971) 40 million acres of the state, and committed payments of $925.5 million over twelve years, to Alaska's 70,000 native people. In return, native Alaskans gave up all claims to the rest of the land and to oil, natural gas, and mineral rights. This was a monumental agreement. According to the state's preeminent historian, it "framed and influenced" virtually "every aspect of Alaska's history" after 1971 (Haycox 2002b, 100).

It financed an economic boom, because the expansion of the oil industry provided tens of thousands of jobs and good wages; oil-taxation revenues underwrote the activities of the Alaskan state; and federal funds poured into the region. It also engineered a "civil rights revolution" by giving native peoples the bases of economic and political parity, and thus near social and racial equality, with other residents of Alaska. Third, in allowing the U.S. Secretary of the Interior to designate up to eighty million acres of Alaska for environmental conservation, it had a profound impact upon the course of Alaskan development. Reflecting the lobbying efforts of a dozen prominent American conservation groups, from the Sierra Club through Zero Population Growth and the National Rifle Association to Friends of the Earth, that came together as the Alaska Coalition (later the Alaska Public Interest Coalition), it brought Alaska into the frame of national discussion. Many coalition members believed that Americans should be more concerned to protect Alaska's wilderness than to use its oil. In their view, the mountains and valleys, and the tundra and forests of the north were far more valuable than the fraction (perhaps 20 percent) of American oil consumption that might be drawn from Prudhoe Bay: "Oil, they argued, was ephemeral; wilderness was forever, until sullied by human imprint . . ." (Haycox 2002b, 103).

In a climate of considerable uncertainty, marked by the environmentally disastrous failure of a badly built supply road bulldozed over 400 miles of permafrost between Fairbanks and Prudhoe Bay in 1969 and a challenge to plans for the development of a trans-Alaska Pipeline System (or Alyeska Pipeline) under the National Environmental Policy Act, the Canadian Government mooted the possibility of constructing an oil pipeline south from Alaska through Canadian territory. There were obvious potential economic benefits to Canada in this, but the government was also concerned about the risk of shipping accidents and oil spills on the coast of British Columbia. Despite a series of high-profile oil spills (including Santa Barbara, from offshore wells; the *Torrey Canyon*, which foundered off the coast of southwest England; and the *Arrow*, which went down off the coast of Nova Scotia in 1970) and continuing opposition from environmental groups, U.S. interests insisted that it was cheaper to ship oil by tanker than to pump it through a pipeline. Washington first explored the idea of sending supertankers through the northwest passage into Prudhoe Bay (thus laying implicit challenge to Canada's claimed sovereignty over northern waters) and then opted to construct a trans-Alaska pipeline (from Prudhoe Bay to Valdez) to facilitate tanker shipments from Alaska to California.

The 800-mile-long Alyeska Pipeline project was beset with difficulties. Legal challenges might have held up its approval even longer had not the OPEC oil embargo of 1973 led to severe gasoline shortages and persuaded Americans

Alaska pipeline. *The Alyeska Pipeline, from Prudhoe Bay to Valdez, was a vastly expensive engineering triumph over a long series of environmental and logistical difficulties. Snaking across frozen ground (permafrost), the elevated pipeline brings north slope oil to year-round shipping lanes for movement into markets in the conterminous United States. (Corel)*

of their vital need for Alaskan oil. Still, the technical challenges of construction, logistical and organizational hitches, and extreme weather conditions meant that the pipeline was not opened until 1977 and that it cost almost nine times the sizable $900 million projected in 1969. By that time, increased oil royalties levied by Alaska (which generated almost three-quarters of its funds from oil and gas revenues) had reduced profit levels. Although they stood at 9 percent, this was considered inadequate on the $14 billion plus invested in exploration, exploitation, and pipeline construction. Moreover, a glut of oil in the California market meant that some Prudhoe Bay oil had to be shipped through the Panama Canal in smaller vessels at reduced rates of return. For all that, twenty-four oil fields linked by pipelines and by 600 miles of roads and trails were exploited in the Prudhoe Bay area in 2003.

The Prudhoe Bay discoveries triggered a frenzy of exploration in Canadian sections of the Beaufort Sea. Initially, drilling proceeded from artificial islands created by dredging up bottom material and building a platform that would become stable in winter ice. Then, in 1976, reinforced drill ships were deployed for ten to twelve weeks in the summer—but the short season meant that at least two years were required to develop a test well. Finally, strong steel or concrete structures were built to serve as caissons that could be filled and sunk onto human-engineered undersea berms to provide a platform capable of withstanding early winter ice pressures. Each of these efforts was a triumph of human engineering over exceedingly difficult environmental circumstances. All quickly made evident the costs and difficulties of commercial exploitation of remote northern energy reserves. Accidental gas discharges from some of these endeavors led the U.S. State Department to protest that an oil strike followed by a blowout would foul the shores of Alaska. Anticipating such problems, the Canadian Government proclaimed the Arctic Waters Pollution Prevention Act (as well as new guidelines for the construction of northern pipelines) in 1972. Prime Minister Pierre Trudeau described the former as "an assertion of the importance of the environment, [and] of the sanctity of life on this planet . . ." (Page 1986, 294).

Meanwhile, the question of Prudhoe Bay gas occupied center stage. The economics of transporting gas are quite different from those pertaining to oil. Shipping gas requires that it be chilled into liquid form. This is far more expensive than moving it in a gaseous state through high-pressure pipelines. American and Canadian interests formulated plans for the development of pipelines on different routes out of the north to link with existing networks and southern markets. The most enduring of these proposals, for a 48-inch high-pressure pipeline from Prudhoe Bay to the Mackenzie Delta and south through the Mackenzie Valley, was designed to serve markets in both the American Midwest and California, and was touted as possibly "the largest project privately financed in the history of free enterprise capitalism" (Page 1986, 1). In December 1973, with oil prices rising and hopes of further oil and gas discoveries in the north running high, the Prime Minister indicated that "it would be in the public interest to facilitate early construction" of a northern pipeline. But well aware of the rising political voice of native people in the north and the growing concern over environmental pollution among residents of the south, he added the caveat that this should be done only by "means which do not require the lowering of environmental standards or the neglect of Indian rights and interests" (Page 1986, 83).

Three months later, in March 1974, Thomas Berger, a British Columbia lawyer, was appointed to head the Mackenzie Valley Pipeline Inquiry and given a mandate to investigate the potential impacts of pipeline development. Opposition was immediate. Pipeline interests feared that delays generated by commission hearings would favor development of an alternative delivery system, the so-called El Paso project, entailing the use of liquid natural gas (LNG) tankers sailing the west coast. Native groups disputed the legality of the inquiry, refusing to accept that their territories were Crown Lands and thus questioning the government's right to determine whether pipeline construction should go ahead.

For all that, Berger was able to find a way forward. From the outset, he insisted on the broad mandate of his inquiry—"I take no narrow view of my terms of reference," he said in October 1974—and made it clear that his efforts would move beyond specific consideration of the Arctic Gas Pipeline application to consider "the whole future of the north" (quoted in Page 1986, 101–102). This meant, among other things, that the commission would hear evidence about traditional patterns of indigenous land use and thus by implication about native land claims. After several months of hearings conducted through dozens of small northern communities, hearings in which residents of all socioeconomic groups and dozens of northern communities were encouraged to speak, the commission concluded its work and produced its report in May 1977. Soon it became common to claim that this document was a landmark in northern development and in the evolution of environmental concern in Canada.

The central recommendations from the Berger Inquiry were that pipelines should not be built across the north slope of Yukon Territory because of the fragility of the environment there, and that there should be a ten-year moratorium on pipeline construction in the Mackenzie basin to allow the settlement of indigenous land claims. The calving grounds of the northern caribou, the traditional economies of native peoples, and the claims of those peoples to land, Berger insisted, were ultimately much more important to Canadians than immediate access to northern gas. Such conclusions had revolutionary implications. First, and in essence, they gave environmental concerns a veto over development. "There is a myth," Berger asserted in words that were an arrow into the heart of prevailing attitudes and practices, "that terms and conditions that will protect the environment can be imposed, no matter how large a project is proposed" (Berger 1977, 1:xi). In the delicate environment of the North, it was wrong to believe that appropriate efforts to mitigate impacts could allow development to proceed. Second, Berger's conclusions stressed the rights of native peoples to self-determination and offered a direct criticism of the long-standing inclination to impose alien political models and expectations upon them.

Reactions were immediate, and criticisms came from many quarters. The mayor of Calgary was particularly outspoken. Contemplating the loss of economic returns to his city and Alberta more generally, he denigrated the inquiry as a "pooling of ignorance" that had provided "a platform for troublemakers" and wondered about the wisdom of efforts to protect the traditional ways of those who endured a "primitive life of insecurity and hardship." The north belonged to Canada and this meant, in his view, that it should be open to development in the interest of all Canadians. Moreover, Berger's recommendations were a slap in the face of "private enterprise in a country which depends upon private enterprise for survival." The Arctic Gas Consortium (which one observer noted always "seemed intent on confronting nature rather than adapting to the physical forces at work") was affronted by the commission's insistence that corporate interests should take a backseat to social and environmental concerns (Page 1986, 162). The assembly of the Northwest Territories voted against the recommendations. Northern Metis communities opined that implementation of the report would doom them to "a welfare economy for a long time." And Alberta Member of Parliament Jack Horner blustered, "If Berger had been around a hundred years ago, we would still have the buffalo herds in the West and the CPR wouldn't be built" (Page 1986, 120).

Many—southern Canadians, members of environmental interest groups, nationalists who disliked the overwhelming American financial interest in the Arctic Gas Consortium, those sympathetic to indigenous rights—hailed the report as a breakthrough. Yet others, cognizant of their society's dependence upon oil and gas and aware of shrinking southern Canadian reserves of both sources of energy, wondered (to paraphrase a telling observation by journalist Richard Gwyn, [quoted in Page 1986, 120]) whether Canadians could afford Berger's ideals. Northern development, as the title of Robert Page's book on these matters has it, framed the Canadian dilemma.

In the end, the outcome was a Canadian compromise. There would be no gas pipeline (at least in the twentieth century) from Prudhoe Bay across the Mackenzie Delta and down the Mackenzie Valley. But the National Energy Board gave conditional approval for the development of a pipeline along the Alaska Highway corridor in 1977 (a project stillborn as a consequence of weakening oil prices and the difficulties of financing such megaprojects under the fiscal arrangements prevalent in the 1980s). Resolve also weakened over time. By 1985 there were plans to construct port facilities on the Yukon coast to service exploration offshore. The Berger Commission's ten-year moratorium was also eroded. The land claims process unfolded very slowly and uncertainly after 1977.

Then, in the early 1980s, Imperial Oil proposed to increase production from its Norman Wells field and to carry the output to southern markets through a new twelve-inch Interprovincial Pipeline. After citing Berger's insistence upon the importance of settling indigenous land claims, the National Energy Board approved the project. The Dene, in whose traditional territory these developments would occur, objected. The Federal Government mediated a settlement that delayed the start of construction for two years and offered a package of economic inducements (from jobs and retraining programs to start-up funds for native entrepreneurs) to northern, native, and Metis interests. Progress toward a land claim settlement quickened, the Dene people softened their formerly intransigent attitude toward nonrenewable resource development and the free-enterprise economy, and in April 1985 crude oil began to flow down the Norman Wells Oil Pipeline to Zama Lake, Alberta. Still, native people doubted the wisdom of these concessions. "[We] are concerned that all is not well with the environment and the processes involved in environmental management," wrote Dene president Stephen Kakfwi to the Toronto *Globe and Mail* on May 15, 1985. "We have concerns with contaminated fish downstream of the oil-field development, inadequate water quality standards and insufficient and untested oil-spill contingency planning" (see also Notzke 1994). For all its radical achievement in placing "environmental and social values on an equal plane with . . . economic considerations," no "significant policy or process changes" can be attributed directly to the Berger Report. It was a rhetorical instrument. It sharpened awareness of the issues involved in northern development, but it did not succeed in creating "a new environmental regime in the North" (Page 1986, 319).

In Alaska, debate over the set-aside of conservation lands continued, and heightened after Jimmy Carter, who had campaigned in favor of conservation in the northern state, was elected president. When a bill introduced in the U.S. House of Representatives in 1977 called for the designation of 115 million acres, "the crown jewels" of Alaska, as conservation reserves, local opposition grew both vociferous and hostile. Sixty percent of American mineral wealth lay in Alaska, said a former governor of the state, and the nation's development would be held back if environmentalists succeeded in defining large areas as conservation lands and wilderness: "We can't," he said, "just let nature run wild!" (Haycox 2002b, 110, 112). The bill received overwhelming support in the House but died in the Senate in the face of opposition from Alaska's senators, whereupon the secretary of the interior and the president exercised their authority under existing acts to withdraw over 150 million acres. Political machinations continued but culminated with the signing of the Alaska National Interest Lands Conservation Act (ANILCA) in 1980, shortly before the Carter administration was succeeded by Ronald Reagan and a Republican majority in the Senate.

The ANILCA was a compromise, and many thought it remarkable. Slightly more than 100 million acres of Alaska were designated conservation units; over half of these (56.4 million) were described as wilderness. Ten new national parks were established and three were enlarged. Some 1.3 million acres were added to the Tongass Forest, and 5.4 million acres of that forest were designated wilderness. The south flank of the Brooks Range was protected from mining

Alaska: Arctic National Wildlife Refuge. *Declining domestic production within the United States and rising world prices for oil have brought strong pressures for oil field development in the spectacular Arctic National Wildlife Refuge. Environmentalists have (thus far successfully) resisted efforts to open the area to oil exploration, but debate continues, political winds shift, and polls taken in late 2005 suggested increasing support among the American public for expanded drilling on the north slope of Alaska. (U.S. Fish and Wildlife Service)*

and other forms of exploitation. The Arctic National Wildlife Range was doubled in size and declared a Refuge. But closer scrutiny revealed that parks and refuges in Alaska would be open to uses prohibited in similar areas in the conterminous United States, that snowmobiles and float planes would be allowed into "wilderness" areas, that many boundaries fragmented rather than encompassed ecosystems, that prospecting was allowed in many designated areas, and that boundaries had been drawn to leave land with economic potential beyond conservation units.

Within the Tongass Forest, large subsidies toward the construction of logging roads and other infrastructure were paid to offset the designation of part of the area as wilderness, and the annual cut was increased by a third over levels that had prevailed through the previous decades to supply the pulp mills, in

Ketchikan and Sitka, operated under the provisions of the Tongass Timber Act of 1947. None of this did much to restrain or deflate the "bonanza" mentality, fueled by revenues from rising oil prices in the 1970s and early 1980s. For a "golden decade," expansion and development proceeded, countless building and construction projects were initiated, and both the state and large parts of its environment were substantially transformed in the drive to establish a modern, materialistic American consumer culture in the north.

Yet resource booms all too often end in busts. Plunging oil prices in the late 1980s almost brought the Alaskan economy to its knees. The state population declined by 60,000 people (15 percent) between 1985 and 1989. At the same time battles over the environment were rejoined. Led by the Sierra Club, environmental and conservation groups took a stand against the exploitation of "old growth" in the Tongass Forest. The struggle mirrored countless others across northern North America in the last two decades of the twentieth century. A remote locale was made the focus of national attention through the publication of a stunning picture book. Scientific reports pointed to environmental abuses, in this case to logging practices that destroyed salmon habitat and affected deer populations, and to the pollution of coastal waters by pulp-mill effluent.

Foresters under attack argued in turn that their scientific management practices were careful and benign. Loggers claimed that they were the "real environmentalists" as a result of their daily work in nature. Families dependent on the industry asserted their rights to jobs and lives in the communities that were their homes. And journalists shaped public opinion with telling "sound bites" and striking turns of phrase, such as the observation by a Seattle writer for the *New York Times* that 500-year-old Tongass trees were being sold, by the U.S. Forest Service, for the price of a hamburger (quoted in Haycox 2002b, 137). In 1990 The Tongass Timber Reform Act went some way to meet environmentalists' concerns by ending subsidies to facilitate extraction, removing the harvest "target," directing the Forest Service to sell logs profitably, and designating 300,000 acres of old growth as protected wilderness.

A year or so earlier, on March 24, 1989, in an event of no little consequence to public concern about the environment of Alaska, the *Exxon Valdez*, carrying 53 million gallons of Alaska crude, went aground on Bligh Reef in Prince William Sound. Nearly 11 million gallons of oil were spilled into the ocean. This was the largest oil spill recorded in the United States (and the thirty-fourth largest in the world). Containment and cleanup efforts were hampered by remoteness and the lack of preparedness for such an emergency, as well as by a storm that blew into the Sound and moved the oil away from the foundered vessel. By some estimates, 1,900 kilometers of coastline and appalling numbers of birds, fish, seals, and other creatures were damaged by the spill. In circumstances such as those in Prince William Sound, accurate assessments of the oil spill's impacts are extremely difficult. Nonetheless, widely accepted counts indicate that 250,000 seabirds, 2,800 sea otters, 300 harbor seals, 250 eagles, twenty-two orcas, and billions of salmon and herring eggs were killed. Cleanup costs amounted to well over two billion dollars (and brought an ironic bonanza to those who flocked into Valdez to benefit from the avails of disaster), but even this level of expenditure failed to return beaches and shorelines to their pristine state within five years of the spill.

When the inquiry convened to investigate the incident heard of the casual, even careless, procedures on the navigation bridge of the *Exxon Valdez* and of the minimal scrutiny over tanker traffic exercised by the U.S. Coast Guard and the Alaska Department of Environmental Conservation, it was easy to conclude that corporate greed and government complacency lay behind the disaster. That was the general view. Exxon reported a decline in gas sales across the United States, and new legislation governing the shipment of oil by tanker was hurried through federal and state legislatures. But few went beyond this, to consider the *Exxon Valdez* disaster as less an unfortunate accident than a symptom of a deep affliction—North America's voracious appetite for oil. Thousands of credit cards were sent back to Exxon in protest at the despoliation of Prince William Sound, but as one commentator observed rather wryly, most of those who took this action probably drove to the post office to mail their letter.

Two decades on, memories have faded, and the outrage at what happened three hours out of Valdez has dissipated. Magazine and newspaper advertisements and television infomercials have steadily and skillfully reiterated the messages that oil and forestry companies are "good citizens" who are "sensitive to the environment" and important contributors to community well-being. Corporate interests in the resource industries continue to insist that resource exploitation is compatible with environmental integrity, particularly with the benefit of "ever-improving" technologies. They find support among scholars, who embrace the promise of sustainable development to suggest that even wilderness areas might be developed without seriously compromising their ecological integrity and that development might even enhance "ecosystem health." As oil prices climb, pro-exploration, pro-development interests look again to the prospects and profits of northern development. In 1999 leases were sold in the naval reserve west of Prudhoe Bay.

Early in the twenty-first century, the gaze of newly merged and rationalized oil interests centered once-again on the ANWR. Recognizing that efforts to open the refuge to oil development were building in 2000, former President Jimmy Carter called on President Bill Clinton to declare it a National Monument before he left office. Alaskan officials unleashed a storm of

Exxon Valdez *oil spill.* *On March 24, 1989, the* Exxon Valdez *tanker ran aground in Prince William Sound, spilling 11 million gallons of crude oil into the Gulf of Alaska. The largest oil spill in U.S. history and a major environmental disaster, the resulting slick covered more than 1,000 miles of the Alaska coastline. Here, the* Exxon Baton Rouge, *the smaller vessel, attempts to offload the remaining oil from the* Exxon Valdez. *(AP/Wide World Photos)*

protest—by some reckonings the coastal plain of the Refuge may contain ten billion barrels of oil—and nothing was done. Four years later, and three days before American forces began to bomb Baghdad in March 2003, Republican senators sought to open the Refuge to oil development by attaching a proposal to do so to a budget resolution. This failed narrowly. Two years later, on March 16, 2005, the U.S. Senate voted narrowly in favor of opening the ANWR to oil drilling. Describing the decision as "a crucial step in President [George W.] Bush's plan for reducing America's dependence on foreign sources of energy through conservation, development of renewable energy sources and increased domestic production of traditional energy," Interior Secretary Gale Norton projected that oil and natural gas would flow from the area within a decade (Pope 2005). In November of that year, however, Republican leaders in the House of Representatives agreed to strip the provisions for oil exploitation in ANWR and the offshore continental shelf from the budget bill. This, reported the *Washington Post,* was "a huge victory for environmentalists, who have made the House their last stand in the decades-long fight to keep oil firms out of the region" (Weisman 2005). Such rapidly fluctuating fortunes suggest that the battle will only continue. The combatants define their positions by the names that they apply to the area and the descriptions they attach to it. Pro-development forces call it "ten-oh-two" (after the section of ANILCA that recommended further study of the coastal plain and gave Congress final authority to open the area to drilling) and portray it (in the words of Secretary Norton) as "a flat white nothingness." Indigenous people (many of whom favor development) have an intimate knowledge of the territory, know it by names that tend to reflect human activities on the land, and resist the designation of this area as "wilderness" because that seems to imply a lack of habitation and human history. Recognizing this, environmentalists prefer to call the area what it is, the coastal plain, or refer to it more grandiosely as "America's Serengeti" in an allusion to the wealth of wildlife and migratory birds dependent upon the region. In Senate debates and general discussion they tend to use images such as the luminously colored and stunningly beautiful pictures made by photographer Subhanker Bannerjee to refute claims (with all they imply about worthlessness) that this is a "snow desert." Here as elsewhere, naming and representing the landscape are strategies for controlling it (Bannerjee 2003; Cronon 2001; Hildebrand 2003). As the debate continues, it behooves all who attend to it to recognize, with political scientist Charles Konigsberg, that "the economics of how we make our living" cannot be separated "from the health and integrity of the natural world" (quoted in Haycox 2002b, 144; see also Ross 2000).

· · ·

As the waters of the St. Lawrence rose to fill locks and shipping channels, they also began to turn the turbines of hydroelectric generators, because the Seaway was as much a power project as it was a navigation system. Some claimed that the generators in the Barnhart Powerhouse near the Long Sault Dam and downriver at the Beauharnois Power Dam were the second- and third-largest hydro plants in the world, behind only the monumental Kuibyshev Dam (that Paul Josephson has called Stalin's concrete pyramid) and its associated generating station on the Volga River. No other part of the entire Seaway, wrote Lowell Thomas, encompassed "more of the elemental things that go into great engineering" than the Barnhart Powerhouse, also known as the Moses-Saunders generating station, or (with an affection typical of the era's infatuation with large projects) "Big Mo."

Here, 1.9 million kilowatts were produced by thirty-two generators, and carried, from this "titan of engineering," "by cables to switchyards, the points of departure for radiation out over New York and Ontario" via pre-existing grids of high-tension transmission lines (Thomas 1957). Divided among Ontario, New York, and Vermont, the power from this site added about a fifth, a ninth, and two-thirds to current supply levels in those jurisdictions. Downstream, the Beauharnois Station ("Big Beau" to its admirers) produced 1.7 million kilowatts for the Québec grid. Together, "Mo," "Beau," and earlier generating plants near Niagara Falls fueled a rapid expansion in the use of electricity for industrial and domestic purposes. By 1961 almost 300 firms employed over 50,000 people manufacturing electrical products in Ontario alone. Between 1983 and 1989 electricity consumption in Québec increased by 7 percent per year, and a few years thereafter it was reported that almost two-thirds of Québec homes were "all electric."

Following the lead provided by the St. Lawrence developments, province after province sought to develop its hydroelectric potential, as consumer demand for electrical appliances soared and energy-intensive industries such as aluminum and magnesium smelting promised to locate where they could be guaranteed access to abundant cheap electrical power. By 1991 half of the provinces, as well as the territories, derived over 70 percent of their electricity requirements from hydropower and (according to the International Energy Agency) Canada was uniquely dependent upon hydropower as a source of electricity generation among the world's leading consumers of electrical energy. Third behind the United States and Japan in electricity production, in 2000 Canada derived 58 percent of its electrical capacity from hydraulic generators (compared with 12 percent for the United States and 19 percent for Japan). To be sure, Norway (99 percent), Australia (66 percent), and New Zealand (62 percent) derive a greater share of their electrical capacity from hydro stations than Canada, but their combined generating capacity is little more than a third of that in Canada.

In sum, the contribution of hydroelectric power to the generation of electricity in Canada is 3.5 times greater than the average for all member states of the International Energy Agency. In 1991 hydro accounted for over 95 percent of electricity generated in Manitoba (22,554 gigawatthours), British Columbia (47,880), Québec (117,040), and Newfoundland and Labrador (34,107). In Ontario, where large nuclear-powered generators operate alongside coal-fired plants, hydroelectric power generation accounted for less than a third of the total output of 135,000 gigawatthours in 1991, but approximately matched that of Newfoundland and Labrador.

Often described as a clean, green, and renewable source of energy that does not produce greenhouse gas emissions, almost all Canadian hydroelectricity comes from enormous generating plants, construction of which has entailed massive geoengineering projects and a wide range of other environmental impacts. In many parts of southern Canada, whether along the Saint John River in New Brunswick, where the Mactaquac dam flooded large areas of good farmland, or in the Arrow Lakes area of British Columbia, where storage dams helped turn the Columbia River into what environmental historian Richard White has aptly called an "organic machine" and displaced thousands of residents, the consequences of hydroelectric power development have been similar to those experienced along the St. Lawrence during the construction of the Seaway and its associated power schemes. By and large, however, it has been in the north, particularly in Manitoba, Québec, British Columbia, and Labrador, that the most gigantic projects and thus perhaps both the fullest manifestations of high-modernist hubris and the most extensive environmental impacts have been concentrated.

The 1960s, it has been said, ushered in a new era of "provincial continental modernization" in Manitoba history, a new era epitomized by the "extraordinary efforts at ecological re-engineering" involved in the closely related Churchill River Diversion and Lake Winnipeg Regulation (CRD/LWR) projects. Brilliantly conceived and executed as an engineering venture, this development confronted and overcame the challenges to hydroelectricity generation presented by rivers flowing northward and subject to periods of low-flow when demand for electricity is greatest, during the winter months. Most basically, the CRD/LWR entailed the construction of a dam across the low-gradient Churchill River at Missi Falls some 400 kilometers above its mouth in Hudson Bay. This raised the level of Southern Indian Lake by about 3 meters, flooded about 1,500 square kilometers, and cut the river's flow below the dam by as much as 85 percent. Water from Southern Indian Lake, now essentially a reservoir, was then directed through a 180-kilometer-long network of lakes, artificial channels, and smaller rivers into the steeper Nelson River, supplementing the flow of that river by as much as 900 cumecs (cubic meters per second) and allowing the construction of four generating stations with a capacity of 3,600 megawatts.

Because these rivers peak in the spring and summer, additional engineering was required to hold back Lake Winnipeg and coordinate its outflow into the Nelson River with the seasonal demand for electricity. To this end, a control structure was erected at Jenpeg, itself capable of generating approximately 128 megawatts of power. In essence, the CRD/LWR turned the Nelson River into a "power corridor" and made Lake Winnipeg a "storage battery." In the process, "it irreversibly altered the hydrological and ecological characteristics of 50,000 square miles of northern boreal forest and rivers" (Hoffman 2002). Peter Kulchyski, sometime head of the Department of Native Studies in the University of Manitoba, recalls "watching propaganda films in the sixties boasting that the new hydroelectric energy would have no negative environmental consequences." But the Nelson River, "once a pristine source of life, became silty and dangerous," and native people found their traditional territories radically altered: "logs blocking access to shores; undrinkable water; water levels that fluctuated according to no locally known logic, making travel unsafe; interred bodies exposed; islands slowly washed away" (Kulchyski 2004).

In hope of kick-starting economic growth in the relatively underdeveloped province of Newfoundland (admitted to the Canadian Confederation only in 1949, after years of economic hardship), in 1953 a consortium of financial and industrial firms known as the British Newfoundland Corporation was granted exclusive rights to the minerals and waters of an enormous area of Newfoundland and Labrador. Early in the 1960s "Brinco" developed a 225-megawatt hydroelectric power plant at Twin Falls on a tributary of what was then known as Hamilton River to supply the developing iron-mining industry of Labrador. A few years later, after the river had been renamed in honor of former British Prime Minister Winston Churchill and as North American demand for electricity soared and enthusiasm for hydroelectric power development seized the imaginations of engineers and politicians alike, the corporation and the Newfoundland Government initiated further development of the 245-foot falls (and the river's 1,000-foot drop in less than 20 miles). An enormous underground powerhouse, designed to include eleven generators, was excavated hundreds of feet below the Labrador plateau.

Almost 1,000 feet long and fifteen stories high, this powerhouse was the hub of a plant capable of producing 5.2 million kilowatts of electricity, enough to supply three cities the size of Montréal. To drive its turbines, the earlier Twin Falls plant was closed and the water in its Ossukmanuan Reservoir was diverted into the massive Smallwood Reservoir on the Churchill River. Five years of round-the-clock work by some 6,000 workers and almost a billion dollars in construction costs went into this powerhouse project. Almost all of the electricity it produced was sold on contract to Hydro-Québec (at rates that have been a source of contention and bitter resentment in Newfoundland and Labrador for decades since, as most of the power is effectively sold on from Québec to New York State at enormous profit). To carry the power from this remote location to market, a 710-foot-wide clearing was opened across the swamps and over the muskeg of Labrador. Within it three high-voltage transmission lines (735 kilovolts) were carried on approximately 400 huge towers, some of them 170 feet high, to link with the Québec electricity grid, 126 miles away (http://ieee.ca/millennium/churchill/cf_home.html, accessed July 25, 2006).

"No list of Bechtel megaprojects," announces that engineering giant's website, "would be complete without the James Bay Hydroelectric Project—one of the largest undertakings ever mounted." Not surprisingly, the rest of this brief account offers a litany of superlatives and achievements: a $13.8 billion project, "appropriately named La Grande," encompassed an area "larger than the state of New York." By "diverting several rivers, Bechtel engineers" all but doubled the power potential of the watershed, giving it the "capacity to generate a whopping 10,300 megawatts." All of this required—and both the emphasis given such numbers and the materials enumerated are revealing and symptomatic—"203 million cubic yards of fill, 138,000 tons of steel, 550,000 tons of cement, nearly 70,000 tons of explosives, and an enormous amount of determination." There were 12,000 workers employed, over two hundred dikes and dams had to be built to serve the four gigantic powerhouses, and getting the power to market in Montréal and Québec "required building a 3,000-mile network of 735-kilovolt transmission lines."

These "great works" were set in motion in April 1971, when Québec Premier Robert Bourassa revealed the essence of the La Grande project as a source of jobs, investment, and economic growth—and thus economic autonomy—for his province. Construction of access roads began within months, before environmental impact statements were completed, and without consultation with the 5,000 Cree and 3,500 Inuit who lived in the area. Work on the first phase of the project (that described by Bechtel) began in May 1972 and continued until December 1985. In 1975, in the face of rising protests against the venture from many quarters, the James Bay and Northern Québec Agreement was reached with indigenous peoples, awarding them 1.3 percent of the land, $500 million in compensation, and participation in a regional government.

But the project remained immensely controversial, the subject of protests by environmental groups, documentary films by people sympathetic to the plight of the indigenes, and countless newspaper reports and stories in other media. Although $250 million was spent on environmental remediation during the thirteen years of construction that went into La Grande, critics of the Bechtel project regarded it as "the most massive destructive engineering and river-replumbing scheme in history" (Sierra Club n.d.) and directed enormous effort, largely successfully, to stopping the second (the so-called Great Whale) and third phases of Bourassa's grand James Bay scheme. Late in the 1980s, the Sierra Club, the James Bay Defense Coalition, the New England Energy Efficiency Coalition, the Grand Council of the Cree, and even northern U.S. coal-mining interests were united in opposition against further development in the James Bay area.

By their accounts, the La Grande scheme was an environmental and social disaster. By one account it submerged 83,000 kilometers of stream and lake banks; by another it flooded 11,341 square kilometers of forested land. Trees rotting under water increased the amount of nutrients in the water and produced an efflorescence of algal blooms. Flooding of peat beds, wetlands, and other vegetation also released considerable quantities of methane, a greenhouse gas. Naturally occurring methyl-mercury also concentrated behind the dams, rising to levels six times those in undisturbed lakes, and causing the collapse of native fisheries through the process of bio-accumulation. Indeed, some reports suggest that almost two-thirds of the people of Chisabi at the mouth of the La Grande drainage suffer from mercury contamination or Minimata disease. Fluctuating water levels in reservoirs have led to a loss of wetland habitat. Changing flow regimes below dams have produced erosion, increased the sediment load of rivers, and damaged and destroyed habitat. Beaver, muskrat, and sturgeon populations have been adversely affected. In September 1984, 10,000 caribou drowned when Hydro-Québec opened flood gates above the point at which they were crossing the Caniapiscau River. Native peoples have seen campsites, burial grounds, hunting areas, and trapping territories destroyed. According to sometime Grand Chief of the Cree, Matthew Coon Come, "our land is our memory" but—he lamented—it had begun to look like a battlefield after a bombing raid:

> Wildlife habitats are flooded. Rivers and lakes are poisoned by mercury. We can no longer eat the fish. Animals are dying by the thousands. Our values are oriented to nature. If you destroy the land you destroy the Cree people. Parents can no longer teach children out on the land. We're losing our way of life. We don't want your money. We want these projects stopped. Where can you buy a wilderness so vast and beautiful? (James Bay Coalition 1991; see also Grand Council of the Crees, 1993)

For all that, dam construction and northern development have been central to the story of Canada in the last half of the twentieth century. From the Peace,

Kemano, and Revelstoke dam projects in British Columbia, through plans for future development of hydro sites in Newfoundland and Labrador (Gull Island, Muskrat Falls), Manitoba (Wukswatim, Conawapa), and elsewhere, the promise of hydroelectric power generation has proven a powerful political tool, as well as a potent instrument of development. With almost all southern sites and rivers exploited to capacity (or regarded as off-limits for main stem dam development—as in the case of the Fraser River where salmon have trumped power to date)—Canadians must perforce look north for future expansion of the country's hydroelectric power capacity, whether this is for domestic or export purposes. If they follow the lead of many politicians of the last half-century, they will have little difficulty doing so.

23

NORTHERN VISIONS

For the Conservative government of John Diefenbaker (1957–1963) the north was integral to a "new vision" of Canada. Plans called variously for road building and railroad construction to open access to the north, an increase in scientific research in arctic environments, and an ambitious scheme for town development at Frobisher Bay on Baffin Island. Delivering these grand designs proved more difficult than formulating them. But the logic behind the rhetoric was encapsulated by Alvin Hamilton, Minister of Northern Affairs and National Resources in the late 1950s, when he described the north as a "new world to conquer . . . a great vault, holding in its recesses treasures to maintain and increase . . . material living standards" (Coates 1985, 199). This was no surprise. The north (Fig. 16) had long been regarded as Canada's frontier, as a "vast warehouse to be exploited vigorously and in some cases ruthlessly, to provide riches for the outside world" (Wonders 1972, 138).

Reading the political wind, and in full accord with the modernist tendencies of the period, several of the country's leading geographers alluded to the need and opportunities for grand-scale master plans for northern development in presidential addresses to their professional association in the late 1950s and early 1960s (Lloyd 1959a; Bird 1960; Wonders 1962; Hare 1964). Late in the latter decade, the idea was picked up and elaborated upon by the private sector. Fostered by Richard Rohmer and Acres Research and Planning Consultants, the idea of the Mid-Canada Development Corridor (a broad arc across the "middle North" of the country, from northern Newfoundland through Schefferville and Noranda, Flin Flon and Hay River to Whitehorse and Tuktoyaktuk) became a prominent feature of national discourse. Mining and mineral wealth were generally at the center of these discussions—as they were at "the plethora of northern development conferences . . . [that] erupted across Canada" in the 1960s (Wonders 1972, 143). By the early 1970s "northern development" was virtually a political mantra. If there was still ample room for debate about precisely what that term implied, and for discussion over how whatever it meant might proceed, few would have disagreed with northern specialist William C. Wonders when he wrote, in 1972, that "in many ways the future of the North is the future of Canada" (137).

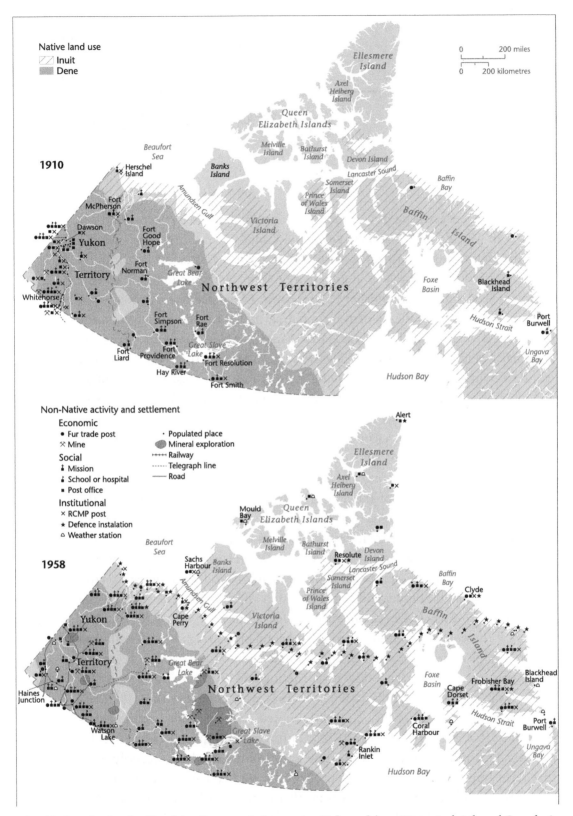

Fig. 16. *Developing the North in the twentieth century. (Adapted from* Historical Atlas of Canada 3: pl. 58)

For military defense strategists during the Cold War era, the north also held the key to the future of North America. No sooner was the Second World War over than the United States began to worry about the threat of air attack by long-range bombers from the Soviet Union. Between 1949 and 1954, the United States and Canada joined to finance and build a line of radar stations across the north to provide warning of any Soviet attack. The so-called Pinetree Line included thirty-three stations arrayed across the country, from Vancouver Island to the coast of Labrador, and was complemented by a similar line of radar monitoring establishments along the American border. When the Soviets upgraded their bomber force early in the 1950s and exploded their first hydrogen bomb in August 1953, the Cold War chill deepened.

American officials mooted the development of radar stations in the far north, but seeking to avoid the buildup of American personnel and interests on Canadian soil, the government opted to build almost a hundred, mostly unmanned, radar stations through the mid-north, forming what became known as the McGill Fence along the fifty-fifth parallel. Even as construction began, there was concern that the Soviets might outflank the Pinetree and McGill warning lines, and late in 1954, Canada and the United States agreed to establish a Distant Early Warning (DEW) Line of radar stations along the Arctic Coast. In three years, twenty-two radar stations were built between Alaska and Baffin Island. All costs (approximately $400 million) were underwritten by the United States, but work on the Canadian sites utilized Canadian companies, employed some native labor, and brought both cash and southern practices and expectations into the north. Within a few years, however, the military importance of the DEW Line was devalued by the development of intercontinental and submarine-launched missiles and advances in satellite tracking systems.

For the 9,000 or so Inuit peoples of the Canadian Arctic in the 1950s, the southern enthusiasm for development, planning, and social experimentation in the north was probably as perplexing as it was tragic. Changes (ranging from the decline of the fur trade to the introduction of diseases associated with increasing numbers of southern white people in the north in the 1940s and early 1950s) affected Inuit populations and economies in a variety of consequential ways in the second quarter of the century. By the early 1950s, internal reports had begun to document and detail the challenges facing Inuit society. By this time, ready cash was flowing into traditional economies from old-age pensions, government relief, aid to the blind, and family allowances (that the Inuit called *kakkalaanituq*, "something for babies"). To Royal Canadian Mounted Police Inspector Henry Larsen the results were all too obvious and disturbing. "The Eskimos generally," he reported in 1951, "have drifted into a state of lack of initiative and confusion." Destitution, filth, and squalor were widespread. "The

***Documentary film: scene from* Nanook of the North.** *In Robert Flaherty's film* Nanook of the North—*often cited by film historians as "the first feature-length documentary"—the lead character (played by an Inuit named Allakariallak) is portrayed in simplistic, romantic terms, and the reality of Inuit lives is much distorted. Walrus and seal hunts and other episodes were staged in an effort to "capture the true spirit" of Inuit life as it was believed to exist prior to European contact. (Library of Congress)*

once healthful and resourceful Eskimo," he wrote to his superiors, "has been exploited to such a degree that he now lives a life comparable to that of a dog" (quoted in Marcus 1995, 20–21).

Such observations were potential political dynamite. In the years since 1922, impressions of the "Happy-Go-Lucky Eskimo" had been embedded in the popular imagination by Robert Flaherty's widely screened film, *Nanook of the North.* Based on the film's carefully constructed portrayal of Nanook as the epitome of simplicity and virtue, the Inuit had come to be perceived as charming, northern "noble savages" (see also Fienup-Riordan 1990). The release of the National Film Board's *Land of the Long Day* in 1952, which presented a happy Inuit family following traditional subsistence activities with the aid of introduced firearms, animal traps, and telescopes, only strengthened the popular sense that the Inuit were resilient and successful in their ability to integrate modern technologies into their age-old ways of life. Not surprisingly, Larsen's reports were diluted and buried in the Ottawa bureaucracy. They were "well-meaning" but they contained exaggerations and came from someone who looked at the Inuit question from a wholly "humanitarian point of view" (James Cantley, quoted in Marcus 1995, 24).

Two books released by American publishers at much the same time as Doug Wilkinson's 1952 film soon lit the fuse of controversy, however. Farley Mowat's *People of the Deer* (1952) told a tragic story with what one reviewer described as "passionate sympathy and indignation" (quoted in Marcus 1995, 16). Serialized in *The Atlantic Monthly*, condensed in the *Reader's Digest,* and vigorously attacked as a "dangerous" and "inaccurate" book by several who had worked in the north, Mowat's account indicted the Hudson's Bay Company, the Royal Canadian Mounted Police, and the Roman Catholic and Anglican churches for their roles in reducing northern peoples such as the Ahiarmiut (who were the focus of the work) to starvation and lamented the government's failure to address their plight. Mowat's basic claim, that the Inuit were confronting shattering difficulties, was given compelling support by Richard Harrington's *The Face of the Arctic,* also published in 1952. An account of five journeys through the Arctic, it contained graphic descriptions and heart-wrenching images of Inuit people in desperate circumstances; one of its chapters is simply yet accurately titled "Portrait of Famine: Padlei, 1950."

Faced with a virtual firestorm of outrage and criticism, the Canadian state sought to address the growing problems of marginalization and relative impoverishment affecting the Inuit. Officials with responsibility for the north had attempted to avoid what they perceived as the development of dependency that had flowed from the establishment of Indian Reserves in the south. In its first manifestations, this approach had produced what might, at best, be called the benign neglect of the 1940s. In the face of growing evidence and rising public concern that such strategies were ineffective, a more interventionist attitude was adopted. If a hunting-gathering people were starving, the problem was overpopulation: Why not relocate them "to areas not presently occupied or where the natural resources could support a greater number of people"? (1952 Conference on Eskimo Affairs, quoted in Marcus 1995, 36). This would reinvigorate traditional economies in the face of the increasing southern presence in the north. It would restore self-reliance. And it would eliminate the squalid conditions in the camps that had grown up around trading posts and that led the author of one magazine article to characterize the Inuit as "slum dwellers of the wide-open spaces" (Phillips 1959). Relocation was seen as a remedy for social welfare problems.

Thus Inuit families from the eastern Arctic (including Port Harrison—Inukjuak—where Flaherty had filmed *Nanook*) were relocated to northern Arctic locations such as Resolute Bay and Grise Fiord in 1953. Years later Inuit would claim that their relocation to Ellesmere Island was undertaken, at least in part, to substantiate Canada's claim to the High Arctic islands, but the government has never formally conceded this point. In 1957 Ahiarmiut people from Ennadai Lake were moved north to Henik Lake, in a tragic venture (subsequently the subject of another book by Farley Mowat, *The Desperate People*, 1959) that led to the deaths of several relocatees by starvation and murder. A year later the continuing rapid decline of the caribou herd brought some 600 Keewatin Inuit to the point of crisis and starvation, and produced another relocation scheme focused on Rankin Inlet and Whale Cove.

All of these initiatives were beset by difficulties. Although plans generally included provision for the supply of medical and others services to the relocated Inuit, the local skills and traditional knowledge of people moved from interior to coast, or to settings hundreds of miles north of the tree line within which their original settlement lay, were substantially devalued. Much had to be learned anew. Kinship bonds were severed and weakened. Networks of reciprocal exchange were fractured. Remoteness, the small size of the resettlement communities, and the lack of suitable clothing and so on also made conditions difficult for the relocated Inuit.

In the end, the logic of economic and administrative efficiency, a logic quite contrary to that of the indigenous communities involved, underpinned Canadian policy during these relocation experiments. Bureaucrats spurred by crises and humanitarian sentiments considered a wide range of possible solutions to the "problems" they faced: Relocation to the High Arctic was weighed alongside the idea of moving Inuit to rehabilitation centers in southern Canada; adjustments to social welfare programs and administration were mooted; eventually, as a consequence of Inuit resistance to some of the policies to which they were subject, consideration was given to the idea of giving Inuit a larger role in policy making. Through all of this, observed Diamond Jenness, the great anthropologist of northern peoples, officials were "steering without a compass" (Jenness 1964, 90).

The results, as Frank Tester and Peter Kulchyski signal in the title of their book on this period, were all too often *Tammarniit* (mistakes). For all the insistence that official interventions were building new gardens of Eden in the far north, conditions were far from utopian. More sensitive policies, community development, and comanagement initiatives and the transfer of land and political authority to Inuit peoples later in the century would begin to redeem some of these ills. But writing on *Eskimo Administration* in 1964, Jenness might well have cast his mind back to the question he posed in *People of the Twilight*, his 1928 account of his sojourn in the Arctic fourteen years before. "Were we," he asked in words that continue to have relevance to northern development in the twenty-first century, "the harbingers of a brighter dawn, or only the messengers of ill-omen, portending disaster?" (quoted in Cook 1987, 436).

24

RAPACIOUS HARVESTS I

From Forests and Mines

Concerns about the state of Canada's forests reached something of a crescendo in the years before World War II. Despite early twentieth-century expressions of interest in "forest conservation," observers began to recognize that exploitation often prevailed over management of the resource. But wartime demands for people and resources undermined concern for the long-term health of the forest. Soon after the war, however, officials recognized that Canada's forests were far from inexhaustible, and that they would need to be "handled as a crop and not as a mine." In 1947 British Columbia implemented the recommendations of a Royal Commission on Forestry that were intended to sustain communities by sustaining yields. The province's forests were divided into extensive "working circles." Some of these were allocated to large operators who undertook to manage them, under Forest Service scrutiny, on a sustained-yield basis by undertaking inventories, developing management plans, calculating an appropriate annual cut, and reforesting logged lands. Others were to be managed by the Forest Service and made available to smaller operators. In both cases exploitation was limited (at least in theory) to levels that would ensure stable, long-term production. Similar arrangements were soon implemented in most provinces.

In essence, these arrangements turned upon a close partnership between corporate interests and the state, and they envisaged forests designed and managed for the efficient production of lumber and fiber. Over time "natural forests" with a mix of species and trees of different ages and sizes would be replaced by plantations composed of even-aged stands of single species. To ensure this progression, and to maximize their efficiency, loggers armed with newly developed, more powerful technologies cut down all the trees in seemingly ever-larger plots laid out according to the dictates of professional foresters. The following summer bulldozers might bare the ground of these clearcuts—a technique known as scarification—to encourage the growth of spruce and pine. Alternatively, the logged ground might be subject to a "controlled burn" before planting crews worked through it placing seedlings in the earth. Intended to mimic the natural cycle of fire in the forest, to open up seed cones, release nutrients from the logging debris, and expose the soil, this practice as often robbed as replenished the area by reducing the organic matter to ash rather than carbon, much of which washed away with the rains. Planted, thinned, and weeded (mechanically or chemically) of competing species, the reseeded plots were conceived of as forest fields, growing crops of trees (in British Columbia, for example, creating a patchwork of spruce here, Douglas fir there, lodgepole pine in another block) to be harvested when they attained marketable dimensions.

The regulator at the heart of the new management regime was the "allowable annual cut" (AAC). This was set each year to ensure that no more wood was cut than was replaced by new growth. Continued productivity—sustained yields—would be realized by limiting exploitation according to the composition of the forest, its expected rate of growth, and the quality of its regeneration. Other things being equal, therefore, effective silviculture—particularly tree planting—could increase the allowable cut. So, too, could improved protection (avoiding the loss of timber to fires, infestation, and so on) and better harvesting techniques. All received a good deal of attention in the third quarter of the century. Late in the 1970s, many felt that unrestrained exploitation had given way to forest management. Although fundamental decisions about the resource were still being made, for the most part, according to economic, political, and social—rather than ecological—criteria, there were many reasons to believe that the industry had "made a leap out of the dark ages" (Swift 1983, 105). The new administrative regime seemed to assure Canadians that the country would have forests forever.

But this promise was only as good as the information upon which the AAC was determined, and the quality of many forest inventories left much to be desired. Late in the 1970s, forestry consultants F. L. C. Reed and Associates concluded that provincial inventories consistently overestimated the extent and value of the forest resource. National estimates that placed the softwood reserve at almost 70 million cubic meters included some 36 million cubic meters of wood in forests that were economically inaccessible; moreover, much of this inaccessible timber was of sizes and species unsuitable for use with existing technologies. Even inventories of the remaining, accessible timber likely overstated the supply. Some claimed that too much money went into sustaining revenue from the forest—into fire protection, road construction, and maintenance, inventory, and planning—and not enough into sustaining the forest itself.

Despite the rhetoric of sustained yields, many students and critics of forest use felt that the companies and provincial governments were "applying little more than a thin veneer of management to the resource" (Swift 1983, 91–123).

Production was being maintained by exploiting new areas of older timber. Exaggerated estimates of the size of the resource effectively discouraged expenditures on intensive forest management. Because regeneration had not matched exploitation, and because much potentially useful wood (in small logs, low-volume stands, difficult locations, and of less-valuable species) was left behind by logging operators, Canadian forests were marked, paradoxically, by over-exploitation and underutilization.

At the same time, smaller operators protested the drift to corporate concentration in the industry. In British Columbia, where the top fifty-eight operators together held barely 50 percent of the area in provincial licenses in 1940, four influential companies held two-thirds of the licensed timberland in the province by 1965. Small businessmen railed against their marginalization and blamed the license system for the advantages it gave their larger rivals. Government officials attributed the changes to a worldwide shift in economic conditions. They admitted no more responsibility for the difficulties faced by small operators than they exhibited capacity to resist the rise of enormous integrated forest products corporations such as MacMillan-Bloedel in the west and Abitibi-Price in the east. By 1982 Canada's leading wood products companies were major players on the world stage. They led a $23 billion industry.

Changes in the organization of the forest industry and in the administration of Canadian forests were paralleled and complemented by changes in woodlands operations. The number of men willing and able to spend a winter working in the woods declined after 1945. Faced with rising labor costs, woods managers sought to mechanize logging and made a virtue of necessity by arguing that new technologies would "improve" utilization of the forest by making it profitable to exploit small trees currently left as waste. Tapping into the rhetoric of sustained yield, they also argued that mechanization would allow year-round operations, the employment of a "permanent" workforce, and the development of stable resource communities. Much effort went into advancing an industrial revolution in the woods. In the 1950s a Québec inventor developed a machine that could cut and stack more cordwood in an hour than traditional operations averaged per man-day. A few years later, two Ontario men designed a "harvester." Both were brought into commercial production, but they were cumbersome and unreliable and neither was a practical success.

Improvements in chainsaw technology were far more important in the early mechanization of logging. Tried on an experimental basis between the wars, early chainsaws were unwieldy and temperamental. But improvements came quickly after 1945. By 1952 some 3,000 gasoline-powered chainsaws were being used in the forests of Ontario, and a few years later they were almost ubiquitous. In the hands of skillful, experienced operators they could double

The Great Lakes Forest Products Company mill, northern Ontario. *Pulp and paper mills were sites of major investment in the forestry sector of the northern North American economy in the twentieth century. They provided a great deal of employment, in both mill and woodland operations, but they were also major sources of environmental (especially water) pollution. (Courtesy of J. Lewis Robinson)*

productivity. If men worked on piece rates, their earnings were increased. And chainsaws undoubtedly made the work of loggers and pulp cutters easier. Coupled with the introduction of mechanical wheeled skidders, chainsaws led to a dramatic decline in the labor requirements of the industry, especially in eastern Canadian pulpwood operations. The Abitibi company reported a 66 percent decline in the labor requirements of cordwood production in the 1950s.

Still the quest for efficiency continued. In the 1970s powerful new machines entered the forest, especially the vast, relatively flat terrain of northern Ontario and Québec. Full-tree logging systems, such as the enormous Koehring Feller Forwarder, with 12-foot-diameter tires, and the capacity to carry up to 12 cords of wood, cut enormous swaths through the forest, shearing trees at a rate of 100 per hour, and delivering them to roadside piles without being touched by human hands. Short-wood harvesters had an equally dramatic impact. Hydraulic booms allowed their operators to cut trees in a broad arc left and right as

The mechanization of logging, near Timmins, northern Ontario, in the 1970s. *Mechanization proceeded apace after 1950 with the development of harvesters, feller-forwarders, and bunchers, all of which allowed a quickened assault on the boreal forest. Here heavy equipment hauls pulp logs through a timber limit. (Courtesy of J. Lewis Robinson)*

they moved through the forest, while other components of the machine limbed and topped and cut the wood into sections before carrying it to a landing for carriage to the mill. These machines had a huge impact on productivity: The manufacturers of one Short-Wood Harvester claimed that two of their machines, operated by an eight-person crew, round-the-clock, six days a week for nine months, could produce as much wood as three camps of a hundred men each would have done through seven months of winter toil three decades earlier.

Heavy machines opened the way for a concerted onslaught on the forest. In addition, their sheer bulk raised concerns about soil disturbance, compaction, and erosion—especially in fragile northern ecosystems—as well as the destruction of undergrowth and seedlings in their path. Some worried that full-tree systems (which delivered limbs and tops of trees to the mill for processing) removed both nutrients and seed cones from logged areas. Studies of these questions by the Forest Engineering Research Institute of Canada were revealing but hardly reassuring. Sites and situations varied immensely, and assessments necessarily rested on simulation models rather than long-term observational data. In general, they found that neither soil compaction nor accelerated erosion could be attributed to the use of the new machinery. But there were grounds for concern about regeneration and the quality of future stands. Hardwood and scrub species were likely to increase, and it seemed probable that, half a century on, yields would be lower because softwoods would be smaller than those cut in the 1970s and would comprise a smaller proportion of the forest.

Through these years, Canadian forest companies were caught up in an emerging, rapidly changing, and highly competitive global market for wood products. Sweeping adjustments—including downsizing of both management and production staff; mill closures; corporate takeovers; construction of state-of-the-art mills; and a movement away from established, integrated modes of production for standardized markets to new, more flexible arrangements predicated on adaptability, and the contracting out of many tasks—were evident across the country. The changes were often dramatic and harrowing. But as industrial restructuring reshaped the forest industry in older established areas, new initiatives spawned large-scale exploitation of far northern forests.

From Québec to British Columbia, the boreal forest came to be seen as the key "to stav[ing] off the day," perhaps for half a century or more, "when Brazil and Indonesia will dominate the [world's] pulp markets." In Alberta, in particular, the worldwide economic downturn of the 1980s, technical advances in milling, a broad-based desire for economic diversification, and a government well disposed to the sale and promotion of its forests encouraged a massive expansion of pulp and paper production. Marketing its northern forests as "the last remaining basket of timber in the developed world," the province provided enormous tax and other incentives to the establishment of exceptionally large, vertically integrated pulp and paper mills. Of the half-dozen major projects initiated in the 1990s, two were dominated by Japanese investors, Daishowa and Mitsubishi. The larger of these, the Alberta Pacific (Al-Pac) mill, can produce over half a million tons of bleached kraft pulp a year; it holds twenty-year renewable leases to almost 75,000 square kilometers of aspen forests, and consumes over 3 million cubic meters of wood annually. The forest, to borrow a revealing phrase from Paul Josephson (2002, 69), had become a "cellulose factory."

As these adjustments created flux and uncertainty in the Canadian forest industries, changing public attitudes toward the environment posed a wide range of new challenges. In Canada, as in the United States, growing concern with questions of beauty, health, and permanence brought forests and forestry under new scrutiny in the last quarter of the century. Public anxieties focused variously on the destruction of "wilderness," on environmental pollution, and on the reduction of forest resources. Debate, protest, and civil disobedience became almost commonplace as citizens' groups sought to preserve "nature" from despoliation by "industry." Generally focused on specific and often spectacularly beautiful sites by new organizations committed to saving wilderness for aesthetic, ecological, and spiritual reasons, these confrontations were nowhere as common or as pointed as in British Columbia. There extensive areas of uninhabited land, breathtaking scenery, and a bountiful habitat seemed to raise the stakes of environmental protest. From the Stein Valley through Haida Gwaii (the Queen Charlotte Islands) to the remote Tatshenshini area near the Alaska boundary, there were bitter conflicts over resource use (M'Gonigle and Wickwire 1988; Islands Protection Society (B.C.) 1984; Budd et al. 1993). Barricades, arrests, worldwide publicity, and international boycotts of provincial and corporate wood products were only the most visible elements of a struggle that divided people and communities. Clear across the country, however, similar battle lines were drawn between citizens worried about a dwindling wilderness and those concerned about the consequences—for companies, employment, communities, families—of excluding large areas of forest from exploitation (Hodgins and Benidickson 1989; Bray and Thomson 1990).

Wilderness preservation was not the only cause of protest, however. As Rachel Carson demonstrated in *Silent Spring*, published in 1962, environmental pollution was insidious. Her case was argued, in part, by reference to the decimation of salmon stocks in the Miramichi River attributable to the use of DDT against spruce budworm infestation in the forests of New Brunswick. When DDT was banned, the antibudworm program was continued with a compound called fenitrothion. Years later, fears that children, inadvertently exposed to this insecticide and the emulsifiers used to spray it, were being seriously harmed led to bitter conflict between concerned citizens and official and corporate interests in both Nova Scotia and New Brunswick. From northern Ontario there were shocking claims that the discharge of pulp mill effluent had produced high concentrations of mercury in river fish and poisoned native peoples.

In light of all this, environmentalists opposed the enormous clearcuts projected by the Al-Pac development and voiced loud concern about pollution of the Athabaska River by its mill waste. Their anxieties were only heightened by the government's rejection of an independent environmental impact assessment that would have delayed construction of the mill until questions about the sustainability of the project and the downstream effects of its effluent discharge had been addressed. The company's own study of the impact of effluent on fish showed that four out of every five survived immersion in tainted water for ninety-six hours. To drive home the point, the company magazine *Forest*

Landscape even carried a photo-essay about an aquarium in the Al-Pac administration building, in which goldfish swam in "raw, untreated effluent from Alberta-Pacific." But this failed to persuade the opposition. Using goldfish as indicator species and "death as an end-point indicator of environmental harm" was hardly reassuring, argued local activists. They were concerned about the trout indigenous to the Athabasca River system (not aquarium species), and they wanted to know whether wild fish would be severely debilitated, fit to eat, or able to reproduce after weeks, rather than days, in the effluent-laden river (Sherman and Gismondi 1997; Richardson, Sherman, and Gismondi 1993, chapter 3).

Researchers with the Science Council of Canada in the 1980s further complicated the lives of provincial foresters and industry leaders when they concluded that Canada's forests still suffered from years of rather heedless exploitation. By their account, much of the country's "high-quality old growth forest" had been harvested and much that remained accessible was "overmature and defective." Moreover, "fires, insects, disease and wind" were destroying "two-thirds as much timber as [was] harvested annually," and every province had faced "local shortages of commercially suitable wood . . ." An eighth of the productive forest area in Canada had been so degraded that "huge tracts" lay "devastated, unable to regenerate a merchantable crop within the next 60 to 80 years." This was a record of "shameful waste," and each year "200,000 to 400,000 hectares of valuable forest" were being added to it (Science Council of Canada 1983, 5).

Responses were already being framed. In 1977 the Canadian Forestry Association had sponsored a national conference on forest regeneration. In the early 1980s further improvements in forest management were encouraged by the publication of *A Forest Sector Strategy for Canada*, and elected officials met regularly as the Canadian Council of Forest Ministers (CCFM). Shortly after a National Forest Congress in 1986, the country's "first truly national statement" on a long list of important issues appeared as *A National Forest Sector Strategy*. Yet such was the pace of change that ten years after its publication this document seemed more closely related to ideas in circulation at the beginning of the century than to the debates in progress at its end. Those who framed this policy were well aware of the ecological, recreational, and aesthetic values of forests, but they put the production of wood first and assumed that attention to other concerns would follow in time and to some degree; their emphasis was firmly upon sustained-yield timber management. As the ink dried on the pages of the national strategy report, however, Canadian policy makers and the public to whom they answered were absorbing the lead, the rhetoric, and the ideas of the Bruntland Commission Report, *Our Common Future.*

In 1989 the Canada Forestry Act directed the minister to "have regard to the integrated management and sustainable development of Canada's forest resources." At its next meeting, the CCFM sought consensus on a much broader agenda than in the past. Social, cultural, economic, and ecological concerns were brought forward together, and there was wide consultation on the form and content of the report. Even more extensive input was incorporated into the 1992 National Forest Strategy document, *Sustainable Forests: A Canadian Commitment.* This ambitious declaration specified the inclusion of timber and nontimber values in forest management decisions and urged protection of the integrity, health, and diversity of forest ecosystems. It was backed up by the "Canada Forest Accord," intended to ensure follow-through on the commitment. Two years later the CCFM identified four important steps toward further realization of their 1992 goals: accomplishing the ecological classification of forest lands; completing a network of protected areas representative of Canada's forests; conducting forest inventories that encompass a wide range of forest values; and developing a system of national indicators with which to measure the sustainability of forests.

Provincial administrations moved in tandem to proclaim "new paradigms" for forest management. Generally, these turned on "sustainable management," the incorporation of multiple economic and social values in decision making, and the protection of biodiversity. In practical terms most also sought greater public participation in decision making, to replace confrontation with "negotiated, interest-based consensus building." Under these new formulations—exemplified by the Ontario Crown Forest Sustainability Act (1994), the Forest Resources Management Act of Saskatchewan (1996), and the Forest Practices Code (British Columbia, 1996)—First Nations groups, environmentalists, loggers, foresters, and others gained voice as resource stakeholders. In jurisdictions as different as New Brunswick and British Columbia, forest management became more transparent, the rights and responsibilities of different interest groups were identified, and enforcement practices were improved. In British Columbia, at least, the provisions of the Forest Practices Code (which mandated careful attention to the environmental impacts of logging) brought the industry a good deal closer to recognizing the highly variable, biologically intricate character of the forest, and to managing it as a complex ecosystem.

But the widespread embrace of a neoliberal ideology by economists, policy makers, and politicians has substantially neutered the commitment to ecologically sensitive and sustainable resource management implicit in the "biodiversity discourse." Pursuing a strategy aptly characterized as "governance without government" (Peters and Pierre 1998), political administrations across Canada slashed budgets and reduced staffing in forestry and other natural resource and environmental ministries in the last decade or so of the twentieth century. Many agencies lost a third of their staff and a corresponding proportion of their funding in a process that has also become known as the "hollowing out" of the state. Environmental research and protection have been crippled by the withdrawal of funding; monitoring and enforcement have weakened appreciably; and codes and regulations that set enforceable standards of environmental performance have been replaced by "policy instruments from the soft, incentive- and persuasion-based end of the spectrum" (Wilson 2004, 271–272). In this institutional climate, two trends seem evident. On the one hand there is withdrawal, a flight from politics, a disengagement from "big issues," and a retreat into the belief that personal actions can salve the conscience and save the planet. This is what Michael Maniates (2002) has called "Individualization: Plant a Tree, Buy a Bike, Save the World?" On the other hand there is the search for new forms of action and the development of new "weapons of the weak" such as consumer boycotts orchestrated through the implementation of such private regulatory strategies as "eco-labeling" and certification that specifies environmental standards have been met in the creation and harvesting of wood products.

·   ·   ·

Few people in modern society recognize the extent of their everyday dependence upon the mining industry. Scattered across the vastness of northern North America, mines are (often remote) pinpricks on the map of human activity. By one estimate, less than 0.03 percent of Canada's land area has been utilized for metal mining in the last century and a half. Mining interests typically describe their activities as a form of temporary land use that disrupts small areas of land for short periods of time. In sum, they point out, billions of dollars' worth of minerals, the extraction and processing of which accounts for over 4 percent of Canada's gross domestic product, are produced from an area less than half the size of Prince Edward Island. Yet life as it is lived at the turn of the twenty-first century would be impossible without the products of these and other mines. Automobiles to zippers depend upon minerals from aluminum to zirconium. Television sets are said to include thirty-five different metals and minerals, ranging from barite to lead. The manufacture of compact discs requires aluminum, nuclear reactors run on uranium, diamonds edge cutting tools and mark betrothals, plants grow better with potash and phosphate, and zinc is a critical element in the sunscreens increasingly necessary as a consequence of increased UV radiation attributable to the "hole" in the ozone layer.

Responding to rising world demands, the output and value of Canadian mineral production rose dramatically after the economic Depression of the 1930s, and the locus of Canadian mining shifted decisively west and especially north in the years after 1945 as float-equipped planes opened up remote areas to exploration. In 1931 Canada's mineral production was valued at $172 million and six important minerals (copper, gold, lead, nickel, silver, and zinc) accounted for almost $120 million of this. Ten years later the equivalent figures were $560 million and $394 million; in 1951 they were $1,245 million and $715 million, and iron production, worth only a million dollars in 1941, had increased thirtyfold in value. By 1964 Canadian mineral production was valued at almost $3,400 million, the six important minerals identified above accounted for less than half of this ($1,300 million), and iron output was worth more than $400 million. Radium, discovered on the east shore of Great Bear Lake in 1930, was exploited for use in the treatment of cancer, but uranium, a by-product, became immensely important for the development of nuclear weapons after 1943. Responding to strong military demand and aided by the German-developed Geiger counter, adapted in Canada as a field instrument for prospecting, new deposits were located late in the 1940s at Beaverlodge Lake, Saskatchewan, and in the Bancroft district of Ontario. In the next decade, the Canadian invention of the scintillometer, and then the development of the gamma-ray spectrometer, greatly improved uranium prospecting, and important finds were made at Elliot Lake in 1953.

Large areas of the north were also surveyed using airborne electromagnetic equipment after the airborne magnetometer (developed during World War II for the detection of submarines) and cumbersome ground electromagnetic survey instruments developed by Canadian geologists in the 1930s were adapted to geophysical exploration after 1945. New and valuable nickel, copper, zinc, and lead sulfide deposits were discovered in overburden-covered areas where older, more conventional forms of prospecting would have had little chance of locating them. Donald Cranstone, Senior Mineral Economist with the Minerals and Metals Sector of the Canadian Department of Natural Resources, has estimated that across some four million square kilometers of northern Canada, outcrop exposures (on which conventional prospecting focused) account for considerably less than 5 percent of the surface area.

But aerial geophysical surveys can detect magnetic anomalies (pointing to the existence of base metal deposits) at depths in excess of 100 meters. Once these areas of potential interest are identified, ground crews carry out further geophysical surveys by drilling and other means. Since 1946 increasingly capital-intensive exploration methods have identified 2,000 metal deposits in which companies have proceeded to assess the tonnage and grade of ore, although not all have proven commercially valuable (Cranstone 2002, 7).

For twenty years or so after 1964, the value of Canada's nonpetroleum mineral production climbed sharply. In constant dollars, calibrated on a year 2000 base, the industry recorded twin peaks in the mid-1970s (approximately $27.5 billion) and the late 1980s ($26 billion). Shifting demand, fluctuations in interest rates (and thus the cost of borrowing capital), competition from other producers, difficulties in mineral extraction, and the inevitable exhaustion of specific deposits affected production levels and the value of Canadian mining output. But new minerals, such as potash and diamonds, have been added to the mix of important products, and early in the twenty-first century the value of Canadian mineral production topped $20 billion. Mining and mineral processing employed almost 400,000 Canadians. Nearly 50,000 worked in mining and almost 60,000 in smelting and refining, with the remainder employed in the manufacture of mineral and metal products.

With growing use of oil and gas for domestic heating, cooking, and automobile fuel, and the replacement of steam locomotives by diesel equipment after 1945, coal faded out of the everyday experience of most Canadians. Indeed, coal's place in the national economy declined sharply; at the end of World War II it supplied about half of Canada's energy needs, and by the 1960s it accounted for only 20 percent of the Canadian energy supply. Production was down to 11 million tons. In the long-established coal-mining areas of Cape Breton, where costs of extraction were relatively high and production was beset with a range of difficulties, mines were closed and operations were consolidated. Jobs were lost and local economies slipped into recession, which meant lower tax revenues, a lessened capacity to ameliorate the environmental effects of mining, and a propensity to seek new sources of economic growth regardless of potential environmental implications. Although coal production continued, Nova Scotia's seven mines produced only about 2 million tons in 1998, and the province was a net importer of coal (about 500,000 tons, primarily for the generation of electricity). When the coal operations of the Cape Breton Development Corporation—which had sustained mining for social and local economic reasons in the face of declining profit margins over many years—were reduced, Nova Scotia's production declined significantly, and early in the new millennium it barely topped a million metric tons. Late in 2001 the island's sole remaining mine of consequence, the Prince Mine, was closed.

When international crude oil prices spiked upward in the 1970s, however, the value ascribed to coal increased. Essential to steel production and now regarded as an important alternative to oil as a source of energy, coal became the focus of renewed mining activity and investment. Encouraged by the efforts of

Japan, Korea, and other major steel-producing countries to diversify their energy supply, new sources of metallurgical coal were developed in British Columbia and Alberta, and new rail and port infrastructures were built to facilitate exports. Production levels began to climb. But export markets proved fickle. With Australia, China (until 2004 when it became a net importer), South Africa, Russia, and other producers also seeking overseas markets after investing heavily in the infrastructure to underpin new mines, Japanese and other purchasers were able to negotiate contracts at rates far below those prevailing when mine developments were begun, and several mines and communities that opened with great fanfare in the 1970s and 1980s were closed before the end of the century.

In 1998 twenty-four Canadian coal mines produced 75 million tons. All but four of these were surface (strip or open-pit) operations, which are generally much more economical (if initially at least much more productive of visible environmental disturbance) than underground operations. The industry employed almost 8,000 people directly and exported approximately half its output. Some 90 percent of domestic consumption is utilized in electricity generation, and another 20 million tons or so are imported each year, mainly from the United States, for this purpose. The twenty-five coal-fired thermal generating plants located in six provinces produce about a fifth of the country's electricity. Continuing adjustments to world demand saw Canadian coal production decline through the last few years of the twentieth century, from 80 million tons in 1997 to 69 million in 2000 before it climbed back above 70 million tons. By 2001, there were only nineteen operating mines, fifteen of them were owned by two companies, and 97 percent of output came from the three western provinces. Alberta (about 30 million tons) generally leads British Columbia (27 million) in output tonnage, but 80 percent of the Alberta total is accounted for by subbituminous coal; Saskatchewan runs a distant third with an output of approximately 11 million tons of lignite.

Mines are of two basic types: surface and underground. Surface (or open-pit) mining is used to work large, near-surface deposits of commodities (such as coal and copper) that have a low value-to-volume ratio. Typically, this requires the removal and disposal of the surface soil and rock (known as overburden) covering the ore or coal deposit, and may also require the stripping of "waste rock" (in which ore concentrations are uneconomically small). Once the ore/coal deposit is uncovered, it is then drilled, blasted, loaded on to trucks (the latest and largest of which have a payload of 400 tons), and hauled to a processing facility. Underground mining is used to extract deposits deep beneath the surface, and depends upon the excavation of a vertical shaft, a horizontal adit, or an inclined passageway, from which subterranean drifts or crosscuts are developed at various levels to open up mining areas known as stopes.

Open-pit mining in hard rock. *Open-pit, or open-cast, mining, as shown here at the Texas-Gulf lead and zinc operation near Timmins, Ontario, in 1976, had massive impacts upon the landscape. A sense of the gargantuan scale of these enterprises is provided by the heavy truck on the road near the bottom of the pit; its wheels were probably five or six feet in circumference. (Courtesy of J. Lewis Robinson)*

Here, too, drills, jackhammers, and explosives loosen the rock before it is hauled to the surface for further processing.

Although all mining involves the extraction of matter from the earth, each and every mine is different. Mines are located in radically different settings, from tundra to prairie, from mountaintop to valley bottom; they employ a range of extraction and processing techniques; and they have markedly divergent environmental footprints. Yet two inescapable themes thread through the story of modern mining: the massive increases in mechanical and technical might deployed in mining, and the huge amounts of capital investment (private and often public) required by modern mining developments. In little more than half a century the scale of mining operations changed almost immeasurably. Late in the twentieth century, the industry depended, as never before, upon brute-force technologies to mount its confined, but nonetheless considerable, assault on the lithosphere.

Generally, mined ore and coal have to be treated initially (in a process known in the mineral industry as milling or beneficiation) to achieve desired levels of size and purity. At its most rudimentary level (as in the coal industry) this entails washing, screening, and drying; at its most complex (as in the processing of copper, gold, lead, silver, and zinc ores) it involves processes of separation and concentration known as amalgamation, flotation, and leaching, all of which entail the use of chemical agents. Amalgamation, which depends upon the capacity of metallic mercury to coalesce after it has attached itself to the surface of certain metals, is used to recover gold. Flotation uses the physical and chemical properties of minerals and reagents to separate commercial material out from the waste of finely crushed ore. The process takes place in an enclosed flotation cell, from which processing liquids and solid waste (tailings) are discarded into holding ponds. Leaching is used to extract minerals from low-grade ore by pouring an acid or cyanide over the mineral-bearing rock.

In perhaps the most environmentally damaging application of the leaching process, dump leaching, enormous piles of low-grade ore (several million tons of rock) are sprayed or washed with an acidic solution that is then collected in a pond. In the similar process of heap leaching, slightly higher-grade ores are crushed, piled in smaller quantities on a specially prepared pad of asphalt, clay, or synthetic material, and treated with cyanide or a strong acid solution. Vat leaching treats crushed ores of higher grade in an enclosed tank or vat. In each of these leaching processes, the leaching solution is pumped from the holding pond to an extraction plant, where dissolved metals are recovered by chemical or electrical processes, and the leaching solution is recycled for repeated use. In theory at least, these are contained processes, but there is always the potential (and sometimes the actuality) of stream, river, and underground water source contamination as a consequence of ruptures in holding ponds or the escape of uncontrolled runoff.

The environmental effects of mining in remote areas begin at the prospecting stage. As researchers for MiningWatch Canada have noted, "although an individual exploration operation may have little effect on its immediate environment, the cumulative effect of thousands of kilometers of geophysical grids cut through vegetation and surface soils can cause considerable erosion, sedimentation and wildlife disturbance" (MiningWatch Canada 2001, 1). The spillage or discard of fuel, drilling fluids, and camp garbage can pollute soils and rivers, and the development of access roads can increase hunting pressure on game. Indeed, the government of Alaska has expressed concern that the access road serving British Columbia's Tulsequah Chief mine, which crosses sixty-nine streams in its 100 kilometers from Atlin, could "open an undisturbed wilderness area to logging and other developments" (for example, other mines) and deleteriously affect salmon runs vital to downstream U.S. fisheries (*Globe and Mail*, 14 July 1998, A7).

Mining claims have been placed on millions of hectares of northern barrens in the Northwest Territories (Nunavut) and Labrador since the announcements of the Ekati diamond and Voisey's Bay nickel discoveries, and the exploration of even a fraction of these would produce significant incursions of people and equipment into thinly populated native space. Airborne geophysical surveys typically involve flying transects (less than 500 meters apart) at altitudes of about 100 feet over the claim, an activity that the Canadian Wildlife Service considers likely to have serious impacts upon caribou and migratory birds. According to material formerly posted on the website of the Innu Nation and reported by MiningWatch Canada (2001), the indigenous inhabitants of Labrador are troubled by illegal hunting by survey workers, the "siting of exploration camps in areas of intensive Innu land use or cultural significance, and increased helicopter and airplane traffic over sensitive wildlife habitat or harvesting areas."

The potential for environmental damage increases significantly once extraction begins. Typically, this leads to the removal of vegetation, the sinking of shafts or the relocation of overburden, and the construction of roads. Heavy equipment generates noise and dust and may compact or otherwise disturb the surface, leading to increased runoff, erosion, and sedimentation downstream. In general terms, however, water pollution is the most widespread and perhaps most significant consequence of mineral extraction. As ores and minerals are exposed, broken up, and crushed, any sulfides that they contain are exposed to the weather. The resulting chemical reaction of oxygen, water, and sulfides produces sulfuric acid. This may, in turn, release potentially harmful minerals before entering groundwater or local rivers and lakes. Such acidic runoff, known as Acid Mine Drainage (AMD), is one of the most serious environmental legacies of mining.

Separate stories in the *Vancouver Sun* in February 1988 pointed to the gravity of the problem. Under the headline "Breaking Up Rock Sets Off Production of Toxic Cocktail," the paper offered the estimate that there were 300 million tons of acid-generating mine waste in British Columbia and claimed that although "at least" five of the sixteen metal mines operating in the province had AMD control programs, "six abandoned mines were acid generating" (*Vancouver Sun*, 12 February 1988, B1). Alongside this, reporters noted that barely a year after Equity Silver Mine began operating in northern British Columbia, "the waste rock dump was already generating acidic drainage." Estimates placed the quantity of toxic liquid escaping from the dump at approximately 1,800 liters a minute. Left unchecked, this would have raised copper concentrations in local drinking water to 750 times the recommended limit and produced arsenic levels twenty times above the allowable threshold. Faced with such outcomes, the company constructed a collection, confinement, and treatment system to deal with the AMD. Long after the mine closed, its operation cost the company over a million dollars a year, and engineers estimated that treatment would have to continue in perpetuity.

Tailings ponds built to contain the slurry of rock particles, water, and chemicals that remain after the target minerals have been extracted from crushed ore are another potential source of environmental concern. Pipes carrying arsenic, cyanide, and other toxic chemicals to tailings ponds have leaked; human errors have led to system malfunctions; and sometimes tailing ponds themselves have leaked, seeped, or failed—with locally calamitous effects. More generally, remote mines have had less disastrous but perhaps no less dreadful impacts as a consequence of their everyday operations. Late in the twentieth century, the U.S. Department of Justice accused Cominco Alaska, Inc., of violating the federal Clean Water Act "by exceeding the allowable limits for metals and pH" at both their Red Dog Mine and their Chuckchi seaport. The company was also charged with repeated violations of sewage disposal requirements in the areas in which it operated. In the end Cominco settled the lawsuit out of court for $4.7 million (MiningWatch Canada 2001, 9; U.S. Department of Justice 1997). By contrast, the environmental concerns associated with coal mining can appear relatively benign. Beyond the tailings piles that are by and large less toxic than those associated with mineral extraction and the pits and processing facilities, they include the generation of carbon dioxide, the production of nitrogen oxides, and the release of fly ash. Wherever underground mining has occurred, however, surface subsidence and ground collapse can be significant and costly problems, even if they only become evident years after mines have been abandoned.

Uranium mining, in which Canada leads the world, has been among the most controversial of modern mining activities. There are, critics say, only two uses for this "deadliest of metals," nuclear weapons and nuclear reactors, so that the end products of its extraction and processing are bombs and radioactive waste (Edwards 1992). But the damaging effects of uranium mining are felt much earlier. Wherever uranium occurs, its slow disintegration produces radioactive by-products, including thorium–230, radium–226, radon–222, and the so-called radon progeny: lead–210 and polonium–210. Where uranium is concentrated in certain hard-rock deposits, background levels of radiation are elevated. But the greatly enhanced release of radon gas during mining quickly pollutes the underground atmosphere with tiny dust particles of "radon progeny." Taken into the miners' lungs, they impart alpha radiation to the sensitive tissue, and produce, in time, exceptionally high rates of lung cancer, fibrosis of the lungs, and other lung diseases. At Elliot Lake, where mining began in 1953, a 1974 study revealed that the incidence of lung cancer among miners was double that in the population at large. A few years later an Atomic Energy Board of Canada study suggested that workers exposed to permissible levels of radiation for thirty years would develop lung cancer at four times the rate of non-miners.

Other significant environmental risks associated with uranium mining include the widespread contamination of air and water by radioactive materials, as well as the pollution of surface and ground waters by such chemicals as heavy metals, acids, ammonia, and salts found in mine tailings. At Elliot Lake, reinvented as a retirement community and described as a vacation playground and the "jewel in the wilderness" since mining ceased in 1996, a ton of ore had to be pulverized to produce two pounds of uranium. Estimates suggest that 10,000 times as much radon gas escapes from these pulverized tailings as from the undisturbed ore, and that despite the extraction of uranium, the tailings remain highly radioactive due to the presence of thorium–230, radium–226, and other uranium by-products. Such findings have prompted zealous critics of nuclear power and weapons systems to portray uranium mining in almost apocalyptic terms. Each mine, writes Gordon Edwards, is

> a "slow bomb"—spreading deadly radioactive poisons over vast areas of the earth, as surely as the Chernobyl disaster did, as surely as atmospheric tests of nuclear weapons have done, but at an insidiously slower rate. Radon gas can travel a thousand miles in just a few days, with a light breeze. As it travels low to the ground (it is much heavier than air) it deposits its "daughters"—solid radioactive fallout—on the vegetation, soil, and water below; the resulting radioactive materials enter the food chain, ending up in fruits and berries, the flesh of fish and animals, and ultimately, in the bodies of human beings. (Edwards 1992)

Yet the conservative Saskatchewan government reversed the early 1990s efforts of its New Democratic Party predecessor to phase out uranium mining in the province, regarding the jobs and wealth it generated as "too important to be eliminated by doctrinaire considerations" and downplaying the environmental impacts of uranium extraction (*Canada's Uranium Production* 2005). Saskatchewan's major producers, the McClean Lake and McArthur River mines, opened in the far north of the province in 1999; both have ISO 14001 environmental certification (an internationally accepted rating system for environmental management, intended to minimize the environmental impacts of products, activities, and services by establishing programs to meet specified objectives and establishing procedures for review and correction of procedures (see ISO 2006).

For all the risks of atmospheric pollution associated with it, uranium mining is potentially a greater threat to groundwater and river systems. Dissolved radioactive materials and AMD wash into the hydrological system in the routine course of mining and milling operations. In the late 1970s fifty-five miles of the Serpent River and more than a dozen lakes were contaminated by pollutants from uranium mines in the Elliot Lake area (MiningWatch Canada 2005, 1) and the International Joint Commission described the drainage system as a major source of radium contamination in the Great Lakes. Chemical pollution from tailings ponds is at least as common as radioactive contamination, and in the near term is almost certainly a greater source of damage. More than thirty tailings pond failures have been recorded at Elliot Lake, and even at the newer Key Lake operations in Saskatchewan there were half a dozen spills within six months of the mine opening in 1985 because the tailings area was too small.

In all cases, major questions surround the long-term implications of uranium-mining operations, and particularly those related to the decommissioning and abandonment of mines and their tailings ponds or piles. In 1979 an Ontario Environmental Assessment Board found reasons to doubt the long-term impermeability of tailings basins in the Elliot Lake area. Fourteen years later a plan to decommission four areas containing 130 million tons of tailings was referred to the provincial environment ministry. The ensuing inquiry concluded that the tailings constituted a perpetual environmental hazard due to the presence of sulfides, thorium, radium, and other heavy metals. Proper containment and permanent water saturation were recommended, although other research has shown that acid mine drainage continues under water cover and that problems associated with drought, low water levels, and bio-accumulation in aquatic vegetation growing over pits and tailings areas continue and will only get worse with time.

In the last months of the Second World War, the National Film Board of Canada released two documentary films, *Coal Face Canada* (1943) and *Coal for Canada* (1944). Filmed in two long-standing centers of Canadian coal production, Cape Breton and Drumheller, Alberta, and intended to reveal the importance of coal to the war effort, they are nonetheless effective in capturing the hard manual work and danger of dynamiting, loading, and hauling coal in underground mining operations—themes reprised in more dramatic and melodramatic tones in the 1995 commercial production *Margaret's Museum*, set in Cape Breton in the immediate postwar years. The 1940s films also provide a marvelous baseline against which to measure changes in the coal-mining industry through the following half-century. Although powerful modern machinery and new workplace regulations had reduced the hard physical edge and perils of coal mining since the early years of the twentieth century, in the 1950s work and life in the mines and their associated communities remained much as they had for decades. By the end of the twentieth century, they were hardly recognizable. Most Cape Breton mines had closed, remaining operations had been consolidated and modernized, and far fewer men worked in the pits. Coal mining had, for the most part, moved away from underground operations, and its new face was most clearly revealed in the open-pit mines of the Cordilleran mountains.

Thirty kilometers southeast of Sparwood, in southeastern British Columbia, the landscape of Coal Mountain appears as some kind of modern-day ziggurat, a monument to the technological might and rapacious character of mining in the new millennium. Bituminous coal was discovered here late in the nineteenth century, and a small underground mine was developed in 1905. In the 1920s open-pit operations were begun, but concerted exploitation and the most dramatic transformation of the landscape occurred in the last quarter century. High in the Rocky Mountains, the Jurassic and Cretaceous sediments of Coal Mountain were once forested, but today the mountaintop rises bald, terraced, and traversed by snaking roadbeds above the surrounding valleys. Here as elsewhere in the Elk Valley, mining has changed the skyline as vegetation and overburden have been stripped away, coal and rock are removed, and waste is either tipped down the mountain side or dumped in the valley.

Two and a half million tons of coal come from the slopes of Coal Mountain every year, and with four other major mines operating nearby, the Elk Valley produces approximately three-quarters (20 million tons) of British Columbia's annual output of coal (about 27 million tons). Once or twice a week, Coal Mountain reverberates to the blasts of ammonium nitrate fuel oil explosives inserted in 31-centimeter-diameter drill holes located to displace the face of 12-meter-high rock benches. Two large hydraulic shovels and a front-end loader capable of lifting 26 cubic meters of rock in a single scoop load the dislodged material into trucks with a payload of 240 tons. Fitted with Geographical Positioning Systems, this equipment is readily monitored and managed, and the actual work of mining employs a mere seventy-four persons. Removed from the pit, the coal is stockpiled before being moved through sorting, crushing, and computerized washing plants that clean and refine the coal by separating it from other sedimentary rock. This employs another twenty-six persons. About seventy-five more work in maintenance, administration, and warehousing areas. Most of the coal is then loaded into dedicated trains that carry it on an 85-hour, 1,175-kilometer journey to the Robert's Bank coal terminal near Vancouver for shipment overseas.

The Coal Mountain operations have received both quality and environmental certification. Mine roads and coal piles are sprayed with water or other special chemical treatments to keep coal and rock dust down, and special ditches surrounding the mine direct runoff to large ponds where the fine sediment can settle out. Since the 1980s, about 100 hectares of waste rock piles have been re-shaped and revegetated. More generally, however, highly mechanized mining on this scale has significant impacts on air and water. In some localities, waste rock is dumped atop streams, creating so-called rock drains, filling in sections of the valley and destroying habitat. Selenium, residues from blasting, and floc-culants used both to coagulate sediments in settling ponds and to reduce dust emissions from railcars have been detected in some streams near mining opera-tions. And the destruction and fragmentation of wildlife habitat can be of con-siderable ecological concern in locations where disturbance proceeds at the scale and pace typical of today's surface-mining operations.

According to Gavin Hilson, the author of a master's thesis completed at the University of Toronto, the environmental performance of the North American gold-mining industry has improved considerably since 1970. Driven by more stringent environmental assessment requirements, facilitated by improvements in mining and treatment technologies, and reflecting, according to company of-ficials surveyed, a corporate desire to comply with legislation and to improve relations with stakeholder groups, mining operations have become more "eco-efficient"—although the operations of many junior mining companies fall short of the standards reached by larger, more established concerns.

What these changes amounted to, in practical terms, is suggested by the de-velopment of the Aquarius Gold Mine, near Timmins in northern Ontario at the turn of the millennium (http://www.ceaa-acee.gc.ca/010/0003/0023/sum-mary_e.htm, accessed August 3, 2006). Focused on a 12.7-million-ton ore body that was expected to yield 2.54 grams of gold per ton over a five-year period, this project was fairly typical of modern gold mining in its commitment of enormous resources to moving vast quantities of earth and rock for relatively small amounts (about 30,000 kgs, or 30 tons) of precious metal. When the open-pit venture centered on a former underground mine was proposed, plans were subject to a review under the Canadian Environmental Assessment Act, and a baseline study was completed in 1997. Approval was given for the development of a 3,500-meter-circumference excavation that would strip off 60 to 80 meters of overburden and extend to a depth of approximately 165 meters, but not with-out careful attention to the short- and long-term impacts of this activity.

Because the 55 million tons of overburden and the 18.9 million tons of waste rock to be removed from the pit would be stripped away from the flank of an esker (a remnant glacial depositional feature) that is part of an important re-gional aquifer, the company was obliged to avoid groundwater contamination. This required construction of a freeze-wall cutoff barrier around the entire perimeter of the pit to separate mine workings from the aquifer. Some 2,200 freeze pipes were laid, one to three meters apart, through the overburden to the bedrock and connected by a network of header pipes to two refrigeration plants that circulated brine chilled to minus 20 degrees Celsius through the closed-circuit freezing system. In addition, stringent site-drainage control requirements were imposed. The overburden and waste rock were to be stockpiled adjacent to the pit, but because they are strongly calcareous and low in sulfides, AMD was not a concern.

According to the environmental impact assessment for the mine, the ore-processing mill just south of the open pit would utilize gravity concentration to recover coarse free gold, and conventional cyanide leaching and carbon-in-pulp recovery to extract the remaining gold. The mill effluent, "containing cyanide and minor associated heavy metals," was to be "treated within the mill using the INCO-SO$_2$/AIR process, to destroy cyanide and to precipitate heavy metals" before the slurry was "discharged to a tailings containment basin for the retention of tailings." Monitoring would ensure that retention was of sufficient duration to ascertain that the final effluent discharge met Ontario Ministry of the Environment standards. In addition, a closure plan was developed before work on the mine began. This envisaged restoration of the site to a "natural or near natural" condition, following the completion of mining activities. All surface equipment, machinery, and infrastructure is to be removed, a new lake "with an associated productive fishery" is to be developed in the pit, and all disturbed areas including the overburden–waste rock stockpile and tailings basin are to be revegetated.

25

RAPACIOUS HARVESTS II

From Land and Sea

East to west across the country, rural life was remade in the second half of the twentieth century. In the most general terms, small farms, using mainly animal power, with uncertain yields and low and variable incomes yielded ground to larger, highly capitalized and mechanized operations, dependent upon external financing and price-support mechanisms, as well as fertilizers and pesticides to return relatively reliable profits. Whereas four in ten Canadians lived and worked on farms before the Great Depression, only four in a hundred did so in the final decades of the century.

Few areas were harder hit by these changes than the Atlantic region, which lost some 60,000 farms and 75,000 jobs in agriculture between 1941 and 1961. In the 1950s alone, employment in agriculture declined by almost 50 percent. Between 1941 and 1981 four in every five farms in the three Maritime Provinces failed, and the area of improved land in Nova Scotia and New Brunswick fell by half. Yet the average size of farms increased as the amount of occupied and improved land declined. Farming operations were mechanized, expenditures on fertilizers doubled, and agricultural labor productivity rose by more than 3.5 percent a year between 1946 and 1965. Basically, agriculture was dichotomized as some farms were consolidated and amalgamated and others became increasingly marginal economically.

Dairy farms in a few favored locales prospered as they benefited from technological improvements in transport and refrigeration and a government quota system that ensured each producer a share of the market and established minimum prices for milk sales. They added milk parlors, large modern barns, refrigeration tanks, and tall storage silos to farmsteads surrounded by extensive limed and fertilized fields devoted to pasture and fodder for sizable pedigree herds. On poorer soils in more remote areas, however, dairy farmers sold milk to processing plants, ice cream factories, creameries, and cheese factories. Those who owned their properties outright and made no substantial capital investments in their holdings could cover cash operating and living expenses.

But they operated on borrowed time. As Atlantic Development Board researchers noted in the late 1960s, "when the owner retires, dies or attempts to sell the enterprise the productivity is often too low to carry the necessary capital and depreciation costs and provide a return to labour as well." Unattractive to buyers, the land was frequently abandoned. Over the years much of it was recolonized by scrub and forest. The scattered regrowth provided admirable habitat for ungulates, and deer populations increased. Eventually, the trees that reclaimed the land into which generations had poured their labor were counted into the country's forest inventory as evidence for claims that careful forest management was increasing rather than depleting Canada's forested area.

As with dairying, so with potato production. In 1950 most New Brunswick farms were diversified, family-owned operations, relatively unmechanized (only one in five had a tractor), substantially dependent upon the knowledge and labor of the farm occupants in all facets of production. Some 20,000 of them produced commercial potato crops, although fewer than 2,000 had more than 7.5 acres in the crop and barely 500 harvested more than a dozen acres. Twenty-five years later, three-quarters of New Brunswick's potato acreage was in a handful of parishes in the upper St. John valley. In the 1980s, 440 farmers cultivated over 52,000 acres of potato fields, 97 percent of the provincial total. New mechanical planting and harvesting methods, the development of chemical agriculture, and the dictates of ever more powerful corporate interests had reshaped every aspect of potato production.

Mechanical harvesters introduced in the 1960s were an essential element of this transformation because they broke the scale limitations imposed by manual picking of the crop from the fields. Capable of harvesting 150 to 200 acres in a three-week period, these machines incurred capital, fuel, and labor costs that forced farmers into new modes of production. In response, potato cultivation concentrated in large gently rolling fields. But harvesters were only the most visible part of a larger packet of changes that transformed agriculture, ecology, and society in the St. John Valley in the second half of the century. New tractors and ploughs quickened and deepened field preparation. Harvesters removed small stones from the soil and changed its structure. The use of heavy machinery increased soil compaction. Erosion from the extensive exposed field surfaces quickened—according to one New Brunswick study spanning the 1945–1980 period, it increased fivefold in the absence of explicit soil conservation measures. Other studies document the loss of twenty tons of soil per acre to sheet erosion, and the consequent withdrawal of land from production.

In the mid-1950s the rising generation of the McCain Produce Company that had bought and shipped potatoes and supplied fertilizers and pesticides to local farmers since the early part of the century brought new vision to the family enterprise. Recognizing the possibilities of new technologies and the market potential of convenience foods in changing urban and domestic environments, they established a plant to process potatoes into frozen French fries. Early in the 1970s, they opened a second. By then the firm dominated the local market for potatoes and was well on the way to its current global preeminence as a producer of frozen French fries. It also expanded its interests into almost all aspects of the farming system, from transportation through the sale of equipment, fertilizers, and fungicides to the ownership of land. Enmeshed in the McCain's corporate web, more and more farmers grew their potatoes under contract to the company and sold their product at a predetermined price. To a considerable degree, farmers became ciphers in the labyrinthine web of corporate agribusiness, their relationship to the land, to their farms, and to the production process changed completely.

To ensure the quality of their frozen fries, McCain's mandated that farmers grow a particular variety of potato, although it "matures late, has only a moderate yield . . . is highly susceptible to disease, and is intolerant of the wet and dry spell weather conditions so characteristic of New Brunswick summers" (Murphy 1987, 28). The company contract also required farmers to accept corporate "instructions and advice" regarding the application of fertilizers and the dates of planting and harvesting. This meant that farmers were required to buy and apply an astonishing range of chemical insecticides, fungicides, and herbicides. Chemical fertilizer application increased almost fivefold in thirty years after 1951. The costs were substantial. One account from the 1980s placed the costs of fertilizer, lime, and chemicals for 100 acres of potato ground at $29,000; specially treated seed cost another $12,000. With potato prices set by the company contract (at $2.50 to $3.00 per hundredweight and "no tolerance for small size"), many farmers found themselves in a cost-price squeeze. Levels of indebtedness rose, and critics began to question the social implications and ecological sustainability of corporate agribusiness, which also dominated provincial hog production (where 13 percent of producers accounted for 85 percent of output) and poultry farming (where 99 percent of meat birds came from 7 percent of producers).

In the mid-1970s, the landscape of the Annapolis Valley was a fascinating mix of old and new. Traces of the past were everywhere. Dozens of weathered houses were "much patched and added on to." Many had sizable kitchen gardens. Derelict barns and other disused farm buildings were common. Abandoned, outdated equipment rusted out back of the "rather sprawling clusters of dilapidated but serviceable outbuildings." Large areas of poorer land were reverting from arable through rough pasture to scrub and forest. Yet there were also "several kinds of newer corrugated iron buildings . . . broiler chicken houses with two or three rows of large ventilating fans spaced along their sides, and piggeries and egg houses . . . with a single row of fans." On the fringes of valley towns and along many a country road there were clusters of nondescript mobile homes and prefabricated houses little different from new buildings elsewhere in North America. On a new highway a few miles from old established towns, a new service center was taking shape. There, automobile dealerships, shopping malls, fast-food franchises, parking lots, mobile home parks, new subdivisions, and apartment blocks sprawled in garish realization of the forces of change transforming time-worn rural landscapes across the continent (Blackmer 1976, 32, 36, 43, 63).

In Annapolis and Kings counties (broadly the Annapolis valley) the farm population fell from more than 50 percent before World War II to approximately 8 percent in the mid-1970s, and the number of farms declined from over 4,000 to fewer than 1,000. On Grand Pré, the heart of eighteenth-century Acadia, mechanization—first hay balers, introduced to the area in 1947, and then larger, cheaper tractors—enabled farmers to cultivate and harvest larger acreages and put up feed for more cattle. Year by year the land was concentrated in fewer hands. In North Grand Pré, where once there had been fifteen farms, only two remained in the 1970s. Similar developments elsewhere on Grand Pré left only twenty-five farms, all highly specialized. Seven dairy operations counted about 800 cattle among them. Two sizable broiler farms, three greenhouse tomato producers, and several pig farms of varying range were also located here.

These trends were accompanied by significant changes in the organization of agriculture. Before 1945 few farms were legally incorporated. Properties were owned and operated as family enterprises. But as former single-family farms were amalgamated into larger operations, as their owners faced the growing capital requirements of farming, and as the tax and inheritance advantages of incorporation were understood, so farms were established as corporate entities. Separate holdings, operated by members of the same family, were sometimes registered as single companies. In other instances, farmers possessed of entrepreneurial vision (and, often, a leaven of good fortune) bought up widely scattered properties over many years and established locally based agricultural "empires."

Even more decisive changes resulted from the entry of national and multinational corporations into regional agriculture. Recognizing that the quota system and price-stabilizing effect of the Provincial Marketing Board worked to sustain small producers of broiler chickens, Canada Packers established a subsidiary to kill and process the birds. Through its feed division it supplied valley farmers with feed and chicks. Producers who contracted with the company were guaranteed a market for their broilers. Still, production and ownership grew more concentrated. Producers either invested in capital-intensive technology—"chicken palaces" designed to maximize the rate of weight gain by 10,000 to 15,000 (or in extreme cases 30,000) birds in an eight-week period—or sold their quota rights. In 1967 Canada Packers owned less than 4 percent of the production quota. By the mid-1970s the company owned or controlled "nearly one-fourth of the total broiler quota, and supplie[d] feed to more than half of the broiler growers" (Blackmer 1976, 74–76, 91–93). Perhaps a quarter of all the chicken consumed in Nova Scotia was sold in franchised fried chicken outlets, most of which had opened since 1970.

In similar fashion, Hostess Foods acquired a small, locally financed potato chip manufacturer in the Annapolis-Kings area in 1960. A few years later they built a new chip factory and began to buy large quantities of potatoes from local growers. Before the end of the decade the company began to buy farmland. By 1974 it was the largest landowner in the Annapolis valley and by far the leading producer of potatoes in Nova Scotia. It cultivated large, well-located farms efficiently with large-scale and expensive machinery, had a guaranteed market in its factory, utilized hired managers, and hired seasonal labor as needed. Such efficiencies kept production costs low and provided the standard against which prices for contract growers were established. Between 1966 and 1971 the number of Kings County farms reporting the sale of potatoes fell by 75 percent, to 150; in 1974 there were a mere 30 commercial potato growers.

Elsewhere there were many farms, on land quite "unsuitable for farming on anything like a full-time basis." Hilly, inaccessible, with small, stony fields, and soils exhausted by years of scourging use, they were "below the competitive margin." Upon them lived an inordinate number of subsistence and part-time farmers. Many of these devoted a small area of land to growing vegetables and fruit, produced "enough cereals to provide the grain ration for a few hens, a couple of pigs, a horse, and one or more cows and young cattle," and used "the remaining and larger part of the total available acreage partly as pasture and partly to grow the hay needed to feed" the stock. Unable or unwilling to change as radically as the new order required, loath to accept unprecedented levels of debt, dubious of advice from "book-larned" experts, these persons were by and large unwilling to enter a new world of quotas, consignments, quality control, and contractual obligation. Their holdings were little more than "sentimental farms," rural residences for an aging population whose children were not interested in working the land.

In Ontario, similar changes transformed the countryside. As rapidly growing cities spilled beyond their official boundaries and growing numbers of people chose to commute relatively long distances by automobile or other means to urban employment, the province's rural population increased, but a smaller and smaller proportion of those who lived in the countryside were involved in agriculture. Farm acreage fell as marginal lands were abandoned and small-scale undertakings were consolidated into large, specialized operations. In the half century after 1951, the number of dairy farms fell significantly as the number of dairy cows and milk output rose. The production of hogs and hens was also increasingly specialized and "industrialized." By contrast, the proportion of Canada's beef cattle herd in Ontario fell by more than half (from over 40 to barely 18 percent) between 1976 and 2000 as that industry concentrated in Alberta.

In Québec, too, the postwar years saw a marked exodus from agriculture. In the 1950s alone, the proportion of the provincial population living on farms fell by half from 20 to slightly more than 10 percent. Subsistence farms were squeezed out of existence as commercial operations claimed an increasing share of investment and output. Expenditures on machinery, buildings, drains, and fertilizers—all of which spelled the modernization and increasing capitalization of farming—were crucial to success in the changing marketplace for agricultural products. Between 1941 and 1961 the proportion of farms with tractors rose from less than one in twenty to almost two-thirds. People left the countryside—by 1981 the agricultural workforce was barely 2.5 percent of the labor force—but agricultural output continued to increase. Where farms had once combined milk, beef, hog, and chicken and egg production by sending older milch cows to the meat market and feeding pigs and poultry with farm-produced milk and grain, now operations were increasingly specialized pork, poultry, or dairy enterprises. Between 1961 and 1981, the average number of pigs on Québec farms keeping hogs rose from 19 to 430. Slightly more than 10 percent of hog farmers kept over 1,100 animals each and accounted for well over half the provincial hog population.

On the prairies, wartime incentives to plant feed grains and increase beef and pork production encouraged the diversification of agriculture and a shift away from wheat. In half a dozen years after 1939, prairie wheat acreage fell by half and hog numbers quadrupled. Beef production also rose, albeit more slowly. These changes affected southern areas less than more northerly ones and Alberta and Manitoba more decisively than Saskatchewan, but their effects were lasting. In the mid-1960s wheat, once the staple of the region, accounted for only a third of farm sales in Manitoba and a quarter of those in Alberta. Here as elsewhere, mechanization proceeded as labor drifted away from the countryside. In 1971 there was $75,000 worth of equipment on the average prairie farm. A year later, geographer Thomas Weir echoed a refrain applicable across rural Canada when he averred that "the [prairie] farmer who lacks managerial skill, who regards farming as a way of life rather than a business, is sure to be eliminated in time" (Weir 1972, 97). In the next few years the region lost an average of over 200 farms a month. In the harsh contemporary assessment of one successful entrepreneurial farmer, cited by historian John Herd Thompson (1998, 178), "A lot of misfits have dropped out of farming. They just couldn't keep up with the challenges."

As market and financial pressures upon prairie farmers intensified, their farms became more and more like factories, designed for the most efficient production of particular commodities, and less and less subject to the unpredictability of the weather, the cycle of the seasons, and the vagaries of nature. New techniques, new organizational strategies, and new technologies quickened and deepened the human modification of the natural environment. In the 1970s pipes, pumps, dams, earth-moving machinery, and the increasing sophistication of electronic control devices facilitated irrigation and doubled the area of former prairie watered artificially. By century's end thousands of hectares were under irrigation.

The development of "feedlot alley" in Alberta neatly epitomizes the massive transformations that remade the nature of farming late in the twentieth century. Long the center of a range cattle industry that shipped most of its stock elsewhere (especially to Ontario) for fattening and slaughter, this very dry region north of Lethbridge was changed almost beyond recognition after the Second World War. First, improvements in refrigerated transportation allowed cattle to be killed and chilled in the west, rather than being shipped as beef on the hoof to processing plants in the major centers of consumer demand. Then adjustments to rail tariffs (with the elimination of the "Crow Rate" in 1983) improved market access; the expansion of irrigation expanded local production of fodder and feed; and improvements in the chilling process (specifically the capacity to produce "boxed hermetically sealed beef") and changes in investment patterns (that shifted ownership of much of Canada's meatpacking capacity to two American interests, each of which operates "a single continental scale beef plant in the heart of Alberta's feedlot country") reshaped the commodity chain linking farm calves to supermarket meat counters (MacLachlan 2002, 329; more generally see also Pollan 2006).

Suddenly, feedlots that had operated for years to carry cattle through the winter became year-round operations. At the turn of the millennium there were some 400 feedlot operations "finishing" cattle purchased from some 15,000 breeders (or "calf-producers") and selling them on to approximately a dozen meatpacking plants. In this system, feedlots are fattening pens. Land use is intense. One typical operation holds about 15,000 cattle in outdoor corrals that cover some twenty acres. Before they became a feedlot, these twenty acres might have provided sufficient spring-summer range for three or four cows.

About 1,600 acres of nearby land, watered by center-pivot irrigation systems, yield corn silage and hay (at about six or seven tons per acre with three cuts a year). Other feeds and nutrients are brought in. On all feedlots, cattle weights are carefully monitored, as are the types and amounts of feed given them; input into a "beef cattle economics" spreadsheet, these data allow close monitoring of feed conversion rates, average daily weight gains, and money received per pound of gain.

All of this has significant environmental consequences. Cattle in these numbers produce vast quantities of manure and methane gas. In the hot dry summers the resulting odor is a significant nuisance, and the powdery dry manure dust a potential health hazard. In this water-short environment, some operators sprinkle their feedlots for an hour or so twice a week to keep down the dust. Local codes require that feedlot manure be spread on farmland at specified rates per acre and that it be turned into the ground within 72 hours thereafter. The manure is rich in nitrogen and phosphates and its high pH helps to offset the salination associated with irrigation. But many feedlot operators do not own sufficient land to spread their manure at required rates. In most cases their surplus is trucked to fields miles away. In others, it is composted in specially prepared, berm-bordered sites for several months before being utilized to fertilize irrigated horticultural land or city parks. This is an expensive and laborious process, but completed properly it does kill pathogens and reduce the moisture and nitrogen content of the waste. Yet leakage, seepage, spills, and groundwater contamination are ever-present dangers. In 1997 a feedlot operator dumped 30 million liters of runoff from his cattle pen into a tributary of the Bow River. The $120,000 fine exacted for this "appalling negligence and irresponsibility" may have deterred others from similar actions, but less deliberate transgressions of health and safety requirements are probably endemic.

Liquid manure collected in catchment ponds on farms across the country is often used as fertilizer. Somewhat degraded by anaerobic bacteria and sunlight, this partially treated waste adds nutrients to the soil, but if application rates are too high or the ground becomes saturated, bacteria and other microorganisms normally filtered out by the soil can contaminate surface and groundwater. In the Ontario town of Walkerton, surrounded by factory-style hog, dairy, and cattle farms, seven people died and many more were sickened in 2000, when the water supply was contaminated by *E. coli* bacteria believed to have come from animal waste from one or more of the surrounding farms. Cattle densities in the vicinity of Walkerton are among the highest in Ontario, and surrounding Huron County has 400,000 hogs.

The disaster was a local one, but the threat of similar problems is widespread. Some eighteen million hogs are raised in Canada for meat, and estimates suggest that each of them produces almost two tons of feces and urine a year. "By raising animals in high densities," environmentalist David Suzuki has written, "we have in effect created crowded cities of animals, with the waste to match" (Suzuki 2000, 1). Without strict regulation and proper treatment, contamination is only to be expected. Indeed, inspections of all water treatment plants in Ontario after the Walkerton crisis revealed that well over half fell short of provincial clean water and monitoring standards. In the nine months after the Walkerton tragedy, boiled water advisories were issued for 70 Ontario municipalities, and 90 Québec water systems. Not all of these were attributable to agricultural contamination, but the pattern of agricultural intensification is clearly a contributing factor, especially as factory farms create larger volumes of waste than the surrounding environment can absorb.

Early in the twenty-first century scientists once more drew attention to the enormous and insidious consequences of refusing to recognize the implications of human disruptions of ecological systems by referring to Lake Winnipeg as "Canada's sickest body of water" and comparing its condition to that of Lake Erie in the late 1960s when journalists declared it dead. For decades sewage effluent and other pollutants from cottages, recreational activities, and settlements on and near Lake Winnipeg entered the large but relatively shallow lake. By and large it seemed that the assimilative capacity of this, the world's tenth-largest body of fresh water, was sufficient to cope with the inputs, as nutrients were flushed from the lake to Hudson Bay.

However, the combination of a growing population in the lake's million-square-kilometer catchment, the development of huge hog farms in the eastern prairies (hog numbers in Manitoba increased from 870,000 to 5.35 million in the last quarter of the twentieth century), the massive increase of fertilizer and manure use in agriculture, and the restriction of the lake's egress to the sea through the enormous geoengineering project known as the Churchill River Diversion–Lake Winnipeg Regulation scheme has led to rapid accumulation of nutrients in the lake. This has fed algae blooms to the point that a thick blue-green slime now covers about a sixth of the lake's surface (some 4,000 square kilometers). This produces vast amounts of algal toxins, which consume oxygen as they decay, leading under certain conditions to the eutrophication of the bottom of the water column and potential kills of certain fish species. For all that, reflected one of the country's leading ecologists, "We'll probably need major episodes of waterborne disease to smarten up politicians and bureaucrats before there is any action" (David Schindler quoted in *Globe and Mail*, 11 December 2004, F6).

• • •

The postwar years were both invigorating and difficult for the east coast fishery. On the one hand, the industry was the focus of an aggressive set of developmental initiatives; on the other, its oldest and in many respects most important constituent, the cod fishery, was subject to grievous overexploitation, the crash of the stock, and the imposition of an embargo on further utilization. Tragically, the two were not unrelated. Contemplating a struggling, labor-intensive, and relatively impoverished salt-fish or inshore sector and recognizing both the growing market for fresh and frozen fish and the intensifying competition on the offshore Banks, a report on the Atlantic sea fishery for Nova Scotia's Royal Commission on Provincial Development and Rehabilitation submitted toward the end of the war called for the modernization of the inshore fisheries and the industrialization of the corporate (offshore) sector. The former was deemed necessary largely for social reasons, to address the immiseration and deprivation evident in scores of underdeveloped communities long under the yoke of the notorious "truck" system by which merchants effectively exercised control over the fishery by holding fishermen in debt. The latter was regarded as an entirely logical, modern, technological response to changing circumstances.

The effect was the bifurcation of the fishery into two very different sectors, one a decentralized, small-capital community-based affair marked by high levels of independent ownership, the other dominated by large-scale capital, corporate interests, and centralization. Into the last quarter of the century, the former was arguably (and contrary to official and bureaucratic expectations) more successful than the latter, especially in Nova Scotia, but because they depended upon the same resource, their prospects ultimately converged in misfortune.

In midcentury Nova Scotia, the inshore fishery was part of "a special type of economy, in which there is generally an integration of fishing, farming and forestry pursuits" (Bates quoted in Apostle and Barrett 1992, 62), that authorities sought to modernize by introducing new technologies (longlines and seine nets) and developing small, efficient, and widely distributed plants to process a diverse catch. Shifting markets and improvements in mechanical drying equipment transformed the salt-fish industry, driving the half-dozen large processors who dominated the industry in 1945 out of the trade and giving it a subsidiary and complementary place in a fishery producing fresh fish and live lobster for markets in the urban northeast.

New vessels—longliners, stern trawlers, and scallop draggers—new fish-catching and navigation gear, new licensing strategies, and the provision of new social programs made the inshore fishery more capital intensive, more professional, and more productive. New independent, competitive processing plants serving independent fishers were "well adapted to a multi-dimensional and seasonal harvesting sector" (Apostle and Barrett 1992, 108). Occasional and part-time fishers were squeezed out of the industry. But in sum, the scale and structure of the inshore enterprise allowed it to respond relatively quickly and effectively to both seasonal and market fluctuations in demand.

By contrast, the offshore fishery was marked by centralization and the ever-increasing concentration of capital as state policies sought to increase the competitive position of the Canadian fleet. Subsidies supported the construction of larger vessels and the development of large-scale fish-freezing plants. In a dozen years after 1950 the Nova Scotia trawler fleet grew from five to thirty-seven vessels. By 1976 there were sixty-one trawlers over ninety-five feet in length fishing out of Nova Scotia. More than 90 percent of them were company owned. Through these years, National Sea Products (NSP) emerged as the dominant corporate player in the offshore industry and the leading recipient of government grants and subsidies intended to expand the fishery. In 1964 NSP built, in Lunenburg, the largest fish-processing plant on the continent, capable of handling 80 million pounds of fish a year. Several other plants were operated around the region. Some of the company's trawlers, part of the largest fleet in North America, could catch 300,000 pounds of fish each voyage. Trip turnaround times were reduced by half and pressures on stocks increased.

As ground fish—cod and haddock—catches declined after 1968, redfish made up the shortfall—at least until 1976, when the NSP fleet was banned from the Gulf of St. Lawrence in an effort to rebuild redfish stocks. Meanwhile, herring were subject to massively increased fishing pressure as world demand for fishmeal rose after the collapse of the Peruvian anchovy fishery. British Columbia fishing interests moved some of their fleet into Atlantic waters after the west coast herring fishery came close to collapse in 1963, and by 1968 the east coast herring catch was up 360 percent to 524,000 tons. Despite a decline in landings through the 1970s, the expansionary tendencies of the capital-intensive offshore fishery were fueled yet further by Canada's declaration of a 200-mile coastal management zone in 1977.

In Newfoundland the roots of change lay in necessity and in the 1933 British Royal Commission Report by Lord Amulree, which offered a damning indictment of the Newfoundland economy and recommended a massive reorganization of the fishery. Initially, the call was for better practices in the traditional fishery: better organization, better production, better markets. But even these were not easy to achieve. The colony—for it remained such—had been hard hit by the Depression and was heavily indebted; in 1934 its government was suspended and the island was, in effect, placed in receivership under the management of a Commission of Government. War brought temporary improvement, even relative prosperity, and in 1949 Newfoundland became a Canadian province. In a strict sense, this brought few direct economic benefits to the island as its resource base differed little from Canada's, but it did lower tariff barriers, provide access to social security benefits, and bring Newfoundland's fisheries policy into the same frame as mainland Canada's Atlantic fisheries strategy.

By this time, too, commentators in Newfoundland were urging that "as many men as possible should be diverted into the frozen fish trade and other branches of the fishery" as postwar circumstances and an expectation that world markets for Newfoundland's traditional salt-fish product would be over-supplied encouraged more radical recommendations for the industry's problems (quoted in Alexander 1977, 10–11). Still the salt-fish trade held such an important place as a source of employment and export earnings on the island that it would not be easy to shift away from it. A single Canadian strategy for reorganization of the Atlantic fishery was one thing; its implications in Newfoundland, where fishermen were a third of the labor force, were quite different even from those in Nova Scotia, where fishermen counted for only a tenth of all workers. In Nova Scotia planners could assume (with what David Alexander [1977, 14] called their "usual *sang froid*") that workers squeezed from the fishery would readily and happily find alternative employment elsewhere in the region; this was a less realistic supposition for Newfoundland, which was the poorest province in the country with personal incomes barely half the national average in 1950.

In the mid-1960s a Norwegian academic visiting Newfoundland was intrigued and puzzled by what he encountered. To his mind, conditioned by familiarity with the coastal regions of Norway, Iceland, and Scotland, the dualistic character of the Newfoundland economy was arresting: "modern, sophisticated, technologically up-to-date industries" existed alongside "economic practices and techniques . . . that appear to be almost medieval, such as inshore fishing and especially the processing of salt fish" (Brox 1972, 6). This pattern posed a number of questions, centered most strikingly on "the seeming immunity of the outport economy to technical modernism" and on the fishermen's reluctance to leave the traditional open-boat fishery for the modern trawler fleet.

Newfoundland officials were perhaps equally confounded, for they had endeavored to promote modernization and change in the inshore fishery for at least a couple of decades. New means of drying fish, "artificially" rather than by traditional labor-intensive methods of air-and-salt drying on flakes, were promoted. The government structured programs of financial support to move fishermen from the inshore to the near-shore fishery, providing substantial subsidies to "longliners," equipped with longlines, gillnets, ringnets, or Danish seines, but little other than a bait service and salt subsidies to those who used small boats and cod traps. And from 1953 formal government assistance was made available to ease, and encourage, the depopulation of struggling outports.

A dozen years later, when many economists and planners believed that economic development was best achieved by concentrating investment and human resources in "growth poles," federal and provincial governments joined together under the Fisheries Household Resettlement Program to relocate residents from outports to designated growth centers, which were nodes of the trawler-based offshore fishery and, in the minds of those who developed the scheme at least, more viable "economic and social units" than the isolated, underserviced outports. In sum 30,000 people were relocated and some 250 outport settlements were abandoned (Head 1967). Still, some 83,000 Newfoundlanders lived in households at least partly dependent on fishing in the early 1980s, and the industry was especially important in providing work for women, who made up the greatest part of the labor force in fish-packing plants.

Canadian interests were not alone in moving to larger, ever more efficient trawler fleets to exploit the fish of the offshore banks. As Paul Josephson has noted (2002, 197), in the early 1960s the nations of Europe and North America declared war on the fish of the North Atlantic Ocean. Vessels from half a dozen nations joined Canadians on the high seas. They were equipped with sophisticated equipment, "armed with the intelligence of oceanography and marine fisheries science," and harvested as though the fish were inexhaustible. According to estimates developed by fisheries biologists Jeffrey Hutchings and Ransom Myers, the 8 million tons of northern cod harvested between 1960 and 1975 approximately equaled the entire catch taken between 1500 and 1750 (see Fig. 17).

Canadian efforts to extend jurisdiction and management over the ocean were both an effort to limit the rapacious exploitation and an attempt to "save" Canadian fish for Canadian fishers. They did reduce fish mortality, at least temporarily. Yet even the 200-mile zone established in 1977 failed to encompass the entire continental shelf, leaving part of the banks open to exploitation from international waters. Surveillance and prosecution for "illegal" fishing remained both difficult and contentious. And the declaration of Canadian interest in this extended zone only served to encourage an intensification of the Canadian fishing effort as yields declined.

Ultimately, the 1977 initiative was simply too little too late. The brute-force technologies brought to bear, after 1950, on what had historically been among the best fishing grounds in the world were too much for the stock to bear. Between 1962 and 1977, the biomass of northern cod (the stock of the northern Grand Banks, northeastern Newfoundland, and southern Labrador) fell by over 80 percent, and that portion of the stock able to reproduce declined by 94 percent. By one estimate, the adult biomass of cod in this part of the North

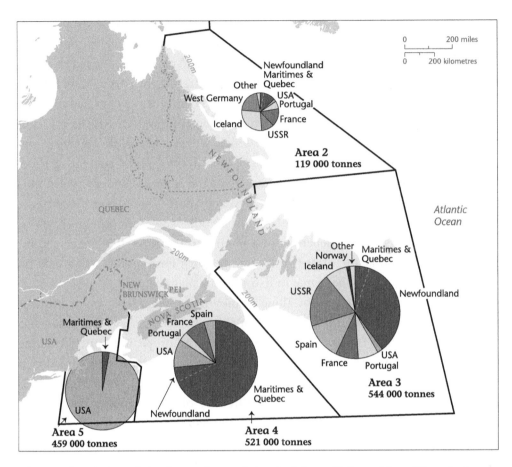

Fig. 17. *The 1958 catch in the northwest Atlantic fishery. (Adapted from* Historical Atlas of Canada 3: pl. 49)

Atlantic fell by 99 percent in little more than three decades. Despite a brief recovery of the stock between 1977 and 1985, further exploitation and further decline led the Canadian Government to impose a moratorium on commercial exploitation of northern cod in July 1992. Little more than a year later, similar moratoria were imposed on five other Atlantic Canadian cod stocks.

They continue. In "Rosie's Cove," a generic evocation of hundreds of Newfoundland communities, one student of the fishery noted in 1996, there are no nets hung to dry on the wharf, or "people leaving the fish plant at the end of a shift." Instead, there are beached and upturned boats on the slipway, and a single boat out catching crab. Now the longliners are used to take "tourists out to watch the whales" in the summer. "Meanwhile, a whole culture, one in which ecology and economy worked hand in hand, is dying before our eyes" (Ommer 1999, 29).

26

URBAN MAPPING

With a view to the future, the Canadian Government established an Advisory Committee on Reconstruction in 1941. Part of its final report, titled "Housing and Community Planning," drew a bleak picture of Canadian cities blighted by slums and surrounded by unsightly, ill-conceived suburbs, before urging the government to develop national programs for social improvement and community development. Almost immediately, the National Housing Act was amended to encourage new housing construction, to support the repair and modernization of existing dwellings, and to improve community environments. A year later the Central (now Canada) Mortgage and Housing Corporation (CMHC) was established to implement the housing agenda. One of its first publications was *Urban Mapping*, a manual predicated on the expectation that an "enormous amount of planning and mapping" would be required to cope with the "on-coming growth of cities," and intended to reduce the "bedlam of misunderstanding and wasted time" that would ensue if the makers and users of maps adopted a variety of graphic languages. Using Brantford, Ontario, as a case study, the manual sought to demonstrate "how every aspect of an urban community with its services, land-uses and densities" could be "described in black-and-white maps." It quickly became the "definitive codification of urban mapping" in Canada (Carver 1975, 118).

The mapping manual defined a language with which to think about and represent urban development. It also, and perhaps inevitably, served to shape the patterns and processes of change in actual places on the ground. Developed to assist planners in their work, it exerted a significant influence upon planning practice. Intended as a tool for the implementation and management of change, it became, in some ways, a driver and determinant of urban form. By simplifying, standardizing, ordering, and making planning procedures broadly intelligible, it offered a way of looking at, and shaping, the world that was entirely in accord with the modernist precepts of the period. The manual was a lens. It clarified the vision of the state and allowed its servants, in the persons of government officials and planners, to respond effectively to the welter of demands and challenges they faced in implementing the postwar agenda.

The influence of this exemplary atlas is reflected in the rise of planning as a profession. Canada's first four degree-granting planning programs were established between 1947 and 1951, but there were only forty-five planning practitioners in the entire nation in 1949. Twenty-five years later, nine more planning programs had been accredited and the Canadian Institute of Planners had some 3,500 members. To a very considerable extent, the ferment of ideas and needs; the social, economic, and political questions; the demographic pressures; and the technological opportunities and challenges that swirled around urban places in the last half of the twentieth century were distilled through the apparatus of professional accreditation, training in fundamental precepts, and manuals of practice into a relatively narrow set of widely implemented principles. Little wonder that historically quite different and distinctive cities across the broad expanse of (northern) North America came to seem increasingly similar, and that Humphrey Carver, one of the leading intellects of the CMHC who envisioned an urban society rather different from that he saw around him, lamented the "uniformity, conformity" of much urban development and urban living in the late 1970s (Carver 1978).

In Canada, as in the United States, a major (and ironic) thread in the story of postwar urbanism was the mass migration of people to outlying suburbs. Suburbs, per se, were hardly new in 1945, of course. But after years of economic deprivation and deferred family formation during the Depression and wartime decades, and rising levels of prosperity in the 1950s, tens of thousands of North American families bought into the suburban dream represented by a new home on the urban periphery. To provide for them, developers built tract housing on a hitherto unimagined scale on greenfield sites that cumulatively encompassed millions of acres across the continent. Typically, these developments were "sub-urban," in the sense that they lacked most of the functions and amenities of true cities. Most (male) residents of these "bedroom communities" commuted to work in relatively distant cities, while their wives and children spent their days in their suburban homes or schools. During these years, the effective area engrossed by already spatially extensive American cities expanded with dizzying speed; it was the age of suburban sprawl and (as American environmental historian Adam Rome has it) of "the bulldozer in the countryside."

The sheer scale of these developments was breathtaking. As Deryck Holdsworth has shown, they were most dramatic in Toronto, which assumed preeminence in the national urban system in the postwar decades, but they worked their effects on the edges of every major city in the country. In Toronto suburban growth was fueled by the rising importance of the city and its role as a major destination of Canada's postwar immigration. The population of the city proper (largely built up and occupied in the century or so since its boundaries were established in 1834) changed little between 1941 and 1961. It remained at approximately 670,000. But numbers in the surrounding municipalities skyrocketed. In the districts incorporated into Metropolitan Toronto in 1953, the population climbed from 242,000 in 1941 to 442,000 in 1951 and 946,000 in 1961.

The demand for new housing was immense, and many municipalities were almost overwhelmed by the combination of opportunities and challenges that it presented. Despite the best efforts of harried employees in local offices that were often understaffed, the growth of many suburban areas seemed less orderly than chaotic. The metropolitan fringe was marked by "tracts of 'strawberry box' bungalows and ranch houses and some new industrial structures [sited to secure good truck access] next to farmland" and the provision of "infrastructure and services had to catch up with" the frenetic building program (*Historical Atlas of Canada* 3: plate 60).

The new suburbs were the product of many things beyond popular demand. The automobile—or more precisely the widespread adoption of the automobile as a mode of private transport—was essential to their growth. Canada lagged behind the United States in this. In 1961, when the population of the country was 18.2 million, there were 5.5 million registered motor vehicles, of which only about 4.3 million were passenger cars. On average then, there were 4.25 persons for every car in Canada. In the United States the ratio was well below four persons per car in 1950 and in 1960 there were even fewer Americans per automobile. Automobiles offered their owners new freedoms by releasing them from the shackles of both timetables and tracks. With their own cars, individuals could come and go as they wished and they could drive wherever roads and wheels would take them. Convenience was combined with opportunity, as the automobile opened the interstices between streetcar lines (that generally ran in a more or less radial pattern from the city center), as well as the cheaper land beyond the outer termini of those lines, to residential development. If "the automobile was the passport to the postwar American dream" (quoted in Meinig 2004, 61), Canadians clamored to acquire the same means of embarking on a journey that massively reshaped the geography of the nation and the shape of its cities.

Improvements in the road network were an equal part of this story of transformation. At the end of the war there were approximately 20,000 miles of paved road in Canada, with less than a quarter of this mileage west of Ontario. Early in the 1960s paving covered 60,000 road miles, and another 225,000 miles or so (up from approximately 125,000 in 1946) were surfaced with gravel and crushed stone. Ontario led the way in road improvement; in 1961 it had more than a third of all the paved roads in the country. Québec accounted for almost a quarter. As suburbs spread and the number of vehicles grew, congestion became a problem. Roads built for fewer and slower conveyances quickly became clogged with cars. "Rush hour" entered the vocabularies and the lives of growing numbers of people.

Traffic engineers joined the expanding army of professionals proffering ways of addressing the problems of expanding cities by designing road systems that would accelerate the flow of traffic. New limited-access highways began to be built. Usually designed with multiple traffic lanes, with relatively gentle curves, and with modest gradients, they were a classic and widespread manifestation of the period's propensity to address problems by resort to technological systems. They cut a wide swath through the countryside, where cuttings, embankments, fills, bridges, and tunnels "smoothed" the topography, and with the great sweeping curves of access and exit ramps, sometimes divided farms and invariably fragmented landscapes and ecologies. In the interests of speed and efficiency, these new highways also typically bypassed smaller towns, whose importance was implicitly downgraded in the new order of things. In response, businesses and services gravitated to locations near the "interchange" that linked these older centers with (and symbolized their place in) a rapidly passing world.

Roads were essential for the automobile revolution, but cars also demanded space. In the suburbs, where their significance quickly imparted symbolic meaning to their size and design, cars required garages, and these soon became integral parts of suburban dwellings, often providing the main connection between the house and the street and almost invariably increasing the size of the standard housing lot. In central cities, commuter vehicles required parking lots where they could stand idle for most of the working day. Selling and supporting automobiles also consumed increasing amounts of space. Competition seemed to spawn gas stations at major intersections, and by the 1960s few sizable towns and no cities were without extensive roadside "strips" devoted to displaying and selling new and used vehicles.

When they were used in the various activities of everyday living—shopping, recreation—cars and the convenience they promised forced the provision of dedicated parking space close to the facilities patronized by their owners. To meet this need, shops moved back from the street front to allow angle parking, and thus more stationary vehicles, on commercial streets. In the competitive scramble to attract customers, businesses built off-street "parkettes," that were soon expanded into larger lots (Harris 2004). Such was the love affair with the automobile, its comfort and convenience and the sense of being "on the move" that it provided, that "fast-food" restaurants began to serve meals to those on wheels (bringing specially designed "tables" replete with burgers, fries, and

Traffic congestion, Toronto freeway. *Growing urban populations, ever-expanding low-density suburban housing, and the public's strong, continuing preference for the private automobile over public transit have crowded freeways in every major city in Canada, although Toronto (shown here) may have the greatest traffic congestion problem. (PhotoDisc, Inc.)*

milkshakes to customers sitting in their vehicles), and moving-picture companies built giant screens in enormous parking lots and called these space-devouring creations drive-in theaters. Soon, automobile-oriented "shopping centers" surrounded by enormous parking lots began to appear on the peripheries of cities. The first of these was built in West Vancouver in 1950, but its counterparts across the country were soon beyond enumeration.

Strategic political choices also shaped postwar suburban growth. Following the 1934 lead of the American Federal Housing Act, which essentially created the modern system of mortgage finance, the Canadian Government passed the Dominion Housing Act in 1935 in an effort to encourage house building and lift the economy out of depression. In effect, the government provided 25 percent of mortgage funds to approved lenders under this scheme, but the small and conservative character of the Canadian loan market largely limited its benefits to the relatively well-to-do. Slightly revised under the National Housing Act in 1938 and later amended and wrapped into the Central Mortgage and Housing Corporation (CMHC), the system of joint loans persisted until 1954, when the CMHC became an insurer rather than a partial-provider of mortgages. All of this made it much easier for individuals to invest in housing: Far more home buyers assumed mortgages, necessary down payments were reduced, and the insurance provisions encouraged lenders to "risk" home loans to those of more modest means.

The expansion of institutional finance spawned regulations that shaped patterns of urban investment By and large, mortgage funding was more readily available for new homes built to certain standards in suburbs provided with piped water and sewers than it was for dwellings in older, established, or poorly serviced areas. The promulgation of a National Building Code (the adoption of which was at the discretion of individual municipalities) also helped establish standards and simplified the administration of mortgage lending. At the same time, several provinces revised or introduced new planning acts, many of which required municipalities to prepare "official plans"—plans that frequently adopted land-use zoning to impose order on the landscape. The effects were striking. Only one Ontario municipality had an official plan in 1946; one other had an all-encompassing zoning bylaw. Ten years later, the corresponding numbers were fifty-seven and forty-eight (Harris 2004, 124). These developments gradually, and more or less effectively, brought growth on the urban periphery "under control," by imposing the planners' systemic, tidy, and rational vision on the landscape. In this new suburban world, everything had its place (housing here, shopping there, offices and industry someplace else) and there was neither room nor use for disorder.

Ultimately, suburban expansion could not have assumed the forms it did without significant changes in building technology and in the pace and scale of land development in the postwar years. Drywall—relatively quick and easy to install—replaced lath and several coats of plaster on interior walls; concrete blocks replaced poured foundations in some areas; and plywood, prefabricated roof joists, and preassembled windows and doors reduced the skills required in construction. Workers utilized a growing arsenal of power tools, including saws, sanders, and drills, on the job. By one estimate, the amount of on-site labor required to build a standard house fell by 60 percent in the two postwar decades. All of this held open a niche for the small builders (constructing about five houses a year) who dominated the early twentieth-century building industry and provided openings for newly arrived immigrant workers. But it also paved the way for a radical shift in the scale of suburban building on greenfield sites where corporate interests increasingly assumed responsibility for land development and housing construction. Here again, Canadian trends followed those in the United States where William Levitt, "the Ford of housing," turned "the assembly line . . . inside out" and produced gargantuan suburban tract developments, where everything from surveys and sewers through design and construction to the marketing and selling of "homes" was integrated in the operation of a single enterprise (Rome 2001).

Canadian ventures into mass construction and salesmanship were on a far smaller scale than Long Island's Levittown, with its 17,500 houses, but in the 1960s they could be seen on the outskirts of every Canadian city. From Westdale in Hamilton (1,700 homes) to West Lynn Park near Vancouver (800 dwellings) and from the outskirts of Ajax to Isle Jésus near Montréal (2,200 units), corporate developers and builders working to CMHC guidelines and broadly supported by that institution accounted for an increasing proportion of new Canadian housing construction. Development on this scale typically followed the Levittown model, in which extensive tracts of land were cleared of their vegetation, leveled, and contoured by heavy equipment and then surveyed and serviced, subdivided, and filled (probably over several years) with mass-produced housing.

Sometimes development adopted the grid, but more commonly—as in Don Mills near Toronto, the largest and most famous corporate suburb in Canada, with more than 8,000 dwellings on 2,000 acres—curvilinear streets and culs-de-sac provided the basic template for the landscape. The number of firms in the development business declined through the 1950s and 1960s, and the leading enterprises erected an ever-increasing share of the country's new housing stock. By one estimate, three-quarters of all homes financed under the

National Housing Act in 1961 were erected by merchant builders (Bacher quoted in Harris 2004, 143).

The results, reflected Humphrey Carver almost twenty years later, were "horrible" expanses of "impersonal, synthetic, exchangeable" dwellings (Carver 1978, 41). Such sentiments were neither new nor uncommon. Writing in *Saturday Night* magazine (18 July 1950), Mary Ross lamented the uniformity of the new suburban landscapes and adumbrated the critique of modernist conformity given popular voice in Malvina Reynolds's famous song about "Little boxes on the hillside, Little boxes made of tickytacky / Little boxes on the hillside, little boxes all the same," when she described "Every home, every family, every person exactly the same. Every single detail the same except for the house numbers." A few years later, Ontario planner Norman Pearson decried the plight of those whose lives were lived in places devoid of the benefits of the countryside and lacking the appeal of cities: "Hell," he proclaimed with stunning conviction, "is a suburb" (Pearson 1957). For all that he (and a small group of like-minded others) worried that the suburbs had established "special compound[s]" in which women were to all intents and purposes "locked away," thousands of others—whose aspirations and expectations were shaped by widely read magazines such as *Chatelaine* and programs such as *I Love Lucy* and the immensely successful Québec *téléroman La famille Plouffe* (first broadcast in the fall of 1953 and subsequently adapted for English-speaking viewers), brought into their living rooms by newly acquired televisions—were appreciative of the space, amenities, privacy, and security offered by their (relatively affordable) new suburban homes.

Generally, critics of the suburbs pursued an intellectual rather than an activist, and a social rather than an environmental, agenda throughout the 1950s and 1960s. Like Carver and Pearson and their counterparts in the United States, they defined the crack in the picture window of suburbia in terms of the anomie, isolation, and monotony they associated with life in these ordered, sterile landscapes that, they insisted, wrapped their residents into illusory communities in which difference and diversity were willfully denied as residents purified identities to conform with pervasive public expectations of life in these broad and barren acres.

All the while, however, suburbs and their inhabitants were playing a consequential part in the transformation of the environment. As cities expanded, they generally absorbed productive agricultural and horticultural land on their peripheries. According to Statistics Canada, Canadian cities and towns engrossed 7,400 square kilometers of former agricultural land between 1971 and 2001. Although the area may seem relatively small, it represents fully 7 percent of Canada's best and most productive agricultural land. In Alberta "less than 2 percent of Class 1 land . . . was urbanized in 1971"; in 2001 the proportion was 6 percent. For Ontario the equivalent figures are 6 and 11 percent (*Vancouver Sun*, 1 February 2005, A4). Cities and the golf courses, gravel pits, and recreational areas they spawn have also filled swamps and drained wetlands. Those who live in these growing centers have grown increasingly dependent on ever-extending transport networks and commodity chains for their daily sustenance. One measure of this is the estimate that the food on a typical North American dinner plate travels an aggregate 1,500 miles or so (with all that implies for fossil fuel consumption and atmospheric pollution) before it is consumed.

As cities grew and impermeable surfaces (roofs, roads, parking lots, and paving stones) replaced vegetation and culverts replaced creeks and streams, hydrological cycles were changed dramatically. Groundwater levels dropped, evapotranspiration decreased, and rainfall flowed more quickly into rivers, whose levels rose and fell more sharply. In the hot weeks of summer, thousands of gallons of water were poured onto suburban lots to keep lawns green and flower gardens blooming. The rate at which incoming solar radiation was reflected or absorbed by the earth's surface was altered significantly as trees and grass gave way to roofs and asphalt. As downtown office towers crowded together and climbed skyward, they markedly changed microclimates in city cores, intensifying winds, reflecting radiation, and almost completely shading some areas of the surface. All of this transformed habitats available to birds and animals. New niches were opened for scavenger (or what David Quammen has called weed [or generalist] species—rats, starlings, pigeons, coyotes—as others closed (for several species of birds, in particular).

Much of this went unremarked through two decades of massive urban expansion across Canada. Then scientists began to investigate and report on the various consequences of growth, mapping and naming the existence of "urban heat islands" in which nighttime temperatures were sometimes elevated several degrees Celsius above those prevailing in the surrounding countryside; charting the hydrographic effects of development; and, perhaps most significantly, identifying the increase, and potentially detrimental health effects, of atmospheric pollution as a consequence of particle discharges from automobiles, furnaces, and factories. By this time, the public at large was also growing concerned about the environmental effects of virtually unrestrained urban growth.

Following the leads provided by such important American critics as Rachel Carson, as well as the arguments of an expanding cadre of concerned fellow citizens, and responding to the growing evidence of environmental despoliation before them, Canadians became increasingly aware of the interconnectedness of ecosystems and of the potentially deleterious and very human consequences of chemical pollution of the earth, air, and water. Joined with a revitalized critique of the social effects of automobile-centered urban development—inspired in part by the work of Jane Jacobs, who left New York City for Toronto late in the 1960s—this spawned significant pockets of resistance to the prevailing course of postwar growth.

Economics, local politics, and the play of personalities affected the salience and effectiveness of radical reform initiatives, but in Toronto and Vancouver—if not to the same extent in Saint John, Calgary, and Montréal—citizens' coalitions affected the course of urban development in important ways. In both cities, major freeway (and associated "urban renewal") projects were stopped after considerable political and capital investment in their development. In both centers, reformers elected to councils worked to redefine the futures of their cities. In neither case were they entirely able to carry the day. Development pressures, aspirations for "world-class" status, the "logic" of the capitalist marketplace (that continued in the main to regard land as a free-market commodity rather than as a community resource and thus to devalue ecological considerations), and other forces (including the sheer press of numbers) combined to ensure that established trajectories of growth were, at best, constrained rather than deflected. But in Vancouver, at least, the unique authority afforded the city under its charter has allowed a visionary group of planners to embrace broad ideas such as "the livable city" and "sustainability" to shape a vital inner core that many regard as a salutary and distinct achievement for its unusual and by-and-large successful attention to social and environmental values.

PART 6

Reflections on the Remaking
of Northern North America

L
ooking backward across the wide sweep of time and space encompassed in
the pages of this book, one might well feel impelled to cry: "Enough!" Ten,
fifteen, thirty, even forty thousand years of human history; hundreds of
millions of years of geological time; from a cold continent covered by enormous ice
sheets to newspaper headlines about the problems of global warming and water
quality at the beginning of the twenty-first century—these are not the usual
parameters of historical narratives. Nor are geographers much accustomed to works
that span such vast areas, studies that stretch across the globe from the western coast
of Greenland to the outermost Aleutian Islands and range from the midlatitudes of rich
Carolinian forests (Point Pelee, Ontario, lies at latitude 42 degrees north) to the sand
plains of Siberia and the sparse scrub and barrens of High Arctic locations.

Prosaic though the measure may be, it is worth remark that these territories
include well over ten million square kilometers of land. Although the focus of this
work is squarely upon Canada and Alaska, plotting the outer bounds of its concerns
on a map of contemporary political jurisdictions carries its trace across the territories
of four nations, ten provinces, several states, and two territories—and this says nothing
of the shifting grid of imperial and commercial claims and administrative units laid
across this space, or of the links that have tied this extensive area to more distant
localities through the movements of people, goods, and ideas over time. In this view,

Canada and Arctic North America is surely and simply too big and too complex and too storied a place to be wrestled within the covers of a single book.

But this is an evasion. No one really expects that studies taking on continents or works treating entire countries will afford every square kilometer, or every inhabitant, of such territories equally detailed attention. Case studies, microscale inquiries, and the identification of representative groups and individuals typically serve to narrow the compass of inquiry while they ground analyses at scales commensurate with human experience and in ways that illuminate and extend understanding. Moreover, the world is infinitely complicated, and any attempt to make sense of it, now or in the past, at large scale or small, is doomed to incompleteness and partiality. There are limits to what we can know, and those who investigate the ways in which nature and society intertwine bring different yardsticks, or measures of significance, to their endeavors. Choices are made and emphases are placed in reflection of these things.

Some will wish that these pages told a different set of stories or perhaps another tale entirely. So be it. The canvas is full and the picture it contains is an expression of what an individual author was able to produce at a particular time. Now it is done, bar the task of offering a few general reflections on the broader implications of this account of the encounters between humans and nature across half a continent and its adjacent islands and seas.

27

TRADE, TECHNOLOGY, AND THE TRANSFORMATION OF THE ENVIRONMENT

The territories that became Canada were among many parts of the world encompassed by European expansion overseas, an expansion that began, effectively, in the fifteenth century and ran into the twentieth. The areas comprehended by this process are sometimes classified as either imperial territories, in which large numbers of indigenous peoples were administered and engaged in trade by relatively few expatriates, or colonies of settlement, occupied by large numbers of European migrants. The British settlement colonies included Australia, New Zealand, and South Africa, as well as Canada and the thirteen colonies that became the United States (and other territories, including Alaska, incorporated into that expanding political domain over time), and it is as well, toward the end of this sweeping survey, to consider northern North America within the broader frame defined by these developments.

Immigrants who came to the settler colonies everywhere initiated a sustained and thoroughgoing transformation of local environments. They were not the first to occupy and change these territories that they tended to think of, generically and revealingly, as parts of the "New World," because all of them had long sustained indigenous peoples. But the newcomers were immensely effective agents of change. Everywhere they displaced original inhabitants and disrupted their traditional patterns of subsistence (although, importantly and for the most part, native peoples and societies were resilient enough to persist). They gave places, plants, and animals new names; brought new species among the indigenous biota; claimed the land of native peoples; and replaced existing land uses with forms of their own devising for both practical and aesthetic purposes. From one perspective, they destroyed much of what they encountered. From another, for the most part their own, they improved it. At the very least they substantially remade the new worlds over which they extended their dominion and they did so at an increasingly rapid rate. According to Thomas Dunlap, the author of a wide-ranging study titled *Nature and the English Diaspora*, the settlers of these various territories spoke a common language of "conquest and command" as they encountered and extended their influence over the new lands (1999, 73).

None of this is to say that settlers never found beauty in colonial scenes or that colonists as a group lacked interest in the nature of the New World. Settler attitudes toward the flora and fauna, the landscapes, and the scenery of the new realms were complicated, multifaceted, and variable. Delight and promise mixed with revulsion and dread. From the first, explorers and newcomers evinced a great deal of curiosity about the natural wonders of the places in which they found themselves, and in the nineteenth century significant numbers of colonists embraced the study of natural history in the colonies. By and large, however, "conquest" and "command" were the dominant refrains in the environmental discourse of those who came to the land and seas of Canada and Arctic North America from the sixteenth century onward.

At some point, however, settlers encountered limits. This happened in different ways at different times in each of the settler dominions and across the broad expanse of northern North America. In the latter case, newcomers were inhibited, from the first, by the unfamiliarity of the New World, the challenges presented by its climate and soils, and by the always significant impediments to the developments of an agrarian empire in this fundamentally niggardly northern realm. With time and in one way or another, they overcame many of these difficulties. But eventually their march of conquest faltered. By and large, this happened earlier in the east, when and where agricultural settlement ran up onto the marginal backlands or into the thin acidic soils of the Precambrian Shield for example, later in the prairies as new strains of wheat pushed back the former climatic margins of agriculture, and perhaps hardly at all in Alaska (where many still see "the frontier" as a source of potential wealth).

In Dunlap's terms it was at such conjunctures that settlers everywhere "learned to obey" nature's constraints. Then they began to consider nature in a new light, moving to keep and protect it in various ways (through the introduction of new policies—"conservation"—or the setting aside of particular areas as parks, for example). The drive to conquer was moderated even if the wish to command persisted. Yet the pendulum of changing attitudes toward the environment was never still. Late in the twentieth century the triumph of high-modernist conviction spawned renewed and concerted efforts to subdue nature. The oscillation of public attitudes toward and national discourses about the environment, so often represented as a pendulum-like motion, continues, but as a contest (now as ever) among competing visions shaded in many and complex ways, rather than as a simple tug-of-war between opposing views.

This book says relatively little of such matters. Its main focus is on the material transformation of the environment by human actions, on the ways in which people, especially immigrant newcomers and their descendants, changed the face of the earth across the northern half of a vast continent. Limited attention is given to considering how settlers "constructed" or conceived of "nature"; there is relatively little discussion of the intricate, competing, and sometimes contradictory sets of ideas that lay behind the actions of those who transformed this territory (whether farmers, engineers, politicians, suburban dwellers, or forest workers). Political debates, the activities of natural history societies, and the civil disobedience campaigns of "new social movement" environmental groups are largely ignored. And virtually no notice is given such cultural activities as landscape painting, nature writing, and the broader body of imaginative prose and poetry in which Canadian literary scholars have identified distinctive attitudes toward the northern environment. This is not because any of these subjects is unimportant. Rather it comes down to the judgment that the main drivers of the story of environmental change unfolded in these pages are trade (which might be seen more broadly in this context as an element of capitalism) and technology (in its broadest sense), rather than the swirl of ideas that conceived of nature as something other than a commodity or resource.

Trade, says a 1994 publication of the Organization for Economic Cooperation and Development, has few direct effects upon the environment. But this is a recent and limited perspective, framed in the language of economists committed to advancing the global exchange of commodities in the name of economic efficiency. The essence of their argument is simple: "Trade increases the efficiency of international markets through the free flow of goods and services, which allows resources to be allocated to the least-cost and highest-return production activities. In this way, trade allows countries with different resources and advantages to overcome constraints of growth due to their own limited capabilities." Insofar as deleterious environmental effects might be associated with the exchange of commodities and the economic growth that this exchange is calculated to produce, they are largely the consequence, in this view, of "market" and "intervention" failures rather than of trade per se. Reading further, one learns that market failures occur when the market neglects to properly value and allocate environmental resources and when it fails to internalize environmental costs in the price of goods. Intervention failures result when government policies fail to correct, or possibly exacerbate, market failures.

Whatever the immediate merits of these contentions—and they essentially ignore the enormous consumption of fossil fuels entailed in global trading and the environmental consequences of such consumption—they are less than helpful in historical perspective. Taking a longer and broader view, such as that offered in this book, it quickly becomes clear that trade had significant implications for colonial environments. Indeed, the Canadian economic historian Harold Innis recognized, many years ago, that the particular circumstances of new countries such as the settlement colonies rendered them especially dependent for economic growth upon the exploitation of abundant, readily transportable, raw materials (resources) that required minimal processing before being transported to external markets. Furthermore, he made it clear that participation in the staple trades generated by these commodities had significant consequences for the environments of the producing countries. Later economists would formalize these consequences as linkages and multiplier effects, but in Innis's original formulation they were conceived of more broadly as influences shaping the development paths of New World societies. They can still be thought of in this way. In this view, staple trades were instrumental in the transformation of New World environments, although the precise dimensions of change depended upon the particular characteristics of the staple commodity, the patterns of its occurrence, the modes of its capture, the returns to labor generated by its exploitation, the constancy or otherwise of its market value, and its capacity to sustain continued exploitation, as well as the setting in which it occurred.

European, Russian, and East Asian markets for fish, whales, seals, and otters first drew newcomers into engagement with the coastal reaches of northern North America. There they exploited the biota and set in train the cascading series of contacts with native peoples that would ultimately and forever transform those groups' accustomed patterns of existence. On the generally forbidding shores and in the harsh northern climes where these activities concentrated, there were few opportunities and perhaps even fewer incentives for those engaged in the trades to settle and diversify their connections with the land. But their commercial enterprises exacted an environmental price, early depleting stocks of whales and otters, later inducing local shortages of fish, and ultimately driving codfish and bowhead whale populations to the threshold of extinction almost entirely in the service of market demand. Within decades after the opening of cod and whale fisheries on the Atlantic littoral of northern North America, European demands for fur drew traders inland, initiating a trade that would carry newcomers clear across the continent, from its St. Lawrence entry to its far northern deltas and on to its Pacific fiords. Trading posts grew into villages and villages into towns as this trade expanded, and significant numbers of people not directly involved in the fur trade came to be supported by it, providing services and provisions to those so engaged.

The rise of the transatlantic trade in timber from British North America—which some have seen as a staple trade par excellence—had even more consequential impacts upon colonial environments, as hundreds of thousands of pine trees were removed from the forest; fires burned over extensive areas; and rivers, lakes, and fish (in particular) suffered the effects of sawdust accumulation, dam construction, and the disturbance of riverine habitat. More than this, and more markedly than did any of the earlier staple trades, exploitation of British North American forests encouraged and facilitated settlement. It provided economic returns for those engaged in the process of settling the land and brought new immigrants flooding into the colonies, many of them on vessels engaged, on their eastbound journeys, in hauling wood to British markets. As colonial populations increased, so colonial economies diversified; towns grew, transportation links were improved, and local services—from carting and hauling to carriage and equipment manufacturing, from the provisioning of ships to the expansion of mercantile and legal and insurance businesses—expanded in tandem.

Important as the staple trades were to the development of colonial economies, in broad terms, there were limits to the impact of the market nexus on everyday life in the early colonies. Until the nineteenth century (in very approximate terms) many settlers sought to sustain themselves and their families rather than to maximize the profitability of their enterprises. Few were entirely divorced from the market. Almost all traded locally, and in these northern and colonial climes they depended upon imported goods (from tea and rum and sugar, to axes and crockery and fine fabric) to satisfy their day-to-day needs. But in their "wooden worlds" on the fringe of the Atlantic economy, many of those who occupied and developed lands around the Bay of Fundy, along the St. Lawrence, and later even in the backwoods of Canada and the eastern colonies sought a modest competence—the humble means of ordinary existence—rather than great wealth (Vickers 1990).

Technological reasons help to explain this. When life and work depended upon the energy and strength of humans and animals, and the additional motive power that could be derived from the wind and running water, there were real limits to the amount of land that most people could cultivate, the size of enterprise they could develop, and so on. In such settings, many remained somewhat apart from the market; they did not live in precapitalist isolation, but the power of the market, like that of the state, attenuated quickly in these still relatively sparsely developed settings. Neither market mechanisms nor government capacities were sufficiently developed during these years to value and allocate environmental resources properly, to internalize environmental costs in the price of goods, or to correct market failures. The transformation of the environment proceeded because settlers needed to convert it to their purposes in order to survive, and they embarked on this task in the substantial absence of economic theorizing and government concern about environmental costs.

Yet as the population of northern North America expanded and developments in the infrastructure and technology of communication facilitated the integration of space, trade and the market became ever more central to the shaping of land and life across this expanding realm. Rapidly rising American demand led the assault on the northern forest, for sawn lumber and then pulp-wood, in the latter part of the nineteenth and early twentieth centuries. The value of gold that lured prospectors and miners into the Fraser canyon, the Cariboo plateau, and the northern fastness of the Klondike and Nome, as well as dozens of other less-fabled sites, was a social construction, but a construction central to the capitalist economy and its expanding empire of trade. Various motives drew settlers to take up prairie land around the turn of the twentieth century, but few who came to this area were able to remain apart for very long from the burgeoning trade in wheat that underpinned the commercial economy. Farmers in earlier-settled areas of the country were forced to recognize and adapt to shifting circumstances that made the world's consumers their customers and the world's producers their competitors, or slide toward failure and farm abandonment if they did not. Mines in the Shield, the western Cordillera, and faraway Alaska produced minerals for national and international markets.

Through the twentieth century, the world trading economy continued to expand, although its trajectory of growth was interrupted periodically and more or less dramatically by events such as world wars, the stock market crash of 1929, and the OPEC oil crisis of 1973. The economies of Canada and Alaska were increasingly imbricated in this pattern of expanding commercial connection (the latter parts of which are now known as globalization). So, too, were the environments of these places shaped in accord with expanding trade's "role in maximizing allocative efficiency among nations" and "altering the international location and intensity of production and consumption activities," in part at least as a result of its effects upon market prices (Organisation for Economic Co-operation and Development [OECD] 1994, 13–14).

Throughout this period, the state's growing capacity for regulation and surveillance shaped the course of economic development and environmental change. Over the course of two hundred years, from the first flawed and faltering attempts early in the nineteenth century (Wynn 1977) to establish a modicum of authority over the exploitation of New Brunswick's forests (for reasons that had much more to do with law and order and revenue generation than environmental concerns), through to the ratification (in the case of Canada) of the Kyoto Protocol, the scale and scope of state interventions (statutes; policies; regulations; review procedures, such as impact assessments and monitoring requirements; and others) affecting the environment have increased dramatically. Entire departments of government have been assigned environmental mandates, and thousands of individuals toil in support of their directives and requirements.

Still, their success in persuading individuals and corporate entities to value and allocate environmental resources appropriately and to incorporate (and accept) a realistic measure of "environmental costs" in the price of goods has been far less significant than might have been. Nor have government policies often served to offer conspicuous and effective correctives to such "market failures." The centrality of trade to the capitalist economy, capitalism's inherent propensity to expand, and the power of international, multinational, and other corporate and political interests have combined to advance the agenda of expanding trade and done relatively little to ameliorate the environmental implications that have ensued. Early twentieth-century "conservation" strategies that sought to ensure "wise use" rather than the "waste" of resources when their finitude became apparent, the "preservation," as parks, of comparatively remote areas of relatively little value for conventional economic development, and the enthusiasm for strategies of sustainable development are all of a piece, in this light and according to proponents of ecological rather than neoclassical economics, in failing to properly value the environment.

As many have recognized, the human capacity to develop ever more powerful and efficient technologies has also been an important part of the story of environmental transformation through the last several hundred years. From their earliest footfalls in this northern realm, the first human occupants of northern North America moved into areas that were gradually growing warmer, and occupied biotic niches that were in flux, as the continental ice sheets continued their long retreat and climatic regimes, floral provinces, faunal distributions, and soil patterns came to approximate those that we know today. The cultures and societies of these first comers were far from static. They adapted to environmental circumstances by developing patterns of seasonal movement that facilitated sequential exploitation of varied resources and by devising new technologies of movement, shelter, and subsistence. The adoption and adaptation of fire, the development of new weapons (such as the bow and arrow), and the introduction of cultivation increased the first peoples' ability to utilize and to alter environments. These environments also changed in response to the pressures exerted upon them by humans. Still, on balance, and by comparison with what would follow, the native peoples of northern North America trod relatively lightly on the earth.

There is no need to conceive of these people as instinctive ecologists or protoenvironmentalists to account for this. With fire their most powerful tool (until the introduction of the musket and rifle, perhaps), few domesticated animals (the largest exception being the horse that came relatively late among the people of the northern plains), and fairly low population densities in most of the territory beyond Huronia and the Pacific coast, their capacity to effect significant and

Alaska salmon hatchery. *Environmental engineering has extended its endeavors to intervention in the reproductive functions of wild creatures. Responding to the construction of large dams on major salmon rivers, engineers built hatcheries to "replace" fish that could not spawn naturally. Soon this intervention was implemented on undammed rivers, in efforts to "cultivate and breed" large numbers of fish in "enclosed and artificial" environments where fry and smolts (young salmon) face "less risk of disease or being eaten." Market demands and falling stocks of "wild" fish have encouraged further interventions, in the form of fish farms now found at many points along the coast of British Columbia and Alaska. As Pacific salmon are difficult to farm, Atlantic salmon (a different species) have been introduced to Pacific waters, leading to environmental concerns about disease, escapement, and possible cross-breeding. (Corel)*

lasting environmental change was limited. These circumstances were forever altered by European incursions into native space, which generated new markets for the products of the native hunt and introduced new technologies into the territories occupied by native peoples: copper pots, firearms, blankets, nails, and similar items of trade. Carried into northern North America on perhaps the most consequential technology possessed by the early newcomers, the vessels that crossed oceans and connected them to distant homelands, populations, and markets, such goods initiated a technological revolution of enormous consequence for native peoples and the environments they had inhabited for many hundreds of years.

As the imprint of European peoples and influences gathered momentum and ranged across this former native space in the second half of the second millennium AD, both indigenous peoples and northern environments were caught up in an accelerating trajectory of change. As in other New World settings, encounters between newcomers and nature were often driven by the imperatives of survival and inflected by misconceptions about the world, to the point that outcomes were as unanticipated as the actions that produced them were unconsidered. Peoples, landscapes, and ecologies were transformed as once-remote territories were brought within the orbit of European-centered interests. Colonialism dispossessed native peoples of long-standing rights and territories. Forests were turned into fields, rivers into driving forces, coves into harbors. New connections to distant markets and to an expansive capitalist ideology led to the commodification of nature and the exploitation of its constituent elements (now identified as "resources") with a quickening frenzy. Beavers were turned into furs, trees into timber, and seas into protein producers. Few gave much thought to nature. The desire for improvement, for progress, framed "waste" and "wild" things (land, animals, plants) as inputs in the development process, as elements for transformation in the cause of human betterment.

In the end, the onslaught was constrained, for the most part, only by the limitations of the technologies that could be brought to bear. Attempts to regulate human impacts upon colonial environments were limited and late. Most of them were hollow endeavors that miscarried because they were poorly conceived or inadequately implemented. Administrators, almost invariably more interested in deriving revenues than in protecting nature, struggled against a thousand adversities. The public mood was against efforts at regulation and the restrictions on access that it implied, surveillance was difficult, and official power had little reach beyond the circles in which it was framed. In the age of wood, wind, and water, the New World was widely perceived as a field of human exertion, and most believed that their efforts to convert and tame and domesticate and exploit the environment were necessary steps toward fulfillment of their mandate to create a productive earth and to make the country be. Both people and places were altered by the protracted mutual engagements this entailed. By the middle decades of the nineteenth century, a population largely convinced that material progress was desirable and realizable had radically altered the look of the land in British North America.

The pace of environmental change only quickened as the evolving technological might available to the ever-growing army of those who occupied and sought to extend their dominion over northern North America was deployed in the century and a half or so after 1850. New sources of energy and power—from coal and steam through electricity produced by coal and hydraulic and nuclear generators to petroleum products and internal combustion engines—shrank this extensive territory by radically compressing time and space and increasing the human capacity to exploit and transform the natural world. New implements expanded the farmer's capacity to extract value from the land. New machines quickened the onslaught on fish and forests and dozens of other commodities. New processing and production methods suddenly rendered new materials valuable and made precious resources of formerly worthless assets. New knowledge, rooted in science and implemented by laboratory technicians and engineers, produced new forms of nature, exemplified by new seed varieties, hybrid species, and hatchery fish. Through the so-called Paleotechnic and Neotechnic revolutions, northern North American nature was subdued and transformed. Today, impressive engineering feats bring oceangoing vessels into the continental interior, harness rivers to light the dark, transform boreal spruce into urban newspapers, and extract thick crude oil from subterranean depths to convert it into (seemingly indispensable) plastics and gasoline (among other things).

Those who inhabit this realm have become increasingly divorced from daily engagement with and intimate dependence upon the natural world. Fewer and fewer Canadians and Alaskans live and work much beyond the urban centers that continue to concentrate wealth and lives in places that are overwhelmingly human-designed and human-built spaces. Automobiles and airplanes cocoon travelers from the environments through which they transport them. Furnaces and air conditioners operate without human intervention to blunt the extremes of seasonal temperatures. Foodstuffs arrive in neighborhood shops dressed and packaged and are available almost year-round. North American consumers receive (and possibly want) only scant information about the locations and circumstances from which the goods in their shopping baskets derive—the feedlots and poultry batteries, the "nightsoiled" fields on the edges of crowded cities in the global south, the monocultural plantations of the tropics—and lament the occasional reminder (provided by the inflated price of Florida oranges after a severe winter storm, for example) that seasons and weather conditions might affect the availability and quality of such produce (Pollan 2006; Tucker 2000).

The Hibernia oil rig. *Reserves in the Hibernia oil field were proven in the late 1970s, but production did not begin for almost twenty years. In part this was due to the danger of icebergs drifting into the oil rigs necessary for exploitation of resources beneath the Grand Banks. Construction of the Hibernia rig (here seen in Halifax harbor) and other engineering and technological work for offshore oil development brought new economic activity to Halifax, and St. John's in particular. (Corel)*

28

HUBRIS AND HOPE

In the 2004 Massey Lectures—Canada's most prestigious and widely known forum for the dissemination of "original study on subjects of contemporary interest" by "distinguished authorities"—Ronald Wright offered his listeners "A Short History of Progress." He had little difficulty establishing the relevance of his topic. "Our civilization," he said a minute or two into his first lecture, "is a great ship steaming at speed into the future." Traveling faster, further, and more heavily laden than any of its predecessors, this vessel ("not merely the biggest of all time" but "the only one left") was spreading its wake over the shipwrecks of previous efforts to navigate these waters, as it raced toward the incompletely charted reefs and hazards ahead. In this predicament, Wright insisted, "the future of everything we have accomplished since our intelligence evolved will depend upon the wisdom of our actions over the next few years" (Wright 2004, 3).

Finding a course through the maelstrom—figuring out where we are going and getting there successfully—is clearly a hugely important and consequential task, but it is also an immensely difficult challenge. Shoals and narrows may be discernible, but some will likely become evident only at the very last moment, threatening destruction to the unwary. Other dangers, metaphorical icebergs and whirlpools, are fickle; they develop almost at random, they shift, they change, and they loom unexpectedly from the swirling mists of self-assurance and ignorance. In all cases, human perceptions of the immediate peril will affect responses and outcomes. Is it better to turn to port or starboard? Might the *Titanic* have survived by proceeding with her engines on "slow ahead"? The answers are as uncertain as our capacity to know the future. Faced with the need to go on, and the risks of doing so, avers Wright, the only sensible course is to understand tendencies that "can tell us what we are *likely* to do, where we are likely to go from here." It is important, in other words, to understand the past, to know where we come from and who and what we are, if we are to continue to prosper—or even make it through.

For Wright, the lessons of the past are a set of parables that repeat themselves; they tell a recurring story of progress and disaster. His lectures sought to illuminate this contention by ranging widely across time and space. They marked how "the perfection of weapons and techniques" by hunters of the Old Stone Age "led directly to the end of hunting as a way of life (except in a few places where conditions favoured the prey)." They rehearsed the sad plight of Easter Island, denuded until it lacked the resources to sustain its people, and thus reduced to serving as "as a microcosm of more complex systems, including this big island on which we drift through space." They carried listeners to the plains of ancient Sumeria, where irrigation yielded bumper crops, then turned the land barren and white as salination destroyed the productivity of the soil, so that "the desert in which Ur and Uruk stand is a desert of their own making." The fall of Rome in the fourth century and the demise of classic Mayan civilization in the ninth were not as dramatic as the collapse of Sumeria and Rapa Nui, but here, too, Wright argued in a clever double entendre, civilizations were pyramid schemes: In a literal sense, stone pyramids (and steel and glass office towers?) reflected both the hierarchical (pyramid-like) structure of human society and the demands that concentrated populations exerted on the natural capital of the earth through ever-extending food chains; more metaphorically, these civilizations resembled "pyramid" sales schemes, gathering wealth from the center to the periphery and thriving only as they grew (Wright 2004, 55, 64, 79, 83).

The idea that human history reveals a repeated pattern of triumph stalked by disaster is hardly original to Wright. Karl Marx, Oswald Spengler, Arnold Toynbee, and others marked the same cycle, though by and large they attributed its turn to factors other than environmental despoliation. To be sure, the American George Perkins Marsh stressed the impact of human actions upon the "balance of nature," and as early as 1864 he recited several examples of the deleterious social consequences brought about by human modifications of the earth. Others have said as much again at various times since the mid-nineteenth century. Yet both the cyclical view of history and the idea that abuse of the environment has been responsible for the fall of societies and civilizations have gained particular prominence of late. Perhaps this reflects a certain millennial anxiety. Perhaps it marks a growing sense of the need for urgent action to halt the latest in the succession of "runaway trains" that Joseph Tainter identified in a 1988 study of earlier disasters on the long track of human-environment interactions through time.

Certainly it has spawned other books than Wright's. Jared Diamond, the evolutionary biologist–physiologist–geographer whose Pulitzer Prize–winning *Guns, Germs and Steel* (1997) found reasons for the rise of European society to world dominance in the biophysical characteristics of the continents, authored a best-selling 2004 sequel titled *Collapse: How Societies Choose to Fail or Succeed*. Here, he too offers a slate of examples of once-thriving societies reduced to destitution (Rapa Nui, Viking Greenland, Norse Iceland, the Maya of Yucatan) on the way to concluding that the downfall of human societies can be explained by reference to five basic causes. These include war; shifting trading patterns; climate change; the damage that people do to their environments; and the political, economic, and social adaptations that they make in response to these ruptures. Like Wright's, Diamond's basic message is simple: Learn from history, or repeat it at your peril. There is no clearer example of this than the moral he draws from the story of Rapa Nui:

> Easter's isolation makes it the clearest example of a society that destroyed itself by overexploiting its own resources. . . . The parallels between [this island] and the whole modern world are chillingly obvious. . . . [It] was as isolated in the Pacific Ocean as the Earth is today in space. When the Easter Islanders got into difficulties, there was nowhere to which they could flee, nor to which they could turn for help; nor shall we modern Earthlings have recourse elsewhere if our troubles increase. . . . [In the] collapse of Easter Island society [there lies] a metaphor, a worst-case scenario, for what may lie ahead of us in our own future. (Diamond 2005, 118–119)

The eminent American scientists Paul and Anne Ehrlich have their eyes more firmly on the future than the past in their recent contribution to this fervent debate about whether humankind is "really on a collision course with the natural world" (Ehrlich and Ehrlich 2004, 7, 17). Convinced that this is the case, that we live in what the agricultural economist Lester Brown has called an "environmental bubble economy" in which "output is artificially inflated by overconsumption of the earth's natural assets," the Ehrlichs make population growth, conspicuous consumption, the need for a sense of limits, and sustainable governance their principal concerns (ibid., 319). They see these issues as important because of the need to heal the world of wounds in which we live early in the twenty-first century. And they are convinced that "a combination of hubris and ignorance of history has caused people around the world to ignore the environmental degradation that helped usher previous civilizations off the world stage" (ibid., 318).

Reflecting this, they invoke the refrain "Lest we forget" from Rudyard Kipling's poem "Recessional," to stress the importance of looking backward to see forward. More than this, they echo Wright and Diamond in their sense that global civilization stands on the brink of collapse. The title of the Ehrlichs' book, *One with Ninevah*, is taken from Kipling's 1897 verse, in which he uses the decline of that once magnificent Assyrian capital in the heart of the Sumerian plain to suggest that pride and arrogance might yet lead the British Empire (at the very flood of its self-conscious glory in the sixtieth year of Queen Victoria's reign) into oblivion: "Lo, all our pomp of yesterday / Is one with Nineveh and Tyre."

Fully sixty years before Kipling wrote his "Recessional," a far less-famous Titus Smith regaled his audience in the Halifax Mechanics Institute with tales of environmental desecration in Syria and Sumeria. "The foot of man," he informed his listeners, "has not passed over what was once the kingdom of Idumea for ages. A few fishermen's huts are all that remain of ancient Tyre; and large districts, once thickly inhabited, present an appearance which seems to say, they will be cultivated no more." There were moral and "economical" lessons to be drawn from this, insisted Smith, but his purposes in invoking the ancient past were different from Rudyard Kipling's. Hubris was the target of Kipling's verse. Smith aimed his barbs at the capitalist greed that had, in his view, produced an undue expansion of manufacturing in Britain, and ran counter to the principle that "the necessaries of life are drawn principally from the culture of the earth" (Smith 1835).

The mantle of history is a complex, multihued garment, and one that is neither easily understood nor readily cast aside. It presents different shades to different lights, its meanings are open to debate and interpretation, and it lends itself to various uses. We neglect it at our peril. We simplify it to our cost. Reducing the past to a series of expostulatory cycles, in the manner of Wright and Diamond, is immensely useful in making emblematic and political claims. It is a wonderfully effective rhetorical strategy. By focusing on, and highlighting, a single pattern in the past, it brings clarity to events and developments that might otherwise and at best seem devoid of order, reason, and design. By fitting the stories of disparate times and places to a single template—rise and fall, progress and decline, triumph and disaster—it drives home the message that if this has happened before it can happen again and we should pay attention. Every orator knows that simplifying and repeating claims and ideas is far and away the best method of getting them accepted. No one concerned about the future of humankind on earth can justly begrudge the use of the past in this way by those who would do something about "the three elephants in our global living room—rising consumption, increasing world population, and economic inequity" (Ehrlich and Ehrlich 2004, dust jacket). And yet every historical scholar knows that the fascination and, yes, the truth of the past—to say nothing, for the moment, of the devil—is in the details.

This tension—between identifying the putative, uncomplicated lessons that the past holds for the present and grappling with the buzzing, howling factuality of former times in hope of understanding the circumstances in which people found themselves as they wrestled with the challenges and complexities confronting them—is unavoidable, and the pages of this book incline more to the latter than the former purpose. But this does not mean that the stories recounted here are devoid of meaning and significance in the present, or that they cannot be of value to us in contemplating the future. Not at all. Through the long sweep of time and space encompassed in this volume, we have encountered people of widely differing backgrounds, armed with radically different technologies, and holding markedly divergent views of the world and their place in it. We have observed them interacting in a great variety of ways with a strikingly disparate array of environments to shape ever-changing economies, societies, lives, and landscapes across the vast northern half of North America. The result is an immensely rich word tapestry, which like its fabric counterpart in Bayeux, serves both to record events and to reveal much of contemporary circumstances, while providing an object of contemplation and quiet instruction for those examining it in the present.

Triumph and disaster—those two imposters, as Kipling called them in another of his poems—walk through these pages, but unevenly, and often (as it were) in others' shoes, so that their footprints are uncertain and ambiguous. There is much in this book to suggest that progress produced decline, that improvement was followed by decay. Think, for example, of Purdy's mill dam in the township of Ops, of the relocations of Inuit, of the pulp mills that produced jobs and dividends but discharged noxious effluent into rivers where it accumulated in the food chain and poisoned those who ate heavily of the fish that swam in these polluted streams. Or consider, in a slightly different frame, the extinction of the great auk and the passenger pigeon, and the near elimination of northern cod fish stocks. Or think of the decimation and displacement of indigenous populations by diseases and peoples who entered their lands from afar. These were enormously consequential events that had hugely deleterious, and in some cases irreparable, effects upon people and environments.

But serious as they were, they were not (in general, and the terrible exceptions such as the extermination of the Beothuk aside) catastrophes to equal the crash of entire societies and civilizations. Any loss of species and any diminution of the possibilities of existence are rightly counted as a subtraction from the balance sheet of life on earth, and the demise of an entire people at the hands of another (even if it be in substantial part unintended or accidental) is indubitably tragic. So, too, the plight of numerous aboriginal groups who continue to struggle with colonialism's grinding legacy of poverty, addiction, and abuse is an awful charge against the actions and accomplishments of the colonizers. But there are credit as well as debit sides to ledgers, and such withdrawals should be placed into account with the positive consequences of human-induced environmental changes.

On balance and in material terms, at least, the conditions of human existence in modern-day northern North America are overwhelmingly superior to those that prevailed in the past. To be sure, many peoples and environments have been radically altered over the long sweep of time encompassed in these pages. Yet almost all have proven remarkably resilient; pliant and durable, they have been in almost constant flux since the retreat of the Wisconsinan ice sheets. This is not a simple story of people having their way with the earth, or of societies destroying the ecological underpinnings of their existence to the point that they are unable to persist. It is much more complicated than that. If triumph and disaster march through this account they do so in parallel rather than in sequence, with the latter scored, for the most part, in a minor key (see also Beckerman 2002 and Easterbrook 2003).

Of course, the transformations wrought upon the face of the earth through the centuries of quickening human exploitation and domination of the planet have been striking. Radically altered rivers below gigantic dams, drained and polluted aquifers, depleted fisheries, overharvested forests, the replacement of a diverse vegetation adapted to its setting (whether prairie grassland, temperate forest, or sphagnum bog) by the near-monoculture of canola fields, by even-age stands of single, commercial tree species, or even by the paved grid of urban growth, are but a few of the environmental consequences produced by the trajectory of recent development. Yet it is one thing to acknowledge that humans have left their mark on almost every square kilometer of the extensive territory considered in this book. It is quite another to argue, as some have on the basis of this observation, that nature has been destroyed, or that we confront its end. To sustain this view it is necessary to believe that "nature" is a pristine entity, that nature is "natural" and that it persists only so long as it is unaffected (undespoiled) by human activity. This is misleading. Although defenders of wilderness areas and national parks often seem to take such a position for granted, it depends upon particular social constructions of "wilderness" and "nature" that set these things apart from culture when they are in truth deeply imbricated with, and ultimately inseparable from, it (Cronon 1995; Loo 2006).

Insofar as nature can be set apart from culture, as a convenient (but still inescapably anthropocentric) shorthand by which to refer to the nonhuman world, it remains both dynamic and resilient. Altered and ordered, lacerated, scarred, subjugated, and forgotten though large parts of contemporary northern North America may be, it seems unlikely that the environmental changes made by people and societies on their way to achieving immense affluence and great material comfort will prove terminal. None of this is to deny the existence, at the beginning of the twenty-first century, of serious global-scale environmental concerns. Problems of atmospheric warming, melting ice caps, extensive drought—these and other environmental phenomena are frequently brought to public attention by news networks, politicians, scientists, nongovernmental organizations, and others. Less noticed but equally troubling are the dramatic environmental injustices, substantially attributable to consumption patterns in the global north, that set the global south apart from that realm. And we should not dismiss the possibility, currently advanced by proponents of "power laws" and complexity theory, that even large-scale environmental systems may reach a critical state of hypersensitivity and be subject to unpredictable ecological chain reactions.

In the end, however, the questions at issue here are questions of scale and perspective. Portentous though the predictions of global climate models are, and mesmerizing as the image of the earth as a latter-day Easter Island may be, there are good reasons to be cautious of sweeping claims that societies have crumbled because of the willfulness of their members and that people have perished at their own misguided hands. Diamond, the geographer David Demeritt has rightly observed, "is not especially self-conscious about his units of analysis." Dramatic though his stories of collapse among isolated societies are, "it is not entirely clear what, in an age of relentless global flows and interconnection, their present day analogues would be" (Demeritt 2005, S93). Moreover, we should be wary of regarding the future as a giant-screen replay of the past and in extrapolating from the long-ago fate of several thousand inhabitants of an ocean-girt speck of land to the fortunes of millions occupying a planet in orbit through the vastness of space. How, we might ask, will people respond to the frequent repetition of such stories? Although intended as dire warnings and reminders of the need for humans to change the very patterns of their lives, might they not, by the sheer weight of their doleful conclusions, do more to encourage passivity than activism? Fear can induce paralysis. Might these accounts conspire to generate a sense of fatalism rather than to encourage a spirit of optimism? Might they even foster an "enjoy it while you can" attitude?

None of these outcomes would curb spendthrift habits of consumption, limit society's dependence upon fossil fuels, shrink rising mountains of discarded inorganic waste, or limit the pollution of air and water that give everyday shape to the recurrent specter of environmental disaster. Ultimately, the challenge is to act, and to act in ways appropriate to the circumstances that confront us. Doing so will require sound knowledge of "how we got to where we are," an understanding that societies are not fixed, homogeneous entities but complex, ever-changing reflections of shifting power relations among competing political and social factions, and a certain faith that the future can be shaped in ways different from the ominous and inevitable forms implied by tales of environmental bubble economies and disaster parables (Mc Elreath 2005, S96–97; see also Turchin 2003).

Those who settled and developed northern North America through the last three or four centuries were not ecological vandals. Most of them were convinced, as most other members of Western society were persuaded (during the latter part of this period in particular), that material and moral progress could be both achieved and demonstrated by increasingly extensive exploitation of the earth. As the activist eco-theologian Thomas Berry has recognized, even in its current manifestation, "the assault on the natural world . . . [is being] carried out by good persons for the best of purposes, the betterment of life for this generation and especially for our children." The actions of these good people, like those of most who preceded them in settling and developing the continent, were based on a "vision of wonderland to be achieved by exploiting the natural world." Even though this vision was pursued ever more violently in an unreflective quest to "take the greatest possible amount of natural resources, process these resources, [and] put them through the consumer economy as quickly as possible," those who subscribed to it were not inherently bad people (Berry 1996, n.p.). By the lights of their times they were "good people acting for good purposes within the ethical perspectives of . . . [Western] cultural traditions." They also generally operated within a juridical framework largely derived from the English common law, which supports the "despotic dominion" of individual property rights yet affords little legal protection for the natural world (McLaren et al. 2005).

In the first years of the twenty-first century, there may well be an urgent need to change widespread attitudes toward "nature" and to shift deeply embedded cultural patterns (reflected in the incapacity of Western scientific and humanistic traditions to move beyond an anthropocentric privileging of the human and the failure of societies such as those considered in this book to recognize and properly to account for environmental values) in order to develop an ethic that places human and nonhuman concerns, ecology and society, at par. But this will not, it seems to me, be achieved by reducing the histories of humans in nature, of environments, to a single broad and easy formula—profligacy spells destruction—and insisting simply and repeatedly that we are all aboard the same sinking lifeboat. "Global citizenship" is a fine mantra, but it is one that is easier to utter than to practice. Instead of lamenting the hubris of those who steered us down the road along which we have come, and contemplating a future seemingly bound to replicate the blunders and catastrophes of the past, we should aspire to learn enough history to understand both the past and ourselves.

Lest we forget, environmental disasters have a human dimension. Rather than "just happening" to people, they are worked out over time, shaped—by both the day-to-day actions and omissions of individuals, groups, and societies and the (un)predictable "forces of nature"—in ways contingent on the rich social, cultural, and psychological contexts in which human lives are embedded. Ultimately, it seems to me that "whole earth" awareness can be gained only by attending to intricate, contested, local, particular, politico-economic and socio-cultural circumstances and considering them in and against larger regional and global political, ideological, and technological contexts. This is a tall order. But a close and sensitive understanding of the past can help us to meet it, if it brings us toward the realization that the damage wrought by humans in devastating forests, driving species to extinction, and polluting air, soil, and water diminishes us as much as it harms the environments over which we exercise mastery.

IMPORTANT PEOPLE, EVENTS, AND CONCEPTS

Acid Rain   This term is attributed to nineteenth-century British pollution inspector Robert Angus. It refers to the transformation of gases, particularly sulfur and nitrogen oxides, into acidic compounds in the atmosphere and their deposition on land and water surfaces as rain, snow, or dry particles that damage soils, forests, lakes, and rivers. The gases are produced by combustion and disseminated by the wind. Relatively pristine environments in northern Ontario, Québec, and the Maritime Provinces have been affected by acid rain produced in southern Ontario, Québec, and the midwestern United States.

   This became a cause of concern early in the 1970s, when the International Joint Commission discovered that much pollution entered the Great Lakes from the atmosphere. By 1978 precipitation monitoring networks had been established in both Canada and the United States, and in 1979 thermal-generating plants and nonferrous smelters were identified as key contributors to acid rain. The United States was found to produce about three-quarters of transboundary air pollution. Nonetheless, in 1985 Environment Canada persuaded the seven eastern provinces to reduce sulfur dioxide emissions by half. Five years later the United States passed the Clean Air Act. It followed the Clean Air Act of 1963, the Clean Air Act Amendment of 1966, the Clean Air Act Extension of 1970, and the Clean Air Act Amendments of 1977, but it was more stringent in mandating reductions, over a decade and a half, of smog and atmospheric pollution from 1980 levels of 10 million tons of sulfur dioxide and 2.4 million tons of nitrogen oxide emissions. In 1991 the Canada–United States Air Quality Agreement included further provisions for emission reductions and a commitment to establish a research and monitoring network. A 1997 *Acid Rain Assessment* determined that Canadian and U.S. emission levels had fallen by 43 percent and 28 percent between 1980 and 1995 (Environment Canada 1984; McKenzie 2002; Munton 2002).

Alpine Areas   Alpine areas are an important feature of the natural environments of Canada and Alaska and have become iconic landscapes. Canada's alpine mountains trend approximately north-south through Alberta, British Columbia, and Yukon, where the St. Elias Range rises to 5,959 meters in Mount Logan. In Alaska, the Coast, Alaska, and Brooks ranges trend approximately east-west and include North America's highest peak in Mount McKinley (6,194 meters). All of these mountains were massive obstacles for early explorers and settlers. Alpinists from the United States and Europe began climbing Canada's western mountains late in the nineteenth century. The Alpine Club of Canada, established in 1906, set up camps for its members and championed wilderness preservation and the protection of scenic, scientific, and recreational values in Canada's national parks. In recent years environmentalists have become concerned about the impacts of ever-increasing numbers of tourists visiting many of these alpine areas. However, the inaccessible Brooks Range in northern Alaska is heralded as one of the last large wilderness areas remaining in the United States (Reichwein 1996).

Audubon, John James (1785–1851)   Artist and ornithologist John James Audubon traveled by schooner to Labrador and Newfoundland in early summer 1833, passing through Nova Scotia on his return to the United States. He was the first ornithologist to visit Labrador, and recorded twenty-three new species there. He published *The Birds of America* in four volumes between 1827 and 1838 (Lank 1998).

Bateman, Robert (1930–)   Nature painter Robert Bateman was born in Toronto in 1930. A keen artist and naturalist, Bateman produced representational depictions of wildlife and wilderness as a child. As a teenager he used several contemporary styles, including postimpressionism and abstract expressionism, to interpret nature. Early in the 1960s, Bateman returned to realism because of his "reverence for the particularity of every square inch of the biosphere." A dedicated conservationist and a member of organizations such as the Federation of Ontario Naturalists and the Ontario Naturalists Club, Bateman is openly critical of the materialistic attitudes of modern Western societies. He advocates an appreciation of the beauty and individuality of nature found in simple, everyday subjects and scenes. Bateman's work began to receive critical acclaim in Canada in the 1970s and in the United States and Britain in the 1980s. He has been awarded the World Wildlife Fund's Medal of Honour, is an Officer of the Order of Canada, and in 1987 he received the Governor General's Award for his contribution to conservation in Canada (Bateman 2000).

Beaufort Sea   The Beaufort Sea is that section of the Arctic Ocean northeast of Alaska and northwest of Canada. Oil has brought it into the international spotlight. In 1923 the United States government withdrew a huge area of the North Slope, running 200 miles along the coast and 250 miles inland, as Naval Petroleum Reserve No. 4. A number of early drilling operations found evidence of oil and natural gas but were unable to locate a major reserve. The first major oil discovery in the region occurred at Prudhoe Bay in 1967 and 1968. With this discovery major oil companies purchased leases to drill in the area. Environmentalists and developers clashed over oil-related developments. But corporate attention was turning to the continental shelf. In 1979 Alaska sold leases for drilling in offshore territory along the Beaufort Sea. Environmentalists, concerned about bowhead whale stocks, attempted to halt the drilling, claiming that it violated U.S. environmental legislation. Their campaign was mostly unsuccessful, but the Interior Department agreed to a 20-mile buffer zone around Barrow to protect the whales. Offshore lease sales went on into the 1990s, with companies sinking 30 drill holes into the floor of the Beaufort Sea. But by the turn of the millennium these sites had been abandoned as dry or economically unfeasible (Bockstoce 1986; Ross 2000).

Beaver   The beaver, *Castor Canadensis,* is a large rodent. Most common in forested areas, beavers are found in streams surrounded by woody vegetation in most regions of Canada. Beaver pelts were used by native peoples to make robes and in Europe for the production of felt hats that became fashionable toward the end of the sixteenth century. The quest for beaver pelts impelled many early European explorers and traders into the interior of North America. Beaver furs became one of the most important trading commodities of early America and formed, with fish, one of the two great staple trades around which the political economist Harold Innis constructed his well-known thesis of Canadian economic development. The Hudson's Bay Company, founded in 1670, and the North West Company, operating out of Montréal after 1783 as a successor to earlier French-Canadian ventures into the interior, each traded chiefly in beaver furs.

At the turn of the twentieth century beavers were becoming scarce across North America. A beaver conservation movement began in the late 1930s under the leadership of **Grey Owl**. Federal and provincial governments have closed the trapping seasons on beaver over many years. This, combined with the intentional reintroduction of beaver into certain regions, has resulted in a massive increase in Canadian beaver populations. Canadians now celebrate the beaver as an unofficial national symbol.

Its image appears on stamps, coins, and trademarks and emblems, carrying connotations of both wildness and industry.

Big Game Hunting   The big game animals of Canada and Alaska include all species of bear, walruses, cougars, wolves, and ungulates such as deer, goats, moose, elk, buffalo, and musk oxen. These animals were the target of imperial hunting adventures that brought well-to-do British sportsmen into the remote reaches of British North America early in the nineteenth century. They were also increasingly sought after by North American trophy hunters, ranchers, and market hunters. In both Canada and Alaska, the turn of the twentieth century saw changes in wildlife law in response to the decimation caused by large-scale settlement, and, in particular, by market hunters (see **Buffalo Hunt**). Anxious to preserve wilderness and especially the sense of virile manhood that came from killing the creatures who lived in it, the sports-hunting lobby was successful in pushing governments to play a more active role in what would become known as wildlife management. By the beginning of the century, control over hunting in much of North America was being ceded to the state, and a new kind of government bureaucrat—the game warden—was charged with enforcing laws designed to restrict and regulate subsistence and commercial hunting in order to accommodate sports hunters.

Still, the first laws pertaining to game management in Alaska were viewed by many residents of that state as the absurd work of "outsiders" who knew little about the local situation. Officials and wardens generally ignored the statutes, and it took decades for Alaskans to recognize that bears and other big game animals had economic worth as attractions to wealthy nonresident game hunters and wilderness tourists (Loo 2001; Sherwood 1981).

Buffalo Hunt   Buffalo were a major resource of the indigenous people of the grassland plains and of the Metis. The meat of the buffalo was eaten; its bones were used for tools, its hide for clothing and shelter, and its sinews for sewing and binding. In summer native peoples herded stampeding buffalo toward a drop-off and sent them plunging over the edge to be butchered by the hunters waiting below. "Head-Smashed-In Buffalo Jump" in southwestern Alberta is a designated World Heritage Site. In winter native hunters built enclosures at selected sites often visited by buffalo, then funneled herds into these pounds where large numbers could be killed relatively easily. After European contact, buffalo populations started to shrink. There were many reasons for this decline. Among them were the incompatibility of free-ranging herds with agricultural land-use systems, the introduction of the horse and the gun, the development of a market for buffalo robes, and the access to the West provided by transcontinental railroad systems in both the United States and Canada.

Some have suggested that the horse allowed hunters to select heifers and young cows for the kill. This strategy would have had an impact on reproduction rates, but the precipitate nature of the decline in buffalo numbers suggests that the other factors, and perhaps especially commercial hunting, were more significant. In any event, a combination of these factors resulted in the near extermination of the prairie buffalo by the 1880s (Isenberg 1994; Lott 2002).

Canadian Pacific Railway   The Canadian Pacific Railway (CPR) was enormously influential in the development of Canada, in changing the landscape, and in fostering new attitudes to the environment. Railway construction began in 1875 at Fort William, Ontario, and the first train left from Montréal in June 1886. Construction of the railroad met many environmental difficulties, from the mountains of British Columbia to the rock and muskeg of the Precambrian Shield. To encourage completion of the railroad, the Canadian government offered large incentives for its construction. The CPR received $25 million in cash, 25 million acres of land, and a twenty-year monopoly over north-south transportation lines. Thus, the company quickly became involved in the sale and settlement of land.

To discourage speculators, they required that half the land of any block be brought into cultivation within four years of purchase. The CPR invested at the turn of the century in the biggest irrigation project in North America, to encourage settlement in the semiarid district of Alberta. The CPR also promoted Canada's scenery to foreign and domestic tourists, and hence played a key role in the development of Canada's national park system. In 1885, after CPR workers discovered hot springs on what is now known as "Sulphur Mountain," in Alberta, the Canadian Government declared the springs and the ten surrounding miles the "Hot Springs Reserve." In 1887 the reserve was enlarged and its name changed to Rocky Mountain Park. This was the first step in the creation of Canada's national park system.

Rocky Mountain Park was established for economic reasons rather than to conserve nature. The CPR aimed to capitalize on the picturesque scenery of this area and hence to make the line's expensive mountain section pay for itself. Under the direction of William van Horne, the company started a system of hotels that were touted to provide the best views of the picturesque Rockies and Selkirks. The first of these, the Banff Springs Hotel, was completed in 1888. At first these hotels were mere stopovers for passengers en route to the coast. However, later they became destinations in their own right. As automobiles were not permitted within park boundaries until 1916, the tourism attracted to this area was a key component of the success of the railway (Dempsey 1984; Eagle 1989).

Canadian Shield   An area of ancient, Precambrian glacial-scoured rock (particularly granite and gneiss), encompassing 4.6 million square kilometers of lake-dotted territory and constituting 40 percent of the Canadian land mass, stretching across six provinces and underlying most of the North West Territories, forms Canada's largest physiographic region. It has come to be known, interestingly and patriotically, as "the Canadian Shield." The Shield supports extensive boreal forests but offers little opportunity for agriculture. Although native people developed subsistence lifestyles well adapted to this environment, Europeans have never inhabited it in great numbers. Instead, settler society has preferred to exploit the Shield's resources from afar. The focus of this exploitation has shifted over time from a demand for timber in the early nineteenth century to pulp and paper in the early twentieth century, and then hydroelectricity and nonferrous metal (gold, nickel, and copper) extraction in the late twentieth century.

Indeed, geographer Iain Wallace has said that the very "wildness" of the Shield landscape—rendered iconographic by the work of the Toronto-based artists who became known as the *Group of Seven*—is now seen as a resource with aesthetic and spiritual values in need of protection. This has led to clashes between environmentalists, native peoples, and those living in local resource communities. Canadians continue to face the challenges of resolving these conflicts and of mitigating the negative environmental effects of industrial activities such as water pollution caused by pulp-mill effluents and mine tailings, and air pollution resulting from nickel smelting (Wallace 1998).

Canals   The Lachine Canal, built between 1821 and 1825 to bypass rapids on the St. Lawrence River, was British North America's first large-scale canal. In 1832 the Rideau Canal network was opened to provide a secure waterway (avoiding the St. Lawrence route) between Montréal and Kingston (regarded as vulnerable after the War of 1812). In the early 1840s the Chambly Canal bypassed rapids on the Richelieu River, and the Welland Canal between Lakes Erie and Ontario (opened 1829) was enlarged and reinforced in 1845. Each of these initiatives entailed locks, dams, and significant changes to river and lake hydrology. Investment in canals was invariably and rapidly rendered more or less obsolete by technological change, from the development of larger vessels to the construction of railroads. Several early canals were rebuilt to accommodate larger craft. The Welland Canal was again rebuilt between 1913 and 1932, with locks 260 meters long and 24 meters wide, and incorporated into the St. Lawrence Seaway project in the 1950s. By contrast, the Rideau Canal system became a recreational resource, well known for bass fishing in the late nineteenth and early twentieth centuries. Sport-fishing lodges were built on several lakes and with the development of motor launches, marinas and summer cottages began to proliferate. In 1926 the Rideau Canal was declared a National Historic Site, and in 2000 the historical and recreational value of the Rideau Waterway led to its designation as a Canadian Heritage River (Legget 1976).

Canoeing/Canoe Tripping   Canoes have become a potent symbol of Canadians' links to the natural environment and their human heritage. Canada's indigenous peoples developed canoes adapted to local environments and constructed of local materials. The birchbark canoe, strong and light, was perfectly suited for traveling the rivers and lakes of eastern and central Canada. In Arctic regions, Inuit peoples built kayaks of driftwood, stunted trees, and sealskins. The native people of the West Coast used dugout canoes, carved from the massive trees of the temperate rainforest. Europeans quickly recognized the importance of the canoe. Early French explorers such as Samuel de Champlain used native-built canoes to penetrate the Canadian wilderness. Later the canoe became the key mode of transportation in the fur trade, connecting a supply chain that ran across the continent from Montréal to the Pacific Coast and north to the Mackenzie River.

Recreational canoeing remains a popular pastime in Canada, allowing Canadians to reach wilderness areas that might otherwise be inaccessible. Canoe trips are central to the programs of summer youth camps in the Canadian Shield, and many Canadians believe that the canoe provides a sense of freedom and harmony with nature that cannot be achieved using other modes of transport (Jennings et al. 1999).

Clayoquot Sound, British Columbia   Clayoquot Sound on the west coast of Vancouver Island is an area of exceptionally beautiful, narrow ocean passages and large estuaries surrounded by temperate rainforest. Forestry corporations exploited the cedar, spruce, and later hemlock of these forests from the early twentieth century. From midcentury the pace of exploitation quickened and large clearcuts left huge bare patches on hillsides. Local people and environmentalists from afar grew concerned about habitat destruction and aesthetic damage, and clashes between environmental groups and forestry interests escalated after 1970. International organizations such as Greenpeace and the Rainforest Action Network made this one of Canada's most famous sites of environmental protest and one of the clearest examples of the globalization of environmental concerns.

The government of British Columbia responded by sponsoring a number of planning initiatives, aimed at finding a balance between resource utilization and the preservation of ecological, recreational, and scenic values in the area. Various task forces and committees grappled with these problems into the 1990s. The result of their labors, the Clayoquot Sound Land Use Decision, was announced in April 1993. It reduced the annual allowable cut by one-third but failed to satisfy either environmental groups or the local indigenous people, the Nuu-chah-nulth, who felt that their traditional uses of the forests and waterways should be accorded more respect. This led to the establishment of the Scientific Panel for Sustainable Forest Practices in Clayoquot Sound. The panel published five reports between 1993 and 1995 that advocated more ecologically sensitive forest management and greater respect for the knowledge, practices, and lifestyles of local peoples, both indigenous and nonindigenous (Berman et al. 1994; Magnusson and Shaw 2003).

Cod Fishery, Newfoundland   The Grand Banks and other parts of the continental shelf, stretching more than 200 miles into the Atlantic from the coasts of Newfoundland and Nova Scotia, formed one of the richest fishing areas in the world. In the sixteenth century fishermen from France, England, Spain, and Portugal worked these waters. Newfoundland became a European fishing station, where fishers spent summers before returning to Europe with their preserved catch. In the seventeenth century scattered, more permanent settlement began, and the continuing importance of the cod fishery to the northeastern part of North America was recognized by Harold Innis when he characterized it as one of the two great staple trades of early Canada. In the twentieth century the development of sophisticated computer-aided devices and new, larger oceangoing fishing vessels massively increased pressure on fish stocks.

In 1977 scientists from the Canadian Department of Fisheries and Oceans, responsible for monitoring fishing activity, announced that cod stocks were at an all-time low. The Canadian government extended Canada's Exclusive Economic Zone to 200 miles in hopes of better regulating the catch, but they could do little against foreign trawlers that fished just outside this zone. In 1992 the entire northern cod fishery was closed. Initially, this closure was intended to be a two-year moratorium, but in consideration of scientific evidence the government decided to extend it indefinitely. The government identified "severe oceanographic conditions" as the cause of this situation, yet inshore fishermen placed the blame firmly at the feet of the insatiable offshore fishing fleets. The collapse of the cod fishery called into question the ability of Canadian scientists to predict fluctuations in fish stocks and to properly prepare for them with well-documented and well-timed legislation (Innis 1940; Matthews 1995).

Commission of Conservation   The Commission of Conservation was established in 1909 by Wilfried Laurier's Liberal government. Chaired by Clifford Sifton from its inception to 1918, it reflected the influence of the U.S. conservation movement that had been influencing U.S. resource policy since the 1890s. Gifford Pinchot, the sage of conservation and scientific management in the United States, spoke at the first Canadian Forestry Convention, held in Montréal in 1906. Canada then participated in the North American Conservation Conference in 1909, where delegates drew up a declaration of principles regarding the use and management of such resources as forests, water, land, minerals, and wildlife, and called upon each participating country to establish its own commission of conservation. The Canadian commission was composed mostly of federal and provincial politicians, and university professors working in areas relevant to resource management. They were divided into seven working committees—for agricultural land, water and water power, fisheries, game and fur-bearing animals, forests, minerals, and public health—and charged with providing a detailed scientific understanding of Canadian resources. These committees published some 200 books and research papers and studies and provided the first large-scale survey of Canadian resources. The commission had less impact on public policy, and as resource management was integrated into the mandates of government departments, the commission became less relevant and was dissolved in 1921 (Artibise and Stelter 1995; Girard 1994).

Conservation and Environmental Movements   Leaders of the early twentieth-century conservation movement, particularly Gifford Pinchot and U.S. President Theodore Roosevelt, placed particular importance on the protection of Alaska's wilderness. They supported the imposition of strict game laws and the setting aside of Alaska's first national parks. They also influenced certain Canadian politicians and bureaucrats such as Clifford Sifton and James Harkin who advocated the wise use of Canada's natural resources (see **Commission of Conservation**). The official activities of these men were complemented by the more personal public crusades of conservationists such as *Jack Miner* and *Grey Owl*. Miner was heavily involved in Canada's most significant early commitment to conservation principles through the Migratory Bird Convention (1916), established with the United States to protect migratory birds. No major environmental nongovernment organizations operated in Canada or Alaska before the 1960s, although some Canadians and Alaskans belonged to U.S. organizations such as the Sierra Club and the Wilderness Society. This decade saw conservationists broaden their concerns beyond wise resource use, to encompass the impacts of human activities on the natural environment. Local environmental organizations such as the Alaska Conservation Society and the National and Provincial Parks Association of Canada (established in 1960 and 1963, respectively) were formed.

In the 1970s U.S. organizations such as the Sierra Club, the Sierra Legal Defense Fund, and the Audubon Society opened branch offices in Canada and Alaska and helped spawn local groups such as the Canadian Nature Federation and the Alaska Center for the Environment, which shaped local, state/province, and federal/national policies and raised the profile of environmental issues. Many Alaskans and hinterland Canadians resent the efforts of the environmental movement because they often focus on the negative impacts of activities that are deemed to be crucial to local economies (for example, mining, logging, and petroleum exploration). Many North American environmental groups such as Greenpeace (founded in Vancouver in 1970) are now international organizations with branch offices around the globe. As environmental issues are seen increasingly in a global context, these organizations have played a crucial role in the development of international environmental standards and agreements. Canada and the United States (and hence Alaska) are now signatories to numerous multilateral environmental agreements dealing with issues as wide ranging as climate change, forestry, endangered species, and air pollution (Ross 2000; Foster 1978; Loo 2006).

Drought, 1930s Dustbowl   Droughts are recurrent in many parts of the Canadian prairies, and particularly in Palliser's Triangle. As the margins of cultivation moved into drier areas, farmers used summer fallowing and shallow cultivation to preserve soil moisture. These strategies left soils vulnerable to wind erosion. Between 1929 and 1937, unusually cold winters and hot, dry summers significantly reduced available moisture and led to sustained and devastating drought over 7.3 million hectares, or one-quarter, of Canada's arable land.

Dust storms began in June 1930. Tons of topsoil were swept off the dry prairie lands, and large areas came to be thought of as "the dustbowl." Infestations of pests such as grasshoppers, sawfly, army worm, and cutworm compounded the farmers' problems. Plant diseases such as stem rust also followed the return of the rains in 1937. Agricultural science, a relatively new field, came to the fore during the drought years as scientists cooperated with farmers to produce new methods and tools for soil cultivation and poisons for pests. In 1935 the Prairie Farm Rehabilitation Administration was created to provide both financial and technical support for the agricultural community. It established forty-eight experimental substations to teach farmers how to cope with soil drift. Techniques included on-farm dugouts and dams for watering livestock, seeding abandoned land for community pastures, and tree planting to protect the soil from wind erosion. These measures alleviated the effects of drought and prepared farmers for future dry periods. Still, the economic impact of the 1930s drought was severe. In the 1920s the prairie wheat crop averaged 350 million bushels. Between 1933 and 1937 it was 230 million bushels. Between 1929 and 1937 a quarter of a million people left the prairie region, with another 50,000 moving to aspen parkland areas north of the grasslands (Friesen 1984).

Endangered Species   In 2000 the Committee on the Status of Endangered Wildlife in Canada (COSEWIC) listed 12 Canadian species as extinct and 15 as "extirpated" (they no longer exist in the wild in Canada but are present elsewhere). A further 326 species were described as "endangered, threatened, or vulnerable." Until 1960 Canadian wildlife politics revolved around game animals, and regulating hunting was the major concern. The Canadian Wildlife Service was concerned about the declining numbers of a few prominent species of birds and mammals such as trumpeter swans, bison, and whooping cranes. Similarly in Alaska, the U.S. Wildlife Service worried about the effects of hunting on polar bears and caribou, but there was little knowledge of or concern for other orders of the plant and animal kingdoms.

In the 1960s North American environmentalists argued for more attention to endangered species and the threats they faced from newly perceived dangers such as chemical contamination. In 1973 the Canada Wildlife Act encompassed a wider range of species, habitats, and threats, and two years later Canada signed the Convention on International Trade in Endangered Species of Wild Flora and Fauna. This required that Canada identify and classify its endangered species, and COSEWIC was created to do this in 1977. In 1988 RENEW (Recovery of National Endangered Wildlife) was established to develop recovery plans for endangered or threatened species. In 1996 wildlife ministers, directors of the Canadian Wildlife Service, and provincial and territorial wildlife agencies agreed to a National Accord committing them to "a national approach to species at risk." At the same time the government's effort to pass a Canadian Endangered Species Protection Act failed.

Canada's first federal legislation aimed at protecting endangered species came only in 2002, with the Species at Risk Act. Recent studies have identified several stressors of biological populations, including habitat destruction, pressures from non-native species, and the effects of chemical pollutants. Furthermore, certain species with economic value are still declining due to human harvesting. Research is starting to fill in the gaps left by the traditional biases of scientific enquiry. Knowledge of vertebrate mammals continues to be the most complete. Less is known about vascular plants, while nonvascular plants, lichens, invertebrates, and microorganisms remain mostly unstudied (Beazley and Boardman 2001; Burnett et al. 1989).

Federation of Ontario Naturalists   The Federation of Ontario Naturalists was founded in 1931 by members of seven smaller Ontario naturalist organizations. By 2004 it had grown to include 119 member groups and 25,000 members. It is a nonprofit society with a mandate to foster the creation of parks and other protected areas, to preserve wetlands and woodlands, and to protect threatened species and wildlife. It also maintains twenty-one nature reserves in Ontario. Among its identified priorities are the preservation of northern wilderness, the protection of Ontario's Carolinian forests and the species they support, and the designation of 15 to 20 percent of Ontario land as park or protected areas. The federation produces a quarterly magazine entitled *On Nature.* The federation's website can be found at www.ontarionature.org/ (accessed July 10, 2006).

Frontier Thesis   Advanced by American historian Frederick Jackson Turner in a paper entitled "The Significance of the Frontier in American History" (1893), the frontier thesis explained American development by the "existence of an area of free land, its continuous recession, and the advance of American settlement westward." Highly influential at the time, the theory is now regarded as simplistic and mystical. Canadian historians (see **Harold Innis**) argued the centrality of staple trades to Canadian development and saw little evidence of a continuously unfolding frontier of settlement occupied by independent agricultural pioneers north of the U.S. border. The same can be said of Alaska. For all that, the concept of the frontier as a site of societal renewal still holds some sway. Many regard sparsely populated areas such as northern Canada and most of Alaska as North America's last frontiers, offering an environmental safety valve for the overpopulated and overexploited locations farther south (Cross 1970; Kollin 2001).

Gosse, Philip (1810–1888)   Philip Henry Gosse, an Englishman born in Worcester in 1810 who died near Torquay in 1888, was a key figure in the popularization of natural history in Victorian England and British North America. He left England for Carbonear, Newfoundland, in 1827 and decided to devote his life to evangelical Christianity. In 1835 he moved to a farm north of Compton in Lower Canada and returned to England in 1838. Gosse's experiences in Compton informed his first published work, *The Canadian Naturalist* (1840), an important early example of Canadian environmental writing. Here Gosse described—in a manner somewhat reminiscent (if less charming than that) of Gilbert White of Sherborne—the flora and fauna of Canada's eastern townships and questioned "the adequacy of the human point of view as a primary focalizing device in descriptions of nonhuman nature." His later publications mostly considered the natural history of Britain. Yet his position as a key figure in the history of Canadian science was further secured with the 1883 publication of the *Canadian Entomologist* (Thwaite 2002; Irmscher 2001).

Greenpeace   The environmental organization Greenpeace was founded in Vancouver in 1970 by a small group of people opposed to nuclear testing in the Pacific. Their first environmentalist action was to protest U.S. nuclear testing at Amchitka Island, Alaska, but the group extended its oppositional tactics to testing by other nations on other islands, and later engaged in high-profile campaigns against whaling, the seal hunt, and clearcut logging. Today Greenpeace is one of the largest and most influential environmental organizations in the world. It claims an international membership of approximately 2.5 million people and operates in some thirty countries. Greenpeace International is based in Amsterdam. National offices in each country are relatively autonomous. They address local issues and contribute to Greenpeace International's major campaigns, among which those directed against whaling, global warming, and mineral exploitation in Antarctica are perhaps best known.

Greenpeace employs a range of nonviolent protest techniques ranging from petitions and the exertion of market pressures to direct intervention on the part of Greenpeace activists. These actions have received varying responses. While international campaigns such as those opposed to commercial whaling have been widely supported, those with a narrower geographic focus, such as attempts to halt the clearcutting of British Columbia's old-growth forests, have been strongly opposed by local organizations. Greenpeace's role in raising public awareness of environmental issues is unquestionable. No longer merely a protest group, Greenpeace is able to fund its own science and technology research and to participate in international treaty negotiations (Zelko 2004; Weyler 2004; Lahusen 1996).

Grey Owl (1888–1938)   The famous Canadian conservationist Grey Owl was born Archibald Stansfeld Belaney at Hastings, England, in 1888. As a child he dreamed of joining a tribe of North American Indians. When he was seventeen, he traveled to Canada and began to identify himself as the son of a Scotsman and a Jicarilla Apache woman. Under the guidance of the native Ojibway people (one of whom, Angele Egwuna, he married) he learned about the wilderness and became an expert trapper. The Ojibway called him *Wa-Sha-Quon-Asin*, which translates to "he who flies by night"—Grey Owl.

Working as a professional trapper in remote areas of Northern Ontario, Grey Owl met and fell in love with a young part-Iroquois woman, Anahareo, who disliked the cruelty of trapping and influenced him to study and write about, rather than hunt, the beaver.

The author of a number of articles in the Canadian Forestry Association publication *Forests and Outdoors*, Grey Owl published his first book, *The Men of the Last Frontier*, in 1931. He wrote three other books, *Pilgrims of the Wild* (1934), *The Adventures of Sajo and Her Beaver People* (1935), and *Tales of an Empty Cabin* (1936) while leading a beaver conservation program in Riding Mountain, and later Prince Albert National Park. By the late 1930s he was Canada's best-known champion of wilderness values. His delivery of conservation lectures in both Canada and England further boosted his international fame. Grey Owl died of pneumonia in Prince Albert, Saskatchewan, in 1938. Shortly after his death his real identity became known, and people reacted strongly against both the man and his message. Grey Owl's contributions to conservation and wilderness protection were largely ignored until the rise of the environmental movement in the 1960s (Smith 1990).

Group of Seven   The Group of Seven originated in the early 1900s when several artists working in Toronto, Ontario, discovered that their painting style and approach to art were similar. This group included J. E. H. MacDonald (1873–1932), Lawren Harris (1885–1970), A. Y. Jackson (1882–1974), Arthur Lismer (1885–1969), Franklin Carmichael (1890–1945), F. H. Varley (1881–1969), and Frank Johnston (1888–1949). Tom Thomson, said to have provided the initial inspiration that brought the group together, was also involved with the group in the early days but died in a canoeing accident in 1917. Dissatisfied with Canadian art, these men were critical of the use of conservative European artistic styles in Canadian settings and sought new ways of painting the uniqueness of the Canadian landscape. Seeking to make direct contact with nature through their art, they made numerous shared sketching trips to rural and wilderness areas of northern Ontario such as Algonquin Park. The rugged and vibrant northern landscape influenced their ideas and identity, and it is for paintings of this region that the group is best known.

The Group formed officially in 1920, when they first exhibited their works together. As their popularity increased, they traveled more widely across Canada and painted new landscapes, including the Arctic. In 1924 Frank Johnston resigned from the Group. Later the Group was expanded to include A. J. (Alfred Joseph) Casson (1898–1992), Edwin Holgate (1892–1977), and Lionel LeMoine Fitzgerald (1890–1956), the only western Canadian member. The "Group of Seven" was disbanded in 1932 after the death of its founder, J. E. H. MacDonald, but many paintings by group members and Tom Thompson remain iconic depictions of the Canadian "wilderness" (National Gallery of Canada 1995).

Haig-Brown, Roderick (1908–1976)   Author, magistrate, and conservationist, Roderick Haig-Brown was born in England in 1908. He came to North America in 1927 and worked as a logger and trapper in the Pacific Northwest before returning briefly to England and immigrating to Vancouver Island in 1931. A prolific author, Haig-Brown wrote twenty-eight books and made regular contributions to publications such as the *New York Times* and *Atlantic Monthly.* He is perhaps best known for his angling guides. *The Western Angler* (1939) is still considered to be the definitive work on western fly-fishing. His other works, including novels, children's books, and a number of essays, also dealt mostly with topics relating to nature. Widely respected within British Columbian society, Haig-Brown was a provincial magistrate and a Provincial Court judge for many years.

In later life, Haig-Brown spoke to audiences in both Canada and the United States on fishing, writing, and conservation. His speeches and essays revealed an increasing concern about the ethos of "progress" that was guiding the utilization of resources in the West. He was also involved in a number of local conservation battles. The first was an unsuccessful effort to save Buttle Lake in Strathcona Provincial Park from the hydroelectric power development in the 1950s. Later he was a key figure in the successful campaign against the High Moran Dam, a scheme that would have flooded the main channel of the Fraser River and led to the extinction of the Fraser River salmon runs.

Haig-Brown was a "gentleman naturalist" who interacted with nature mostly as a hunter and fisherman and who believed in the necessity of resource development. The struggle to reconcile his love of wilderness with his understanding of the need for development pervaded much of his writing. He was forward thinking for his time, describing nature as a highly interconnected and dynamic entity. Thus, he advocated a wise use of resources, which involved "putting them to use, seeking a valuable return on them, and at the same time ensuring future yields of at least equal value." His position as an accessible, popular author and public speaker allowed him to reach audiences that might not have otherwise considered these ideas. Therefore, he served an important function in provoking a wider contemplation of humanity's place within nature in the mid and late twentieth century (Keeling and McDonald 2001; Metcalfe 1985).

Halibut Treaties   In 1923 Canada and the United States signed a Treaty for the Preservation of Pacific Halibut. It established a three-month winter closed season and called for the creation of an International Fisheries Commission (IFC) to study halibut biology and make recommendations for future regulations. The treaty is significant for two reasons. It was the first-ever treaty aimed at international management of a high-seas fishery and set an important precedent for future fishery management. It was also the first-ever international treaty entered into by the Dominion without a British signatory, and so stands out in the political history of Canada.

A second treaty, signed in 1930, empowered the IFC to divide the Pacific coast between Oregon and Alaska into four zones and set the total catch of halibut in each. Minor adjustments were made to the treaty in 1954, and major changes in 1977, following the extension (by Canada, the United States, and other coastal states) of national jurisdiction to 200 nautical miles. The IFC continued to study the fishery and set national quotas reflecting the calculated Maximum Sustainable Yield, but Canadian and American fishery management organizations allocated the catch to achieve the so-called Maximum Economic Yield (Thistle 2004).

Harkin, James B. (1875–1955)   James B. Harkin, born at Vankleek Hill, Ontario, in 1875, was Canada's first Commissioner of National Parks. He was secretary to successive Ministers of the Interior, Clifford Sifton and Frank Oliver, before assuming office as Parks Commissioner in 1911, when the ministry created a separate park's branch. Harkin placed great importance on the moral value of outdoor education, arguing that wilderness areas should be both protected and made accessible to the public. A leader in the field of wildlife conservation, he played a key role in the development of the Migratory Birds Convention Act (1917). Later he worked tirelessly toward the establishment of a separate National Parks Act, believing that this was a necessary step to protect Canada's parks from the encroachments of industrial developments.

Under his leadership (1911–1936) the parks system expanded significantly with the number of parks tripling to eighteen by 1932. He established the first Canadian standards for wildlife preservation, control, and management. In 1972 the Canadian Parks and Wilderness Society created the annual Harkin Conservation Award to honor the memory and achievements of James Harkin. The award recognizes Canadians who have served the cause of conservation with distinction (Bella 1987; Lothian 1987; MacEachern 2001).

Innis, Harold Adams (1894–1952)   Canada's most influential economic historian, Harold Innis was a member of the University of Toronto political economy department from 1920 to 1952. Innis explored staples theory in two major works, *The Fur Trade in Canada* (1930) and *The Cod Fisheries* (1940). Innis developed a distinctly Canadian approach to economic history. His concern with resources has led some modern scholars to read a conservationist or environmental philosophy into his work. However, Innis was concerned with the economic, rather than the environmental, implications of Canada's situation. He warned that a staples-based economy was inherently vulnerable to market fluctuations, but was not concerned with issues of environmental management or sustainability. Nevertheless, staples theory has much to offer Canadian environmental historians, particularly in understanding the role of economics in resource exploitation and environmental perception (Creighton 1957; Watson 2006).

James Bay Project   Initiated by Québec Premier Robert Bourassa in 1971, the James Bay Project is a massive hydroelectric power scheme on the eastern shores of Hudson Bay in northern Québec. It formed a central component of Bourassa's plans for Québec's economic development. Power generated at the site was to be used in Québec and sold to U.S. states. The potential for negative impacts on native peoples and the environment made this a highly controversial project.

The local Cree and Inuit peoples were the first to challenge the scheme. Despite the fact that the project would cause extensive flooding of their traditional homelands, they were not consulted in any of the initial planning stages. Construction work on the first dam began in 1973, two years before the James Bay and Northern Québec Agreement was signed between the Cree, Inuit, the Province of Québec, the Canadian Government, and Hydro-Québec. The native peoples surrendered their lands under this agreement, receiving hundreds of millions of dollars in compensation.

Scientists also raised concerns about the potential impact of the project on Arctic ecosystems. As this was the first "mega-scale' hydroelectric scheme to be built in an Arctic region there was little scientific data for environmental planning. The first phase of the project involved the diversion of waters from the Eastmain, Opinaca, and Caniapiscau rivers to dammed reservoirs on La Grande Rivière. Completed in 1985, "La Grande Phase I" increased the average flow of La Grande Rivière from 1,700 to 3,300 cubic meters per second and created five enormous reservoirs. The project flooded thousands of kilometers of wilderness lands. Forests were also burned to clear debris and reduce the amount of vegetation rotting in the reservoirs. When not removed, this remnant vegetation created increased levels of mercury in local fish populations, and hence in the diets of local people. Environmental and native rights groups have focused much of their criticism of the project on this particular issue.

Phase II of the project began in 1989. Environmental groups in Canada and the United States opposed it so vigorously that Maine, New York, and Vermont canceled their contracts with Hydro-Québec, which led to the temporary suspension of work on Phase II in 1994. The high profile of the project ensured a proliferation of research on various aspects of its social and environmental impacts. This has, in turn, aided project planners in producing detailed environmental impact assessments for later phases of the project (Salisbury 1986; Hornig 1999; Carlson 2004).

MacKenzie Valley Pipeline   The federal government proposed the Mackenzie Valley Pipeline in 1974 to transport natural gas (and later oil) from the Beaufort Sea to Alberta. Two consortiums presented proposals for the construction of the pipeline. Canadian Arctic Gas Pipeline, Ltd., proposed a lengthy route from Prudhoe Bay in Alaska across the northern Yukon to the Mackenzie Delta and then south to Alberta. Foothills Pipe Lines, Ltd., proposed a shorter route starting in the Mackenzie Delta and running directly to Alberta. The challenge of building a pipeline over permafrost posed a major problem for both proposals. To avoid melting the frozen Arctic soils, a pipeline would have to be both refrigerated and insulated. Justice Thomas Berger of British Columbia led a federal Royal Commission of Inquiry into the pipeline proposals. Appointed in 1974, the commission was particularly concerned with the potential social and environmental impacts of the pipeline.

Berger himself became engrossed with the pipeline issue, making a long personal visit to the western Arctic in the summer before the hearings opened. The commission held formal hearings in large and small communities across the north during 1975 and 1976. In May 1977, its report, entitled "Northern Frontier, Northern Homeland," was released in Ottawa. The report warned against building a pipeline across the fragile environment of the northern Yukon. It found that a pipeline from the Mackenzie Delta down the Mackenzie Valley to Alberta was feasible, but proposed a ten-year delay so that native land claims could be dealt with properly. Plans for pipeline development were shelved. But interest has been renewed since 2000. Currently, the North West Territories government supports construction of a pipeline from Alaska to the Mackenzie River Delta into Alberta and south to the United States. In 2003 the Minister of Indian and Northern Affairs, Robert Nault, announced the creation of a "pipeline readiness office" to coordinate planning, environmental assessment, and regulatory review of a pipeline over the period from 2003 to 2006 (Pearse 1974; Berger 1977).

Macoun, John (1831–1920)   Botanist John Macoun is best known for his re-assessment of the agricultural potential of the Canadian prairies. Born in Dublin in 1831, Macoun compensated for his lack of scientific training with his boundless enthusiasm for the natural world. In 1872 he took part in a government survey of western Canada, and in 1880 he was employed by the Department of the Interior to further investigate the Southern Prairie region. At this time the standard evaluation of Canada's prairie land was based on information produced during the expeditions of John Palliser and Henry Hind. They identified a narrow fertile belt running through the Northern Prairies and an arid area, unsuitable for agriculture, in the Southern Prairies, known as "Palliser's Triangle." Macoun denounced these assessments, claiming that none of this land was desert and that it was well suited to agriculture. Following the presentation of these arguments in an annual report of the Department of the Interior for 1880, Macoun was appointed Dominion botanist to the Geological Survey of Canada in 1882 and became survey naturalist and assistant director in 1887.

Macoun's dismissal of Palliser's Triangle contributed to the choice of a southern route for the Canadian Pacific Railway and influenced patterns of prairie development significantly. However, historians (and some of Macoun's contemporaries) charge that he allowed fervent idealism to override scientific inquiry in serving the interests of the institutions that funded him. Certainly his work was expedient for those promoting agricultural settlement of the prairie regions, and the hardships suffered by the thousands who attempted to farm the Southern Prairies (see **Drought**) suggest that his conclusions about this region were optimistic. Macoun died in Sidney, British Columbia, in 1920. His large collection of flora and fauna is now housed in the Canadian Museum of Nature (Waiser 1989).

McIlwraith, Thomas F. (1824–1903)   Businessman and ornithologist Thomas F. McIlwraith was born at Newton upon Ayr, Scotland, in 1824. In 1851 he emigrated to Upper Canada where he managed the Hamilton Gas Light Company before buying his own business, which dealt in coal and industrial products. McIlwraith was a keen naturalist and spent much of his time in Hamilton observing birds. He shot, skinned, and stuffed his own specimens, and for a period in the 1880s and 1890s traded bird skins with collectors across North America. A longtime member of the Hamilton Association for the Advancement of Literature, Science and Art, McIlwraith was elected president of this association in 1880. Three years later he was invited to serve as a founding member of the American Ornithologists Union and was quickly appointed as coordinator of bird migration records for Ontario. His many years of bird observation in the Hamilton area made him an authority on local bird populations and in 1886 he published the land-mark work *Birds of Ontario* (McIlwraith 1894; Ainley 1985, 53–63).

Migratory Bird Treaty (1916)   The Migratory Bird Convention signed by the United States and Canada in 1916 is North America's oldest international wildlife conservation agreement and binds the two countries to coordinate efforts to protect migratory birds from indiscriminate harvesting and de-struction. The United States and Mexico signed a similar treaty in 1936. These agreements banned all migratory bird hunting in North America be-tween March 10 and September 1. But these arrangements failed to recog-nize traditional harvests of migratory birds by northern indigenous peoples (for whom the birds are a vital source of food) during the spring and summer months. The Parksville Protocol, an amendment to the Convention, which came into force on October 7, 1999, permits regulated spring subsistence hunting for the indigenous peoples of Canada and Alaska (Dorsey 1998).

Miner, Jack (1865–1944)   Born at Westlake, Ohio, in 1865, John Thomas Miner moved with his family to Kingsville in southwestern Ontario in 1878 and lived there until his death in 1944. As a teenager, Miner became a skilled hunter, but he was greatly affected by the extinction of the passenger pi-geon in the early 1900s. In 1904 he bought a number of live geese and cre-ated a pond for them in his backyard, hoping that he could attract wild geese to his property. His home in Kingsville was well positioned for this task, sitting at a cross point of the Mississippi and Atlantic flyways, and he was soon feeding migrating flocks of geese, ducks, doves, and songbirds in a series of artificial ponds. The Kingsville site was one of the first bird sanc-tuaries in North America and is still operating today. Fascinated by the mi-gratory movements of birds, Miner began tagging visitors to his ponds in August 1909. Over the next six years he tagged thousands of birds, provid-ing data that was crucial to the formulation of the 1916 Migratory Bird Treaty between Canada and the United States.

Miner became an authority on conservation, methods of banding, and habitat preservation. He lectured widely and wrote books on these subjects. In 1931 he organized the Jack Miner Migratory Bird Foundation. During the 1920s and 1930s the Jack Miner League spread across Canada. Like the Izaak Walton League in the United States, this organization encouraged children to become interested in nature. During his life Miner was the re-cipient of many awards including, in 1943, the Order of the British Empire conferred by the British monarch for his outstanding work in the field of wildlife conservation. In 1947 the week of April 10 (Miner's birthday) was proclaimed Canadian National Wildlife Week, to be celebrated each year in his honor (Linton 1984; Loo 2006).

National Parks   Canada and Alaska lagged behind the conterminous United States in establishing national parks. A hot springs reserve was established at Banff on the Canadian Pacific Railway line in the southeastern Rockies in 1885 and became the country's first, 673-square kilometer, national park in 1887. The enabling act for "Rocky Mountain Park" (later Banff National Park) explicitly encouraged commercial development and resource exploitation. Railroads also played a part in the creation of Alaska's first national park at Mount McKinley, created in 1917 to protect game from market hunters who supplied the Alaska Railroad Company. Still, the Park Act allowed mining within the park and gave miners and prospectors special privileges to hunt game there.

Other mountain reserves were soon established near Banff, and there were several boundary adjustments: Yoho (1886), Glacier (1886), Waterton Lakes (1895), and Jasper (1907) parks were all established to profit from tourism. With the creation of the Dominion Parks Branch, as a section of the Department of the Interior, in 1911, park policy began to change. The first park's commissioner, James Harkin, created nine new national parks during his tenure (1911–1936) and broadened their mandate beyond mere profit. Parks such as Elk Island Park (Alberta, 1913) and Point Pelee Park (Ontario, 1918) were created to preserve wildlife and vegetation. St. Lawrence Islands (Ontario, 1914) and Prince Albert (Saskatchewan, 1927) brought new geographical areas within the developing park system. In 1930 the National Parks Act presented a new view of Canadian parks as wilderness reserves and specifically cited the need to leave parks "unimpaired for the enjoyment of future generations.'

It took another four decades to establish a systematic, coordinated approach to park creation, however. The National Park System Plan of the 1970s divided Canada into thirty-nine distinctive natural regions and aspired to the creation of national parks representative of each and all of these regions. In Alaska the creation of Mount McKinley National Park was quickly followed by the creation of two national monuments (which later became national parks), Katmai (1918) and Glacier Bay (1925). These were created as wilderness preserves, partly to allow for the scientific study of Alaskan species and their habitats, and partly to allow tourists to experience pristine, undisturbed Alaskan environments.

The lack of treaties between the U.S. government and Alaskan native peoples held up further park development in Alaska until 1981, as the idea of national parks as wilderness areas, undisturbed by human activities, proved incompatible with traditional native land uses. The United States National Park Service (NPS) responded to this situation in various ways.

In Glacier Bay National Monument they attempted to remove native peoples from their lands. In Mount McKinley National Park they permitted native peoples to live within the park boundaries but tried to regulate their activities. Neither approach was successful. Under the Alaskan Native Claims Settlement Act (ANCSA) of 1971, Alaska's native peoples were allowed to select 40 million acres of land, and the U.S. Secretary of the Interior was authorized to withdraw 80 million acres of Alaskan land for new federal conservation units. This enabled passage of the monumental Alaska National Interest Lands Conservation Act (ANILCA) in January 1981. It created ten new national parks and enlarged the Mount McKinley, Glacier, and Katmai National Parks.

The ANILCA signaled a new approach to Alaskan national parks: Parklands serve a range of purposes and are considered both wilderness areas and traditional homes of native peoples. There are now thirty-nine national parks and national park reserves (areas still affected by unresolved native land claims that are designated to become national parks) in Canada, adding up to a total land area of nearly 62 million acres. Alaska's sixteen national parks and reserves encompass approximately 54 million acres (Bella 1987; Lothian 1987; Catton 1997).

Niagara   Although Niagara Falls attracted explorers from the earliest days of European adventuring in the interior of the continent, it was not until the advent of railroads in the late nineteenth century that it became truly popular with visitors from the United States and Europe. In 1885 the Ontario provincial government established the Niagara Parks Commission and Queen Victoria Park, a 62-hectare landscaped area, constructed to beautify an environment much disrupted by settlement, the development of transportation networks, and the availability of comparatively cheap power. Industrial expansion was aided by the opening of the first hydroelectric operations at Niagara Falls in 1905 and the construction of additional large power stations (Adam Beck 1 and 2) in 1922 and 1954.

Massive diversions of water were required to harness the energy of the river. Deep canals were cut into the rock above the falls, tunnels were constructed under the city, the flow of the Welland River was reversed, and the larger region became known for its iron and steel, abrasive products, pulp and paper, chemicals, and automobile parts industries. At the same time, tourist numbers rose as the automobile became the dominant form of transportation. In 1939 Canada's first multilane, divided highway, the Queen Elizabeth Way, was opened between Toronto and Niagara. Agricultural land use peaked in 1911, and spreading residential developments threatened to destroy the scenic rural landscape of the Niagara escarpment.

The Parks Commission responded to these pressures by extending the parklands along the river. The park reached Fort Erie by 1915 and Niagara-on-the-Lake by 1931. Still, the impacts of urban sprawl and industry on the natural environment of the Niagara region have been severe. Both the Niagara River and the Great Lakes are severely polluted. Yet the region receives some twelve to fourteen million visitors per annum.

Conservation, urban quality, and respect for the environment have become major issues in the Niagara area. The Niagara Parks Commission attempts to protect the falls area from unsightly and disruptive developments and continues to extend the Niagara Parkway system. Casinos and sideshows draw vast numbers of visitors to the town of Niagara Falls, and agritourism associated with the peninsula's wineries is also expanding. In many ways, the Niagara region offers a microcosm of wider land use changes and environmental management challenges across much of southern Canada (Gayler 1994).

North/Nordicity   Early Canadian boosters believed that the nation's character would be forged by its northern climate and that the national destiny would turn on utilization of the treasures of "the North." This obsession with nordicity continued into the twentieth century when artists and intellectuals such as the ***Group of Seven*** found inspiration in Canada's northern regions. Despite its unquestionable influence on the development of the Canadian identity *the North* is remarkably difficult to define. It is a geographical region but cannot be defined purely in terms of latitude and temperature. Definitions of nordicity based exclusively on the physical environment center on Canada's arctic and subarctic regions, or upon those territories in which commercially viable agriculture is impossible. But scholars have also argued the importance of isolation, a resource-based economy, political powerlessness, and a high percentage of aboriginal people in the population as markers of northern-ness.

Like Alaska, Canada's North is considered both a last frontier, waiting to be developed, and a last wilderness, in need of protection. The inherent contradiction in these views became especially evident in the late twentieth century when hydroelectric power generation, mineral mining, oil and gas extraction, and transportation megaprojects were proposed for northern regions. The associated real and potential degradation of northern environments became the focus of much attention from environmentalists. Concerns about northern environmental management have also given rise to conflicts between indigenous and nonindigenous groups. Those seeking to preserve the North as wilderness have found certain indigenous practices disturbing. They have therefore attempted to circumscribe these activities, by imposing regulations such as game laws and by reserving areas of land as parks. The indigenous peoples of the North see their land as neither a frontier nor a wilderness, but as a home. The lack of attention paid to this point of view reveals the extent to which the North is understood in southern terms (Mulvihill et al. 2001; Sandlos 2001).

Palliser's Triangle   From 1857 to 1860 John Palliser led an expedition, sponsored by the Royal Geographical Society, to observe the natural and human environment of British North America's western interior. In his 1863 report, Palliser divided the prairie region into two distinct geographical areas. The area bounded by present-day Cartwright (Manitoba), Lloydminster (Saskatchewan), and Calgary and Cardston (Alberta), and subsequently known as Palliser's Triangle, was unsuitable for agriculture due to its desertlike climate and poor, sandy soils, which supported only short stubble grasses and shrubs. It was surrounded to the north and east by a "fertile belt." In the 1870s John Macoun reassessed Palliser's Triangle and claimed that it was ideally suited for settlement and agriculture. This was excessively optimistic, but late in the twentieth century farms and ranches within Palliser's Triangle accounted for a substantial part of Canada's agricultural production. Today, however, there are mounting fears that global climate change will increase the frequency of droughts and reduce the prospects for sustainable agriculture in the region. In 1991 the Geological Survey of Canada began a multidisciplinary research project to evaluate the impacts of climate change on the land and water resources of Palliser's Triangle (Owram 1980; Spry and Palliser 1968).

Provincial Parks   Like the early national parks, Canada's early provincial parks were created at the end of the nineteenth century to provide opportunities for recreation in a natural setting. The first, Queen Victoria Park at Niagara Falls, was established in 1885. Over the next sixty years, seven more provincial parks were created in Ontario. The first provincial parks in Québec, the Laurentides north of Québec City and Mont Tremblant north of Montréal, were created in 1895. In the middle of the twentieth century, the Québec Government created many small parks to provide recreation for those living in urban industrial centers. Strathcona Park on Vancouver Island was the first provincial park in western Canada. Provincial parks were not established in the Prairie Provinces until those provinces received control of their resources in 1930. Alberta passed a Provincial Parks and Protected Areas Act in 1930 and by 1972 had created fifty-one parks. Saskatchewan followed Alberta's lead, creating three provincial parks in 1931. Manitoba and the Atlantic Provinces lagged behind the rest of the country, and did not establish parks until late in the twentieth century.

Territorial parks are relatively recent creations in Yukon and the North West Territories.

This varied pattern of provincial park creation reflects the distinctive purpose of these parks. Unlike the first national parks, the first provincial parks were established relatively close to urban areas. Generally, they were to provide recreational opportunities for local people, although certain parks—such as Ontario's Algonquin Park (1895)—were created expressly to protect natural environments. The preservationist impulse has been a far less prevalent influence upon the creation of provincial than national parks. Indeed, many provincial parks have been significantly modified by human activities, and others have been radically altered after designation. Provincial governments have the power to decide which activities will be permitted within their parks, and resource exploitation in the form of mining and logging continues in many provincial parks today (Killan 1993; Warecki 2000).

Roberts, Charles G. D. (1860–1943)   Born in Douglas, New Brunswick, in 1860, Charles George Douglas Roberts won fame for his poetry and his animal stories. A member of the "Confederation Poets" group, Roberts published his first poetry collection in 1886. His early works showed great promise. In 1897 he left Canada for New York City and Europe, where he stayed until 1925. During this period he published both poetry and a number of works of fiction. By far the most successful of these were his animal stories, which drew on his experiences growing up in the Maritime Provinces. He published over a dozen collections of these stories, the last of which—*Further Animal Stories*—appeared in 1936. Roberts and Ernest Thompson Seton are widely regarded as establishing the animal story as a prose genre, and as the first nonnative purely Canadian art form. At his death in 1943, some considered Roberts to be "Canada's leading man of letters" (Ower 1971; Keith 1969).

Seton, Ernest Thompson (1860–1946)   Nature artist and writer Ernest Thompson Seton is best known for his extraordinary animal stories. Born in Durham, England, in 1860, Seton immigrated to Canada when he was six years old and settled with his family on a farm near Lindsay, Ontario. It was here that he developed an interest in wildlife. A talented young artist, Seton earned a scholarship to study at the Royal Academy of Arts in London, England, in 1881. Upon his return to Canada in 1882, he joined his brother on a homestead near Carberry, Manitoba. There he spent much time taking notes and sketching wildlife. He also began to write. In 1891 he published *The Birds of Manitoba,* which led to his appointment in 1892 as Provincial Naturalist. In 1898 Seton published his first and most famous collection of animal stories, *Wild Animals I Have Known*. Seton chose to write from the animal's point of view and often invested his animals with human qualities such as curiosity, sympathy, and longing, an approach that was criticized by influential members of U.S. society, including President Theodore Roosevelt. Yet, the stories were very well received by North American children, and their protagonists such as "Lobo, the King of Currumpaw" and "Raggylug," the cottontail rabbit, quickly became household names.

In 1902 Seton organized and wrote a manual for a boys' club named the Woodcraft Indians. In 1910 he helped to found the Boy Scouts of America. The Woodcraft Indians merged with the Boy Scouts, but Seton was critical of their militancy and was expelled from that organization in 1915. Early in the twentieth century Seton published two important works: *The Life Histories of Northern Animals: An Account of the Mammals of Manitoba* (2 vols., 1908) and *Lives of Game Animals* (4 vols., 1925–1927). He also worked with Sioux and Pueblo Indians, giving expression to his lifelong admiration of America's native peoples in *Gospel of the Red Man* (1927). He died in Santa Fe, New Mexico, in 1946 (Wadland 1978).

Sifton, Clifford (1861–1929) Politician Clifford Sifton was born in 1861 near Arva, Canada West, and practiced law in Brandon, Manitoba, before entering the provincial legislature, where he served as Attorney-General between 1891 and 1896. He then entered federal politics and was Minister of the Interior and Superintendent of Indian affairs from 1896 to 1905. In 1903 he presented Canada's case to the Alaska Boundary Tribunal, and he chaired Canada's Commission of Conservation between 1909 and 1918. He is best known for his aggressive immigration policies that brought thousands of eastern Europeans to Canada's Prairie Provinces (Hall 1981, 1985).

Society Promoting Environmental Conservation (SPEC) The Society Promoting Environmental Conservation was founded in 1969 in Coquitlam, British Columbia. Originally created because of concerns about the declining quality of life in urban areas due to traffic congestion, poor air and water quality, and inadequate sewage treatment, the society quickly broadened its outlook, campaigning on a broad range of environmental issues (Keeling 2004a).

Spruce Budworm Native to North America, the spruce budworm (*Choristoneura spp*) is a destructive pest that damages spruce, fir, pine, larch, and hemlock trees, although its impact has been most severe on eastern balsam fir. In Alaska budworm infestations have damaged urban ornamental trees; in the 1980s low levels of budworm activity were detected in forests from the Kenai Peninsula to Fairbanks, where they primarily affect white and sitka spruce. Prolonged outbreaks cause severe defoliation, reduced growth, and tree mortality. In Québec a twenty-five-year outbreak that began in 1967 destroyed almost 13 million hectares of commercial forest. The earliest known budworm outbreak occurred in 1704, but the incidence and severity of infestation appears to have increased considerably in the twentieth century as the proportion of fir in eastern forests increased as a consequence of heavy exploitation of pine and spruce. There is now more research on this forest insect than any other. Much of it considers techniques for budworm control. Chemical insecticides such as DDT were used against budworms until their harmful environmental effects were documented. Today chemical and biological controls (some of which remain controversial) are used in conjunction with silvicultural techniques to limit budworm population growth (Schmitt et al. 1984; Sandberg and Clancy 2002).

Staples Theory/Staple Trades   Economic historian Harold Innis argued that Canadian history was best understood through the study of natural resources. He described the natural resources that formed the backbone of the Canadian economy as "staples," arguing that each stage of Canadian history was characterized by a dominant staple and the technologies that were developed to exploit it. He posited that staples shaped the society and political tendencies of those who were involved in their exploitation, and that the exhaustion or decline in demand for a staple caused major social and political upheaval (Neill 1972).

Stewart, Elihu (1844–1935)   Elihu Stewart, a Dominion Land Surveyor, was appointed the first Chief Inspector of Timber and Forestry within the Department of the Interior in August 1889. Stewart set the focus of the Dominion Forestry Branch on conservation and propagation. In 1900 he was instrumental in the establishment of the Canadian Forestry Association, of which he was president in 1907 and 1908. Stewart's conservation strategies focused on fire fighting, tree planting, and forest reserves. He saw a particular need for tree planting on the Canadian prairies, where drought and erosion made soil conservation difficult. From 1901 to 1920 the Dominion Forestry Branch distributed over 50 million seedlings to prairie farmers (Drushka and Burt 2001; Natural Resources Canada n.d.).

St. Lawrence Seaway   In 1909 Canada and the United States established the International Joint Commission with responsibility for studies of power and navigation on the St. Lawrence River. Early plans met opposition from railroads and coastal ports, but after World War II improvements to navigation along the St. Lawrence were required to link iron ore reserves in Labrador with markets in the Great Lakes, and a growing demand for power pushed

Ontario and New York State to convert the kinetic energy of the massive waterway. In 1954 Canada and the United States agreed to joint construction of the Seaway. Locks between Montréal and Lake Ontario lift westbound vessels the 65 meters necessary to bypass various rapids and hydroelectric developments. The Welland Canal raises westbound vessels another 99.4 meters to circumvent Niagara Falls. Locks on the St. Mary's River Canal, between Lakes Huron and Superior, overcome a 6.5-meter difference in elevation. The Seaway locks matched the dimensions of the 1932 locks on the Welland Canal (244 meters by 24 meters) and the Seaway channel is dredged to a depth of 8 meters. The costs of the Seaway's navigation facilities were apportioned between the United States ($130 million) and Canada ($330 million, plus a similar sum for improvements to the Welland Canal). The $600 million spent on hydroelectric development was shared between the two countries (Mabee 1961).

Strathcona Provincial Park   The oldest provincial park in British Columbia was established in 1911 by an act of the provincial legislature that reserved half a million acres of Vancouver Island's mountainous central plateau as a "public park and pleasure ground." However, forests and mineral resources within the park were exploited, and plans for hydroelectric generation at Buttle Lake were mooted from the 1920s. In 1951 a new proposal to dam Buttle Lake sparked the fury of anglers, campers, and nature enthusiasts led by *Roderick Haig-Brown*, who pleaded for protection of the scenic, recreational, and biological values of the park. Although the lake was raised by a dam just beyond the park, in 1955 the campaign raised the profile of conservation issues in British Columbia. The battle between environmentalists and developers was rejoined when the provincial government changed the classification of the park to allow resource extraction and industrial development within its bounds. Continuing conflict led to the appointment of the Strathcona Park Advisory Committee in 1988 and the establishment of permanent park boundaries, as well the cessation of logging and mining activity in the park (Baikie and Philips 1986).

Strong, Maurice (1929–)   A Canadian businessman, bureaucrat, and activist, Maurice Strong is best known for his role as Secretary-General of the 1992 U.N. Conference on Environment and Development in Rio de Janeiro (the "Earth Summit") and his call on that occasion for governments in the developed world to extend over $600 billion in financial aid to the developing world, in reparation for a century of interdependent industrial development and environmental degradation. Earlier, as director of the 1972 U.N. Stockholm conference on the Human Environment, Strong had brought global environmental issues to the fore. Self-described as a "socialist in ideology and a capitalist in methodology," Strong, who was president of Canada's Power Corporation in his thirties, has been head of the Canadian International Development Agency, director of the U.N. Environment Program, head of Petro-Canada, and senior adviser to both the Secretary-General of the United Nations and president of the World Bank (Foreign Affairs and International Trade Canada, n.d.).

Tourism When naturalist and explorer John Muir claimed that Alaska's awe-inspiring mountains, rivers, and wildlife offered visitors a chance to experience American landscapes as they had been before the arrival of Europeans, he tapped into a deep well of appreciation for the wild empty landscapes of northern North America. "Tourists" (and others such as big game hunters and sport fishers) were encouraged to view (and exploit) scenic and other resources. After 1885 the Canadian Pacific Railway promoted tourism through its rail and steamship services and developed hotels and spas in the national parks of the Rocky Mountains. The remoteness of Alaska meant that organized tourism was substantially delayed until air travel opened the area to game hunters in the 1940s. The negative environmental impacts of these activities were limited by the small numbers involved and confined by the technological constraints on movement.

With growing affluence, increased leisure time, and the popularization of the automobile, however, the sheer number, and behaviors, of tourists came to jeopardize the existence of what they came to see. In 2001 Alaska received almost a million visitors; Canada counts approximately twenty million tourist visitors a year. Burgeoning numbers place extra pressures on natural environments. Technologies such as helicopters, all-terrain vehicles, and sophisticated outdoor equipment cause environmental damage. Areas that were once the preserve of wilderness tourists gradually open up to a wider market, a shift associated with a move from authentic to artificial, from low density to high density, and from low-energy consumption to high-energy consumption. The throngs of tourists that now climb Alaska's Mount McKinley each year are testament to this trend.

In response, the U.S. National Park Service and the Canadian National Parks Commission sought to render tourism a "handmaiden for conservation." They promoted recreational and educational programs to encourage support of nature conservation and paved the way for "low impact" or "ecotourism" tailored to visitors who hold conservationist/environmentalist ideals. The number of ecotourism outfits in Canada and Alaska has grown steadily over the past few decades, but this has created conflict between tourism, forestry, mining, and environmental interests. For tourist operators who seek to provide their customers with an authentic wilderness experience, the preservation of pristine wilderness environments is an economic rather than a moral or aesthetic concern (Jasen 1995; Nelson 2000; Cruikshank 2005).

Tundra   Arctic tundra covers about one-tenth of the earth's surface. It forms a distinct circumpolar ecological unit, marked by a lack of trees and the general presence of permanently frozen ground, permafrost. Tundra covers vast areas of northern and western Alaska. In Canada the southern boundary of the tundra zone extends from the Mackenzie Delta to southern Hudson Bay and runs northeast to Labrador. These regions are characterized by short summers with almost continuous daylight and long, dark winters. Tundra plants, mostly lichens, mosses, grasses, and low shrubs, grow close to the ground, and their leaves are typically small, leathery, and hairy, so as to retain as much moisture as possible. Their reproductive cycles tend to be short and they rely upon arctic winds for seed dispersal. Animals and birds of the tundra (reindeer, moose, lemmings, and ptarmigans) have thick layers of fur and feathers to insulate themselves against the northern climate. Tundra ecosystems pose many problems for the construction of roads, buildings, oil drills, and so on. Construction materials must withstand the intense cold climate and cope with seasonal variations in soil stability. Because the biological productivity of tundra areas is very low, they are fragile, and damage to landscapes and ecosystems can be long lasting (Chernov 1985).

Western Canada Wilderness Committee (WCWC)   A nonprofit environmental organization, the Western Canada Wilderness Committee was founded in Vancouver in 1980. Its initial mandate was to provide information about Canadian wildlife and wilderness issues to build broad public support for wilderness preservation. It has been involved in a number of high-profile campaigns to protect western Canada's wilderness areas, such as the Stein and Carmanah valleys and Clayoquot Sound. The WCWC is also involved in trail building, mapping wilderness areas, and public education projects. In 1991 the government department known as Environment Canada judged it to be the most effective environmental group in Canada (Western Canada Wilderness Committee n.d.).

Whaling   North American native peoples have hunted whales for millennia. The Eskimo and Inuit of the Arctic and the Nootka of Vancouver Island hunted whales for subsistence, but estimates suggest that Western Arctic whale hunters took only about a hundred whales a year from a population of 30,000. Basque whalers processed approximately 17,500 whales in southern Labrador between 1545 and 1585. In the seventeenth and eighteenth centuries, European, British, and later American ships hunted in the Davis

Strait region; early in the nineteenth century catches may have exceeded a thousand whales a year.

The depletion of whale populations in the Eastern Arctic region led whalers to new hunting grounds. In 1848 an American whaler, the *Superior*, encountered large numbers of bowhead whales in the Bering Sea. By 1852 more than 220 ships were hunting in the Bering Strait region, and an estimated 2,682 bowheads were killed. The catch dwindled thereafter. The whalers shifted to the Sea of Okhotsk and back to the Bering Strait and equipped their vessels with steam engines, but the catch never again exceeded 600 whales a season. By the early twentieth century whale stocks had been depleted to the point of extinction, and the market for baleen had diminished. By 1909 there were only three vessels in the Western Arctic whaling fleet.

Significant improvements in whaling technology after 1920 sparked international concern, the first International Whaling Agreement in 1937, and the establishment of the International Whaling Commission (IWC) in 1946. Hunting certain species such as bowheads was prohibited (although Alaskan Eskimos were permitted to take fifty bowheads a year) and quotas were placed on the exploitation of other stocks. Greenpeace led a high-profile campaign against whaling on the high seas, but Canada resigned from the IWC in 1982 in disagreement with a proposed whaling moratorium, an action strongly criticized by antiwhaling groups. However, Canada continues to ban commercial whaling and cooperates with the IWC's scientific committee.

Conservation measures have allowed bowhead whale stocks in northern Canada to recover to approximately 700, and sightings of gray whales and orcas are relatively common along the coasts of Alaska and British Columbia. Whale-watching companies now "hunt" their quarry to provide customers with close-up views of whales in their natural environments. Though far less destructive than the hunting trips of the past, these activities have aroused concern among scientists and environmentalists, who believe they may alter whale feeding, mating, and migration patterns (Bockstoce 1986; Allen and Keay 2001, 2004; Kraus et al. 2005).

TIMELINE

The *Canadian Encyclopedia* has a full and excellent timeline with many interactive links at www.thecanadianencyclopedia.com/.

50,000–20,000 BC   First Nations cross Beringia

27,000 BC   Estimated date of bone scraper found at Old Crow, Yukon Territory

13,000 BC   Date of Bluefish Caves, 54 kilometers southwest of Old Crow site

10,000 BC   Emergence of Clovis and Folsom cultures

8,000 BC   Development of Plano Culture

6,000 BC   Archaic Boreal and Shield cultures

5,000 BC   First buffalo jumps

4,000 BC   Laurentian Archaic Culture; development of distinct Northwest Coast Culture

3,000 BC   Emergence of early Paleoeskimo Culture; Maritime Archaic Indians

1,000–500 BC   Huron adopt agricultural techniques

500 BC—AD 1500   Dorset Culture

1000–1600   Development of Thule Culture

1000   Leif Eiriksson reports discovery of Vinland

ca 1400   Development of Iroquois Confederacy

1420   Basque whalers in Gulf of St. Lawrence

1497, June 24   John Cabot reaches and claims Atlantic Coast

1498   Cabot's second voyage

1501   Gaspar Corte-Real coasts Atlantic

1506   First fishing vessel from Normandy to Grand Banks

1520   Joao Fagundes visits Newfoundland

1534, 2 April   Cartier reaches Labrador

1534, 24 July   Cartier lands at Gaspé

1535–1536   Cartier's second voyage

1535, 10 August   First use of name *Saint-Laurent* for river system

1535, 13 August   First European use of name *Canada* for territory

1535, 14 September   Cartier reaches Stadacona

1535   Cartier visits Hochelaga

1541–1542   Cartier's third voyage

1544   Basques in St. Lawrence estuary

1557, 1 September   Death of Cartier

1576, 11 August   Frobisher sights Baffin Island

1577   Frobisher's second voyage

1578   Frobisher's third voyage

1579   Drake on the Pacific, possibly north of 49°

1583, 5 August   Gilbert claims Newfoundland

1585   John Davis, first voyage to Baffin Island, at 66°40′N

1586   Davis's second voyage

1587   Davis's third voyage reaches 72°12′N

1600–1602   Tadoussac established as first official fur-trading post in Canada

1603   Champlain reaches Tadoussac

1603, 4 July   Champlain reaches Montréal

1604   Champlain establishes settlement on St. Croix Island

1605   Founding of Port-Royal, Acadia

1608, 3 July   Founding of Québec

1609   Champlain and native allies engage Iroquois at Lake Champlain

1610   Founding of Cupids Bay settlement, Newfoundland

1610, 3 August   Henry Hudson enters Hudson Bay

1611, 23 June   Hudson's crew mutinies

1613, 9 January   Publication of Champlain's *Voyages*

1615, 1 July   Champlain reaches Huronia

1615, 11 October   Champlain's third battle with the Iroquois

1616   William Baffin and Robert Bylot reach Lancaster Sound and 77°45′N

1619-1620   Jens Munk in Hudson Bay

1623   Seigneurial system instituted

1627   Compagnie des Cent-Associés established

1629, 20 July   Champlain surrenders Québec to English adventurer David Kirke

1632   Beginnings of seigneurial survey system

1632–1672   *Jesuit Relations* describe Jesuits' work in New France

1632, 29 March   Treaty of Saint-Germain-en-Laye

1633   Champlain returns to Québec

1635   Charles de la Tour granted land on St. John River

1635, 25 December   Death of Champlain

1639   Establishment of Ste. Marie among the Hurons

1639   First coal mined at Grand Lake (New Brunswick)

1642   Maisonneuve founds Ville-Marie

1645   Attack on Fort La Tour (Acadia)

1649, 16 March   Jesuits Brébeuf and Lalement killed by Iroquois

1649, April   Huronia destroyed

1650   Population of New France recorded as 705

1651, 25 February   La Tour appointed governor of Acadia

1654, 16 August   Port-Royal captured by New England fleet

1663, 24 February   New France declared a Crown Colony

1664   Coutume de Paris (civil law of France) established in New France

1665   Filles du Roi and Carignan-Salières regiment arrive in New France

1665   Jean Talon becomes intendant of New France

1666   Canada's first census, 3,215 non-native persons in New France

1666   Mohawks defeat French troops

1667   Treaty arranged with Iroquois

1667   First iron mined in Canada

1668, 29 September   *Nonsuch* reaches James Bay

1670, 23 March   Explorers reach Lake Erie and claim it for France

1670, 2 May   Hudson's Bay Company founded

1672   Frontenac appointed governor of New France

1673   Second census of Canada, 6,705 enumerated in New France

1673   Cataraqui founded by La Salle

1673, 15 June   French reach the Mississippi River

1680   First ship on Great Lakes

1682   La Salle reaches mouth of Mississippi River

1682, 9 April   La Salle claims Louisiana

1682, 5 August   Fire destroys settlement at Québec

1686   French expedition to James Bay seizes Hudson's Bay Company posts

1687, 19 March   La Salle assassinated

1690, 22 January   Peace treaty with Iroquois

1690, 19 May   American adventurer William Phips plunders Port-Royal

1690–1692   Henry Kelsey of the Hudson's Bay Company explores from York Factory to Saskatchewan River; first European to see Prairies, grizzly bear, and bison

1692   Caterpillars destroy most of crops in New France

1699   "An Act to Encourage the Trade to Newfoundland" passed in British Parliament

1701   Iroquois agree to neutrality in France-England wars

1702, 15 May   England declares war on France

1710, 5 October   Francis Nicholson and two thousand Englishmen capture Port-Royal

1713, 11 April   Treaty of Utrecht signed

1713   Louisbourg founded

1717   Hudson's Bay Company establishes post at Churchill

1720   Three vessels and French settlers to Isle St. Jean (Prince Edward Island)

1728   Danish explorer Vitus Bering, in employ of tsar of Russia, discovers and names Bering Strait

1730   Blackfoot acquire horses from south

1731, August   Fur trader and explorer Pierre La Vérendrye reaches Grand Portage

1732   La Vérendrye in the Northwest

1734   La Vérendrye at Red River

1734   First road between Québec and Montréal ("Chemin du Roi") opened

1736   Forges du Saint-Maurice established near Trois Rivières

1739   Census of New France returns 42,701

1741   Exports from New France exceed imports for first and only time in French regime

1742   La Vérendrye and sons in Dakota

1743, 1 January   La Vérendryes sight Rockies

1744–1748   War of the Austrian Succession

1745, 17 June   Louisbourg surrenders to English forces

1747   Claude de Ramezay attacks British garrison at Grande Pré, Nova Scotia

1748, 18 October   Treaty of Aix-la-Chapelle; Louisbourg returned to France

1749   English establish settlement at Halifax.

1753   German Protestants settled at Lunenburg

1754   Population of New France placed at 55,000

1754   Anthony Henday explores vicinity of present-day Red Deer

1755, June   Robert Monckton and 2,000 British troops take Fort Beauséjour

1755   Expulsion of the Acadians (Grand Dérangement)

1755   Marquis de Vaudreuil appointed governor of New France

1756–1763   Seven Years' War

1757, 9 August   Montcalm takes Fort William Henry

1758, 26 July   Louisbourg surrenders

1758, 27 August   French Fort Frontenac falls to John Bradsteet

1759, 13 September   Battle of the Plains of Abraham, Generals Montcalm and Wolfe die, British victorious

1759, 18 September   Québec surrenders

1760   French defeat British at Sainte-Foy but British hold Québec

1760, 5 September   Treaty between Huron and British

1760, 8 September   Montréal surrenders

1763, 10 February   Treaty of Paris ends Seven Years' War, France cedes all North American territory except Saint-Pierre and Miquelon and Louisiana to British

1763, 7 October Royal Proclamation sets western boundary of Québec and establishes Indian Territory in interior

1766 British conclude treaty with Ottawa chief Pontiac, ending uprising begun in 1763

1771, 17 July Samuel Hearne and native guide Montonabbe reach the Arctic

1773, 15 September Arrival of the *Hector* in Pictou from Scotland

1774, 22 June Québec Act (effective 1 May 1775) established French civil and English criminal law and freedom of worship in Québec and extended boundaries to the Ohio valley

1774, 15 July Spanish explorer Juan Pérez Hernández sights Haida Gwaii (Queen Charlotte Islands)

1775, 9 September Hurricane hits Newfoundland

1775, 11 November Governor Carleton abandons Montréal to American invaders following outbreak of American Revolution

1775, 5 December–1776, 6 May Québec besieged by American forces

1776, 1 April First Loyalist refugees arrive in Nova Scotia

1778 Peter Pond reaches Athabaska River

1778, 7 March English explorer James Cook reaches Vancouver Island

1779 North West Company established in Montréal to trade in furs

1781–1782 Smallpox epidemic on plains and west coast

1783 British government purchases land north of Lake Ontario from Mississauga people

1783, 20 September Treaty of Paris ends American Revolution, establishes boundary between British North America and the United States along St. Lawrence River and through Great Lakes

1784, 16 August New Brunswick established as separate colony

1784, 26 August Cape Breton Island becomes a separate colony

1784, 11 September Island of St. John becomes part of Nova Scotia

1784 Census of Québec enumerates 113,012 people

1785 French explorer La Pérouse on Pacific Coast

1786 Island of St. John becomes a separate colony

1788 Fort Chipewyan established on Lake Athabasca by North West Company

1789, 5 May Spanish vessels under Esteban Jose Martinez at Nootka plunder British ships, initiate Nootka Sound controversy over claim to Northwest Coast

1789, 10 July Alexander Mackenzie of the North West Company reaches Arctic Ocean

1791, 26 December Constitutional act creating colonies of Upper and Lower Canada goes into effect

1792   George Vancouver explores Strait of Georgia

1793, 20 July   Alexander Mackenzie reaches Pacific at Bella Coola inlet

1794, 11 January   By second Nootka Convention Spain and England acknowledge mutual rights to trade

1795   Hudson's Bay Company establishes Fort Edmonton

1796, 1 February   York (later Toronto) first capital of Upper Canada

1798   Formation of fur-trading XY Company

1799, 3 June   Island of St. John becomes Prince Edward Island

1802   Scottish highlanders settle in Sydney, Cape Breton

1804, 5 November   XY and North West fur-trading companies unite

1808   First stagecoach, Montréal to Kingston

1808, 2 July   Simon Fraser reaches Pacific down river that now bears his name

1809, 30 March   Labrador Act transfers that territory to Newfoundland

1810   John Jacob Astor establishes Pacific Fur Company to trade on Pacific Coast

1811, 12 June   Hudson's Bay Company grants 185,000 square kilometers to Lord Selkirk to establish settlement at Red River

1811, 15 July   David Thompson reaches Pacific via Columbia River

1812   First settlers arrive at Red River

1812, 18 June   Declaration of War of 1812

1812–1814   Several battles between U.S. and British North American troops, York captured and looted

1814, 24 December   Treaty of Ghent ends War of 1812

1816, 19 July   Governor and twenty Red River colonists killed in skirmish with Metis at Seven Oaks

1818   Expedition led by William Parry reaches Ellesmere Island

1818, 20 October   Convention sets boundary between British North America and United States on 49th parallel from northwest point of Lake of the Woods to Rocky Mountains

1819   Locusts devastate crops at Red River

1820, 16 October   Cape Breton Island becomes part of Nova Scotia

1821   John Franklin descends Coppermine River and explores Arctic Coast

1821, 1 June   Hudson's Bay and North West fur-trading companies merge

1822   First gypsum mined near Paris, Ontario

1823, 30 November   Promoter William Hamilton Merritt begins work on Welland Canal, between Lakes Erie and Ontario

1825, 19 March   Hudson's Bay Company establishes Fort Vancouver near mouth of Columbia River

1825, 5 October   Miramichi Fire

1825   Census returns for Lower (479,288) and Upper (157,923) Canada populations

1825–1827   John Franklin continues Arctic explorations

1826, May   "Greatest known flood" of the Red River

1826   Construction of Rideau Canal begins

1827   Horse-drawn railway to haul coal built in Pictou, Nova Scotia

1827   William Parry reaches 82°45′N

1827   Construction of Fort Langley on Fraser River

1829, 6 June   Death of Shawnadithit, the last Beothuk

1829, 30 November   Welland Canal opened

1832   Cholera epidemic in British North America, Grosse Isle established for quarantine

1832, 25 February   Champlain and St. Lawrence Railroad incorporated

1833   *Royal William* completes Atlantic crossing (Pictou to Gravesend) under steam

1836   Coal mined on Vancouver Island

1836, 21 July   Champlain and St. Lawrence Railroad opened

1837   Severe smallpox epidemic decimates native peoples of Prairies

1837   Uprising of Patriotes in Lower Canada and of rebels led by William Lyon Mackenzie in Upper Canada

1839   Hudson's Bay Company granted exclusive trading rights on Vancouver Island

1839   Lumbermen from New Brunswick and Maine clash over undefined border in the "Aroostook War"

1839   Four kilometers of Albion Mines Railway opens in Pictou County, Nova Scotia; second steam railway in British North America

1840   Great auk becomes extinct

1840   First production of hydraulic cement in Canada at Hull, Québec

1840   *Britannia* inaugurates steam-packet mail service across Atlantic

1841, 10 February   Upper and Lower Canada united as Canada (East and West)

1841   Monies provided for geological and natural history survey of Canada

1842   William Logan appointed first director of the Geological Survey of Canada

1842   David Fife develops red fife wheat near Peterborough, Canada West

1842   Webster-Ashburton Treaty defines border between United States and British North America and provides for settlement of Maine–New Brunswick boundary issue

1845, 17 March   St. Lawrence and Atlantic Railroad chartered to build link from Montréal to Portland, Maine

1845   Last sighting of John Franklin expedition in Baffin Bay

1845   Canada's first graphite mine, Grenville, Québec

1846   Abraham Gesner of New Brunswick demonstrates kerosene

1846, 9 June   Fire destroys St. John's, Newfoundland

1846, 15 June   Oregon Boundary Treaty extends border between United States and British North America along 49th parallel to Pacific and awards Vancouver Island to British North America

1847   First zoo in Canada established in Halifax, Nova Scotia

1847   Geological Survey of Canada reports copper ores in Eastern Townships of Québec

1847   Corundum discovered in Lanark County, Ontario

1847   Gypsum mining begins near Hillsborough, New Brunswick

1848   Forsyth iron mine opened at Hull, Québec

1850–1854   Search for Franklin reveals much of Arctic region

1851   Famous clipper ship *Marco Polo* launched in New Brunswick

1854, 27 January   Great Western Railway open from London to Windsor

1855, 18 March   Suspension bridge across Niagara River near falls opens to rail traffic

1856   Railway links Halifax to Truro and Windsor, Nova Scotia

1856   Grand Trunk Railroad open from Montréal to Sarnia

1857, 28 February   World's first commercial oil well, Petrolia, Canada West

1857   Copper discovered at Tilt Cove, Newfoundland

1857–1860   Royal Geographical Society expedition led by John Palliser explores and maps western interior

1857–1858   Canadian government expedition led by Henry Youle Hind explores southern Prairies

1858, April   First gold miners of Fraser River rush arrive in Victoria

1858, 2 August   Establishment of British Columbia as a Crown Colony

1858, 12 August   First telegraph message, Newfoundland to Ireland

1860   Victoria Bridge spans St. Lawrence at Montréal

1860   First mica mined in Ontario

1862   Cariboo wagon road from Yale to Barkerville goldfields in British Columbia initiated

1862   First production of oil in Ontario, at Oil Springs

1866   Great Eastern lays first transatlantic communications cable

1866, 6 August   Vancouver Island and British Columbia united as the single colony of British Columbia

1867, 1 July   Confederation of Nova Scotia, New Brunswick, and Upper and Lower Canada into Dominion of Canada

1868, 31 July   Rupert's Land Act allows Crown to transfer Rupert's Land to Dominion of Canada

1869, 2 November   Red River Insurrection, Louis Riel occupies Fort Garry to protest transfer of Hudson's Bay Company territory to Canada

1869, 1 December   Hudson's Bay Company surrenders Rupert's Land

1869, 23 December   Louis Riel declares his Provisional Government of Red River

1870   Canadian Government acquires North West Territories

1870   Beginning of coal mining at Lethbridge, Alberta

1869–1870   Smallpox epidemic afflicts Blackfoot Confederacy

1870, 15 July   Manitoba Act creates new Canadian province of Manitoba

1870, 15 July   Britain officially transfers Rupert's Land and North West Territories to Canada

1871, 8 May   Treaty of Washington grants Americans fishing rights in Canadian waters and use of canals and St. Lawrence; Canadians allowed to navigate Lake Michigan and Alaska rivers

1871, 20 July   British Columbia joins Confederation

1871, 3 August   Treaty No. 1 with Swampy Crees and Assiniboines

1871, 21 August   Treaty No. 2 with Chippewa of Manitoba

1871   First apatite mine in the Lièvre River district, Québec

1871   Discovery of silver at Eureka Mountain near Hope, British Columbia

1871   First production of soapstone (asbestos) in Brome County, Québec

1872   Dominion Lands Act opens west to settlement

1873   Red fife wheat introduced to Manitoba, lengthens growing season

1873, 1 May   Twenty Assiniboine killed by wolf hunters in Cypress Hills Massacre

1873, 13 May   Coal mine disaster at Westville, Nova Scotia, kills sixty

1873, 23 May   North-West Mounted Police (NWMP) established by act of Parliament

1873, 1 July   Prince Edward Island joins Canada

1873, 25 August   Hurricane strikes Nova Scotia, kills 500, destroys 1,000 vessels, and does much property damage

1873, 3 October   Treaty No. 3 with Salteaux (Chippewa)

1874, 15 September   Treaty No. 4 with Cree, Salteaux, and others

1874, 1 October   Police post established at Fort Macleod on Oldman River in present-day Alberta

1875, 8 April   North-West Territories Act separates North West Territories from Manitoba

1875, 1 June   Construction of Canadian Pacific Railway begins

1875, 20 September   Treaty No. 5 with Chippewa and Swampy Cree

1876, 23 August   Treaty No. 6 with Plains Cree

1876   Intercolonial railway links Montréal and Halifax

1876, 1 October   First wheat shipped from western Canada to Ontario

1877, 20–21 June   Fire destroys Saint John, New Brunswick

1877, 22 September   Treaty No. 7 signed with Blackfoot, Blood, Peigan, Sarsi, and Stoney

1878   Discovery of gold in the Lake of the Woods area of Ontario

1879   Sandford Fleming proposes World Standard Time (adopted 1884)

1879, 9 February   Railway links Montréal and Québec along north shore of St. Lawrence

1879, 14 March   National Policy of Protective Tariffs into effect

1880   New Brunswick legislature burned

1880, 9 October   Canada assumes sovereignty over British Arctic islands

1881–1882   Matthew Cochrane establishes first ranch in Bow River valley with cattle from Montana

1883–1884   Anthropologist Franz Boas visits Inuit

1883, 1 June   Canadian Pacific Railway reaches Medicine Hat, Alberta

1883, 18 November   Canada adopts Standard Time

1884   J. B. Tyrell discovers partial skull of dinosaur *Albertasaurus sarcophagus* near Red Deer River

1885, 19 March   Louis Riel proclaims Provisional Government of the North West at Batoche

1885, 26 March   NWMP defeated by Metis at Battle of Duck Lake, "North-West Rebellion" begins

1885   Several skirmishes between natives, Metis, and NWMP

1885, 7 November   Last spike driven on Canadian Pacific Railway railbed

1885, 16 November   Louis Riel hanged

1885, 25 November   Canada's first park reserve established, 26 square kilometers around Banff Hot Springs

1885   Smallpox claims over 5,000 lives in Montréal

1886–1888   Banff Springs Hotel built

1886   Canadian Pacific Railway reaches Vancouver, city incorporated

1886, 14 December   Yoho National Park established

1887   Mining for nickel and copper begins at Sudbury, Ontario

1887   World's first electric streetcar system in St. Catherine's, Ontario

1888   First hydroelectric power generated in Canada at Georgetown paper mill

1888   First coal mining at Canmore, Alberta

1888   Discovery of natural gas in Essex County, Ontario

1889   Gold and copper discovered at Rossland and lead and zinc at Kimberley, British Columbia

1890   Petroleum discovered in Athabasca River

1891, 21 February   Coal mine explosion in Springhill, Nova Scotia, kills 125

1891   Canadian Pacific Railway reaches Edmonton, Alberta

1892   Area around Lake Louise set aside as a government reserve

1892   Two-thirds of St. John's, Newfoundland, destroyed by fire

1893, 27 May   Algonquin Park, Ontario, established as Canada's first provincial park

1895   First Forest Reserve Act and establishment of Laurentides National Park

1895   Waterton Lakes National Park established

1896   Canadian Government passes the North-West Irrigation Act to control water resources of Prairies

1896, 17 August   News of gold discoveries on Bonanza Creek, Yukon

1897   Baffin Island claimed for Canada

1897   Yukon gold rush begins

1897, 6 September   Crow's Nest Pass Agreement to reduce Canadian Pacific Railway freight charges on grain and flour

1898, 13 June   Yukon becomes a separate territory

1898, 29 July   White Pass and Yukon Railway completed from Skagway to Whitehorse

1898, 11 September   New Westminster, British Columbia, destroyed by fire

1898   Crow's Nest Pass Railway opened

1899   Canadian Northern Railway incorporated

1899, 21 June   Treaty No. 8 with Cree, Beaver, Chipewyan, and Slavey

1900   Niagara Parks Commission agrees to construction of massive hydroelectric power generating facility on Canadian side of Niagara River

1900, September   184-kilometer irrigation canal opened between Kimball and Lethbridge, Alberta

1903, 25 March   Alaska-Canada boundary established as it is today; Canada denied access to sea from Yukon and northern British Columbia

1903   Canada asserts Arctic sovereignty with police post at Herschel Island

1903   Rich silver vein discovered at Cobalt, Ontario

1903   Highest temperature ever recorded in Canada measured at 46°C (115°F) at Gleichen, North West Territory (now Alta)

1903, 29 April   Turtle Mountain or Frank rockslide kills seventy people in coal-mining town of Frank

1905, 1 September   Alberta and Saskatchewan become provinces under Dominion Act, federal government retains control of lands and natural resources

1905, 24 November   Canadian Northern Railroad completed to Edmonton, Alberta

1905   Roald Amundsen claims discovery of Northwest Passage

1906   Prince Rupert (British Columbia) established as Grand Trunk Pacific Railway terminus

1906   Canada lays claim to Arctic islands

1906, 14 May   Hydro Electric Power Commission of Ontario created; world's first publicly owned electric utility

1907   Fast-maturing marquis wheat developed by Charles Edward Saunders; first farmed in 1909; by 1920 it accounts for 90 percent of Prairie wheat crop

1907, 14 September   Jasper Forest Reserve, later Jasper National Park, established

1908   Jack Miner establishes one of first bird sanctuaries in North America at Kingsville, Ontario

1909, 11 January   Boundary Waters Treaty signed by Britain and United States creates International Joint Commission to adjudicate minor disputes

1910   Newfoundland given authority to regulate fisheries used by Americans and Canadians as well as by Newfoundlanders

1910, 11 October   First long-distance electricity transmission in Canada from Niagara to Berlin (Kitchener)

1911   Gold discovered at Kirkland Lake, Ontario

1911, 11 July   Forest fire destroys town of South Porcupine, northern Ontario, kills fifty

1912   Federal government extends northern boundaries of Québec, Ontario, and Manitoba

1913   Elk Island National Park established east of Edmonton, Alberta

1913   Rockslide blocks Fraser River at Hells Gate

1913, 17 November   National transcontinental railroad completed

1914   Oil discovery at Turner Valley, Alberta

1914   Bassano Dam on Bow River, Alberta, completed for irrigation scheme

1914, April   Grand Trunk Pacific Railway completed

1914, 19 June   Methane explosion at Hillcrest Mine, Alberta, kills 189 men

1916   Edmonton, Dunvegan, and British Columbia railway opens Peace River district for settlement

1916, July   Wildfires burn Cochrane and Matheson, northern Ontario, and kill 228

1918, 23 January   Explosion at Stellarton coal mine, Nova Scotia, kills eighty-eight men

1918, September   Outbreak of Spanish influenza kills 50,000 Canadians in next twelve months

1919, 6 June   Canadian National Railway created to consolidate various private, unprofitable lines

1921   World's largest generating station opened at Niagara Falls (Sir Adam Beck No. 1)

1922   American filmmaker Robert Flaherty releases *Nanook of the North*

1923, 2 March   Halibut Treaty signed with United States

1925, 24 February   Canada signs treaty with United States establishing international Lake of the Woods Control Board

1927, 1 March   Labrador boundary determined by Judicial Committee of the Privy Council

1929, October   Stock market crash precipitates economic Depression

1930   By separate acts of Parliament, Canada transfers control over their lands and natural resources to Manitoba, Saskatchewan, and Alberta

1930   Pitchblende, the source of radium and uranium, discovered at Great Bear Lake, North West Territories

1932   Trans-Canada telephone system established

1932   Waterton National Park linked with Montana's Glacier National Park to form world's first international park

1933, 2 December   Due to dire financial circumstances, Newfoundland loses Dominion status and reverts to British Crown

1934, 16 February   Newfoundland placed under British-appointed Commission of Government

1937, 1 September   First regular flight of Trans-Canada Airlines

1940–1942   First west-to-east navigation of Northwest Passage by Henry Larson in RCMP vessel *St. Roch*

1940, 21 June   National Resources Mobilization Act

1941   Wartime Prices and Trade Board established with strong power over the economy

1945, 8 March   United States announces abandonment of Canol Pipeline Project (Norman Wells to Yellowknife)

1947, 13 February   Vern Hunter strikes oil at Leduc, Alberta

1949, 23 March   Newfoundland becomes tenth province of Canada

1950, 31 October   1,770-kilometer gas pipeline from Edmonton to Great Lakes completed

1951, 12 December   St. Lawrence Seaway Authority established

1952, 24 April   First Alberta oil reaches Ontario by pipeline and freighter

1953, 15 October   Transmountain (Edmonton–Vancouver) oil pipeline completed

1954   Pinetree Line of radar early warning stations built

1954, 8 January   Crude oil from Alberta reaches Sarnia, Ontario, by pipeline

1954, 10 August   Ground broken for St. Lawrence Seaway project

1954, 15 October   Hurricane Hazel strikes near Toronto, kills eighty-one, 124 kilometer/hour winds and floods do extensive damage

1955   Dinosaur Provincial Park established in Alberta

1957, 1 January   Mid-Canada Radar Warning Line into operation

1957, 31 July   Distant Early Warning (DEW) Line radar stations brought into operation

1958, 23 October   Springhill, Nova Scotia, mine explosion kills seventy-four

1959, 1 April   St. Lawrence Seaway officially opens

1960   Helge Ingstad announces discovery of Viking settlement site at L'Anse aux Meadows, Newfoundland

1961, 1 September   First Arctic oil-drilling efforts at Melville Island

1962, 30 July   Official opening of Trans-Canada Highway

1963   Beginning of Red River Floodway project in Manitoba, nation's largest earth-moving job

1964   Canada passes the Territorial Sea and Fisheries Zone Act, which establishes a nine-mile fishing zone outside the three-mile territorial limit established by the law of the sea

1964   Canadian-U.S. treaty for collaborative development of Columbia River becomes effective

1969   Pollution Probe established at University of Toronto

1970   Pacific Rim National Park established

1970   Inco builds 381-meter smokestack at Copper Cliff, near Sudbury, to disperse pollutants

1970   Don't Make a Wave Committee formed in Vancouver to protest U.S. nuclear testing in Alaska

1970, 4 February   Liberian tanker *Arrow* goes aground in Chedabucto Bay, Nova Scotia, spills 3.5 million gallons of oil

1970, 17 February   Canadian government bans use of phosphates in laundry detergent

1970, 24 March   Mercury contamination leads to closure of Lake St. Clair fishery

1970, 31 March   Sale of perch and pickerel from Lake Erie prohibited because of mercury contamination

1970, 1 August   Canada announces unilateral extension of the territorial sea to twelve nautical miles

1971   Department of Environment formed in Ottawa

1971, 10 February   Federal Government establishes fines of up to $200,000 for air pollution

1971, 5 April   First CANDU nuclear reactor power generator operated at Gentilly, Québec

1971, 4 October   Oil discovered on Sable Island, Nova Scotia

1971, 2 December   Norway agrees to cease fishing in Canada's territorial waters

1972, 24 February   Panarctic Oils, Ltd., announces first Arctic oil find on Ellesmere Island

1972, 25 February   Pickering nuclear power plant goes into operation

1972, 27 February   Several European nations agree to reduce fishing in eastern Canadian waters

1972, 15 April   Great Lakes Water Quality Agreement signed by Canada and United States

1972, May   First Canadian Arctic Resources Committee Conference, Ottawa

1972   Governments ban fishing off Gaspé, New Brunswick, and southwest Newfoundland due to low stocks

1972, 16 June   Churchill Falls power project goes into operation

1972, 2 August   Arctic Waters Pollution Prevention Act proclaimed

1973, 25 January   Freighter *Irish Stardust* goes aground north of Vancouver Island, spills 378,000 liters of fuel

1973, 31 January   Supreme Court of Canada recognizes aboriginal land title

1975   Syncrude project to develop Athabasca oil sands near Fort McMurray, Alberta

1975, 3 March   Mackenzie Valley Pipeline Inquiry hearings begin

1976, 15 April   Dome Petroleum gains approval to drill for oil in Beaufort Sea

1977, 1 January   Canada extends territorial waters to 200 miles

1977, 11 April   Justice Thomas Berger issues final report of Mackenzie Valley Pipeline Inquiry

1978   Committee on the Status of Endangered Wildlife begins development of species-at-risk list

1979   Syncrude plant extracts first oil from tar sands at Fort McMurray

1979, 27 October   First hydroelectric power from James Bay project

1980, 6 October   British Newfoundland Corporation agrees to sell Churchill Falls power to Hydro Québec for forty-year period

1980, 28 October   National Energy Program aims to increase Canada's energy self-sufficiency

1982   Supreme Court rules that resources of the continental shelf rest with federal not provincial governments ("Hibernia decision")

1982   Canada charges that acid rain attributable to air pollution originating in United States killed all fish in 147 Ontario lakes

1982, 15 February   *Ocean Ranger,* world's largest semi-submersible oil drilling rig, capsizes and sinks during a storm on Grand Banks, all eighty-four workers are killed

1987   In Montréal, fifty-three industrial nations agree to eliminate chlorofluorocarbons, a cause of ozone depletion

1989   *Exxon Valdez* grounds on reef in Prince William Sound, hundreds of kilometers of coastline fouled

1994, 14 June   Canada and other industrial nations agree, in Oslo, Norway, to reduce acid-rain-producing sulfur emissions

1995   Federal-provincial agreement on tighter emission standards for automobiles sold by 2001

1996, 12 March   Vancouver City Council bans smoking in restaurants

1997, 17 November   Hibernia field off Newfoundland produces first oil

1998, January   Severe ice storm in northeastern North America brings down electricity transmission lines, leaving a million households without power and causing at least twenty-five deaths. Millions of trees were destroyed as well as 120,000 kilometers of power lines and telephone cables, 130 major transmission towers, and about 30,000 wooden utility poles

1999, 3 June   Canadian-U.S. treaty governs conservation and sharing of Pacific salmon

1999, 14 September   Canadian Environmental Protection Act, 1999, approved: "An Act respecting pollution prevention and the protection of the environment and human health in order to contribute to sustainable development"

1999, 30 November   Western provinces and Québec reject national initiative to prohibit freshwater exports

2000, 11 April   Canadian Government introduces Species at Risk Act reflecting recognition of the Committee on the Status of Endangered Wildlife in Canada view that some 340 species endangered

2000, May–June   *E-coli* contamination of water supply in Walkerton, Ontario, kills nine and leaves hundreds ill

2002, January   Report of inquiry into Walkerton water tragedy released: lays some blame on the provincial government

2002, October   The government of Canada announces its intention to create ten new parks and five new national marine conservation areas over the next five years

2002, December 16   Canada signs the Kyoto Accord, limiting greenhouse gas emissions

2003, March   Ontario declares a public health emergency as a result of SARS

2003, May   Mad-cow disease (bovine spongiform encephalopathy) discovered in a northern Alberta farm, threatening the $7.6 billion beef industry

2004, 17 March   Canadian Food Inspection Agency creates a program for routine testing of poultry for avian influenza (bird flu), after British Columbia had to destroy 57,000 chickens

2004, 22 March   Federal government announces an aid package worth almost $1 billion to farmers hurt by mad cow disease

2004, 24 March   Canadian Food Inspection Agency orders slaughter of 275,000 chickens and turkeys in British Columbia to fight avian flu outbreak

2004, 6 April   Canada orders slaughter of 19 million British Columbia poultry due to avian influenza

2004, 10 September   The federal government announces $500 million to help cattle farmers hurt by the restricted trade of cattle stemming from one case of mad cow disease in 2003

2005, September   Almost ten years after the introduction of genetically engineered farm crops in Canada, Prince Edward Island Premier Pat Binns initiates a process to examine prospects for a GE-free Prince Edward Island

2005, December   World governments reach a historic agreement at the Montréal Climate Summit, committing to a stronger Kyoto Protocol

2005   By year's end five provinces/territories and at least seventy-five municipalities have legislation or regulations limiting or banning smoking in public spaces indoors

2006, 7 February   British Columbia Government announces plan for environmental protection of two million hectares of the central coast in the Great Bear Rainforest Agreement

REFERENCES AND
FURTHER READING

In a work of this scope, it is impossible to cite the source of every idea, claim, and number in the text. Even to attempt this would seriously compromise the "user-friendliness" that I have sought in these pages. I have, therefore, compromised. Quotations are acknowledged (although occasionally only by a single blanket reference at the end of a passage). I have also tried to note those works that have had a major influence on my thinking about particular topics, and to point to others that extend, amplify, or in some cases challenge, a particular line of argument; together these references are intended as pointers and stimuli for further exploration rather than as comprehensive bibliographies. There are library shelves of important writing on dozens of topics treated en passant in these pages.

Readers interested in pursuing details beyond those caught in the fairly tight net of references in the text (there are over 650 items in the bibliography below), should consult the following sources. The three volumes of *The Historical Atlas of Canada* (from which many of the maps in this volume have been adapted and redrawn) are indispensable, and offer much fine-grained detail about the evolving country. The on-line edition of *Historical Statistics of Canada*, edited by F. H. Leacy (www .statcan.ca/english/freepub/11-516-XIE/sectiona/intro.htm), contains more than a thousand tables of data pertaining to the economic, demographic, institutional, and social circumstances of the country between 1867 and the 1970s. There is a very useful website, Censusfinder. Canada Census Records Online (www.censusfinder .com/canada-census-records.htm), with links to dozens of early provincial and colonial censuses and other local enumerations. For information about individuals there is no better starting point than the *Dictionary of Canadian Biography*, available online at www.biographi.ca/EN/index.html, which includes biographies long and short of many thousands of those who shaped the country and died between approximately 1000 and 1930, as well as a number of extended, more general essays.

Beyond this, earlier surveys treating the history, geography, or historical geography of all or part of northern North America generally contain a wealth of specific detail: Among the most accessible of these items are the following, listed in full below: Harris and Warkentin 1974; McIlwraith 2001; Wynn 1987a; and the several contributions (by Ray, Moore, Wynn, Waite, and Morton) to R. C. Brown, ed. 1987, *The Illustrated History of Canada* (Toronto: Lester and Orpen Dennys). Those wishing to burrow deeper might begin with the Integrated Database in the Canadian Environmental History Project, developed and coordinated by Stephane Castonguay, Canada Research Chair in the Environmental History of Québec, at Université du Québec à Trois Rivières. This source currently offers convenient access to over 24,000 partially indexed bibliographic references and is available at www.cieq .uqtr.ca/fci_hec/index_en.php (accessed 10 July 2006).

A

Adney, Tappan. 1900. *The Klondike Stampede.* Vancouver: University of British Columbia Press; reprint 1994.

Ahlgren, Isabel F., and Clifford E. Ahlgren. 1960. "Ecological Effects of Forest Fires." *Botanical Review* 26, 4: 483–533.

Ainley, Marianne G. 1985. "From Natural History to Avian Biology: Canadian Ornithology, 1860–1950." Ph.D. diss., McGill University, Montréal.

AIS. 2006. Aquatic Invasive Species (AIS) and the Great Lakes: Simple Questions, Complex Answers. Available at: www.glerl.noaa.gov/pubs/brochures/invasive/ais .html (accessed 10 July 2006).

Akenson, Donald H. 1984. *The Irish in Ontario: A Study of Rural History.* Montréal and Kingston, ON: McGill–Queen's University Press.

Alexander, David. 1976. "Newfoundland's Traditional Economy and Development to 1934." *Acadiensis* 5, 2: 56–78.

Alexander, David G. 1977. *The Decay of Trade: An Economic History of the Newfoundland Saltfish Trade, 1935–1965.* St. John's: Memorial University of Newfoundland.

Alexander, David G. 1983. *Atlantic Canada and Confederation: Essays in Canadian Political Economy.* Toronto: University of Toronto Press.

Alexander, Sir James. 1849. *L'Acadie, or Seven Years' Explorations in British America.* London: H. Colburn.

Allardyce, Gilbert. 1972. "The Vexed Question of Sawdust: River Pollution in Nineteenth-Century New Brunswick." *Dalhousie Review* 52: 177–190.

Allen, John L. 1992. "From Cabot to Cartier: The Early Exploration of Eastern North America, 1497–1543." *Annals of the Association of American Geographers* 82, 3: 500–521.

Allen, Robert C., and Ian Keay. 2001. "The First Great Whale Extinction: The End of the Bowhead Whale in the Eastern Arctic." *Explorations in Economic History* 38: 448–477.

Allen, Robert C., and Ian Keay. 2004. "Saving the Whales: Lessons from the Extinction of the Eastern Arctic Bowhead." *Journal of Economic History* 64, 2: 400–432.

Ames, Herbert Brown. 1897. *The City Below the Hill.* Montréal: Bishop.

Anderson, T. W. 1989. "Vegetation Changes over 12,000 Years." *Geos* 18, 3: 39–47.

Ankli, Robert E., and Kenneth J. Duncan. 1984. "Farm Making Costs in Early Ontario." *Canadian Papers in Rural History* 4: 33–49.

Apostle, Richard, and Gene Barrett; with contributions by Pauline Barber et al. 1992. *Emptying Their Nets: Small Capital and Rural Industrialization in the Nova Scotia Fishing Industry.* Toronto: University of Toronto Press.

Armstrong, Christopher, and H. Viv Nelles. 1986. *Monopoly's Moment: The Organization and Regulation of Canadian Utilities, 1830–1930.* Philadelphia: Temple University Press.

Artibise, Alan F. J., and Gilbert A. Stelter. 1995. "Government Planning and Urban Planning: The Canadian Commission of Conservation in Historical Perspective." Pp. 152–169 in Chad Gaffield and Pam Gaffield, *Consuming Canada: Readings in Environmental History.* Toronto: Copp Clark.

Atwood, Margaret. 1970. "The Planters." Pp. 16–17 in Atwood, *The Journals of Susanna Moodie.* Toronto: Oxford University Press.

B

Baikie, Wallace, and Rosemary Philips. 1986. *Strathcona: A History of British Columbia's First Provincial Park.* Campbell River, BC: Ptarmigan.

Bailey, Alfred G. 1937. *The Conflict of European and Eastern Algonkian Cultures, 1504–1700: A Study in Canadian Civilization.* Toronto: University of Toronto Press; reprint 1969.

Balcom, A. B. 1928. "Agriculture in Nova Scotia since 1870." *Dalhousie Review* 8, 1: 37–43.

Bannerjee, Subhanker. 2003. *Seasons of Life and Land.* Seattle: Mountaineers.

Barkham, Selma. 1980. "A Note on the Strait of Belle Isle during the Period of Basque Contact with Indians and Inuit." *Etudes/Inuit/Studies* 4, 1–2: 51–58.

Barkham, Selma. 1982. "The Documentary Evidence for Basque Whaling Ships in the Strait of Belle Isle." Pp. 53–96 in G. M. Story, ed., *Early European Settlement and Exploitation in Atlantic Canada.* St. John's, NL: Memorial University of Newfoundland.

Barkham, Selma. 1984. "The Basque Whaling Establishments in Labrador 1536–1632—A Summary." *Arctic* 37, 4: 515–519.

Barkham, Selma, 2001. "Between Cartier and Cook: The Contribution of Fishermen to the Early Toponymy of Western Newfoundland." Pp. 23–31 in Olaf Janzen, ed., *Northern Seas 1999: Yearbook of the Association for the History of the Northern Seas.* St. John's, NL: Memorial University of Newfoundland.

Baskerville, Peter A. 2002. *Ontario: Image, Identity and Power.* Don Mills, ON: Oxford University Press.

Bateman, Robert. 2000. *Thinking Like a Mountain.* Toronto: Viking.

Beazley, Karen, and Robert Boardman, eds. 2001. *Politics of the Wild, Canada and Endangered Species.* Don Mills, ON: Oxford University Press.

Bechtel Corporation. La Grande project, at www.bechtel.com/spjames.htm (accessed 5 July 2006).

Beckerman, Wilfred. 2002. *A Poverty of Reason: Sustainable Development and Economic Growth.* Oakland, CA: Independent Institute.

Bella, Leslie. 1987. *Parks for Profit.* Montréal: Harvest House.

Bellavance, Claude. 1994. *Shawinigan Water and Power, 1898–1963: Formation et déclin d'un groupe industriel au Québec.* Montréal: Boréal.

Benidickson, Jamie. 2006. *The Culture of Flushing.* Vancouver: University of British Columbia Press.

Berger, Carl Clinton. 1966. "The True North Strong and Free." Pp. 3–26 in P. Russell, ed., *Nationalism in Canada.* Toronto: McGraw Hill.

Berger, Carl Clinton. 1971. *Studies in the Ideas of Canadian Imperialism.* Toronto: University of Toronto Press.

Berger, Thomas. 1977. *Northern Frontier, Northern Homeland: The Report of the Mackenzie Valley Pipeline Inquiry.* 2 vols. Ottawa: Minister of Supply and Services Canada.

Berman, Tzeporah, et al. 1994. *Clayoquot and Dissent.* Vancouver: Ronsdale.

Berry, Lynn. 2001. "The Delights of Nature in This New World: A Seventeenth-Century View of the Environment." Pp. 223–235 in Germaine Warkentin and Carolyn Podruchny, eds., *Decentring the Renaissance: Canada and Europe in Multidisciplinary Perspective, 1500–1700.* Toronto: University of Toronto Press.

Berry, Thomas. 1996. "Ethics and Ecology." Paper delivered to the Harvard Seminar on Environmental Values, Harvard University, April 9. Available at http://ecoethics.net/ops/eth&ecol.htm (accessed 5 July 2006).

Berton, Pierre. 1972. *Klondike: The Last Great Gold Rush, 1896–1899.* Toronto: McClelland and Stewart.

Berton, Pierre. 1976. *My Country: The Remarkable Past.* Toronto: McClelland and Stewart.

Bilson, Geoffrey. 1980. *A Darkened House: Cholera in Nineteenth-Century Canada* Toronto: University of Toronto Press.

Binnema, Theodore. 2001. *Common and Contested Ground: A Human and Environmental History of the Northwestern Plains.* Norman: University of Oklahoma Press.

Bird, J. Brian. 1960. "The Scenery of Central and Southern Arctic Canada." *Canadian Geographer* 15: 1–11.

Bitterli, Urs. 1989. *Cultures in Conflict. Encounters Between European and Non-European Cultures, 1492–1800.* Trans. Richie Robertson. Stanford, CA: Stanford University Press.

Blackmer, Hugh A. 1976. "Agricultural Transformation in a Regional System: The Annapolis Valley, Nova Scotia." Ph.D. diss., Stanford University.

Bleakney, Sherman. 2004. *Sods, Soils, and Spades: The Acadians at Grand Pre and their Dykeland Legacy.* Montréal and Kingston, ON: McGill–Queen's University Press.

Bockstoce, John R. 1986. *Whales, Ice and Men: The History of Whaling in the Western Arctic.* Seattle: University of Washington Press.

Bonatto, Sandro L., and Francisco M. Salazano. 1997a. "A Single and Early Migration for the Peopling of the Americas Supported by Mitrochondrial DNA Sequence Data." *Proceedings of the National Academy of Sciences of the United States of America* 94, 5 (4 March): 1866–1871.

Bonatto, Sandro L., and Francisco M. Salazano. 1997b. "Diversity and Age of the Four Major mtDNA Haplogroups, and Their Implications for the Peopling of the New World." *American Journal of Human Genetics* 61: 1413–1423.

Bordo, Jonathan. 1992–1993. "Jack Pine: Wilderness Sublime or the Erasure of the Aboriginal Presence from the Landscape." *Journal of Canadian Studies* 27 (Winter): 98–128.

Borneman, Walter R. 2003. *Alaska: Saga of a Bold Land.* New York: Harper Collins.

Bouchard, Gérard. 1996. *Quelques arpents d'Amérique: Population, économie, famille au Saguenay, 1838–1971.* Montréal: Boréal.

Bouchette, J. 1832. *A Topographical Dictionary of the Province of Lower Canada.* London: Longman, Rees, Orme, Brown.

Boyd, Robert. 1996. "Commentary on Early Contact-Era Smallpox in the Pacific Northwest." *Ethnohistory* 43, 2: 307–328.

Boyd, Robert. 1999. *The Coming of the Spirit of Pestilence: Introduced Infectious Diseases and Population Decline among Northwest Coast Indians, 1774–1874.* Seattle: University of Washington Press.

Bray, R. Matthew, and Ashley Thomson, eds. 1990. *Temagami: a Debate on Wilderness* Toronto: Dundurn.

Breen, David H. 1983. *The Canadian Prairie West and the Ranching Frontier 1874–1924.* Toronto: University of Toronto Press.

Brière, Jean-François. 1997. "The French Fishery in the 18th Century." Pp. 47–64 in James E. Candow and Carol Corbin, eds., *How Deep Is the Ocean? Historical Essays on Canada's Atlantic Fishery.* Sydney, NS: University College of Cape Breton Press.

Brown, Jennifer S. H. 1980. *Strangers in Blood: Fur Trade Company Families in Indian Country.* Vancouver: University of British Columbia Press.

Brown, R. M., et al. 1983. "Accelerator C14 Dating of the Taber Child." *Canadian Journal of Archeology* 7: 233–237.

Brown, Robert Craig, and Ramsay Cook. 1974. *Canada 1896–1921: A Nation Transformed.* Toronto: McClelland and Stewart.

Brox, Ottar. 1972. *Newfoundland Fisherman in the Age of Industry: A Sociology of Economic Dualism.* St. John's: Memorial University of Newfoundland.

Bruce, Charles. 1954. *The Channel Shore.* Toronto: Macmillan.

Brundtland, Gro Harlem. 1987. *Our Common Future.* Oxford: World Commission on Environment and Development.

Brunger, Alan G. 1972. "Analysis of Site Factors in Nineteenth-Century Ontario Settlement." Pp. 400–402 in W. P. Adams and F. Helleiner, eds., *International Geography 1972: Papers Submitted to the 22nd International Geographical Congress, Canada.* Toronto: University of Toronto Press.

Buchanan, Mark. 2001. *Ubiquity: The Science of History or Why the World Is Simpler Than We Think.* London: Phoenix.

Budd, Ken, Ric Careless, and Johnny Mikes. 1993. *Tatshenshini: River Wild* Vancouver: Raincoast/SummerWild.

Bumsted, John M. ("Jack"). 1992. *The Peoples of Canada.* Toronto: Oxford University Press.

Burnett, J. Alexander. 2003. *A Passion for Wildlife: A History of the Canadian Wildlife Service.* Vancouver: University of British Columbia Press.

Burnett, J. Alexander, C. T. Dauphine, S. H. McCrindle, and T. Mosquin. 1989. *On the Brink: Endangered Species in Canada.* Saskatoon, SK: Western Producer Prairie.

Busch, Briton C. 1985. *The War against the Seals: A History of the North American Seal Fishery.* Montréal and Kingston, ON: McGill–Queen's University Press.

Butler, William Francis. 1872. *The Great Lone Land.* London: Sampson Low, Marston, Low and Searle.

Butzer, Karl. 1998. "A 'Marginality' Model to Explain Major Spatial and Temporal Gaps in the Old World and New World Pleistocene Settlement Records." *Geoarcheology* 3: 193–203.

Butzer, Karl W. 2002. "French Wetland Agriculture in Atlantic Canada and Its European Roots: Different Avenues to Historical Diffusion." *Annals of the Association of American Geographers* 92, 3: 451–470.

C

Cadigan, Sean T. 1995. *Hope and Deception in Conception Bay: Merchant-Settler Relations in Newfoundland, 1785–1855.* Toronto: University of Toronto Press.

Cadigan, Sean, T. 1999a. "Failed Proposals for Fisheries Management and Conservation in Newfoundland, 1835–1880." Pp. 147–169 in Dianne Newell and Rosemary E. Ommer, eds., *Fishing Places, Fishing People: Traditions and Issues in Canadian Small-Scale Fisheries.* Toronto: University of Toronto Press.

Cadigan, Sean T. 1999b. "The Moral Economy of the Commons: Ecology and Equity in the Newfoundland Cod Fishery, 1815–1855." *Labour/Le Travail* 43: 9–42.

Cadigan, Sean T., and Jeffrey A. Hutchings. 2002. "Nineteenth-Century Expansion of the Newfoundland Fishery for Atlantic Cod: An Exploration of Underlying Causes." Pp. 31–65 in Paul Holm, Tim D. Smith, and David J. Starkey, eds., *The Exploited Seas: New Directions for Marine Environmental History.* Research in Maritime History, no. 21. St. John's, NF: International Maritime Economic History Association.

Cameron, Wendy, S. Haines, and M. M. Maude. 2000a. *English Immigrant Voices: Labourers' Letters from Upper Canada in the 1830s.* Montréal and Kingston, ON: McGill–Queen's University Press

Cameron, Wendy, S. Haines, and M. M. Maude. 2000b. *Assisted Emigration to Upper Canada: The Petworth Project, 1832–1837.* Montréal and Kingston, ON: McGill–Queen's University Press.

Campbell, Claire Elizabeth. 2005. *Shaped by the West Wind: Nature and History in Georgian Bay.* Vancouver: University of British Columbia Press.

Canada's Uranium Production and Nuclear Power. 2005. Nuclear Issues Briefing Paper no. 3. Available at: www.uic.com.au/nip03.htm (accessed 5 July 2006).

Carlson, Catherine C., George J. Armelagos, and Ann L. Magennis. 1992. "Impact of Disease on the Precontact and Early Historic Populations of New England and the Maritimes." Pp. 145–147 in John W. Verano and Douglas H. Ubelaker, eds., *Disease and Demography in the Americas.* Washington, DC: Smithsonian Institution.

Carlson, Hans M. 2004. "A Watershed of Words: Litigating and Negotiating Nature in Eastern James Bay, 1971–75." *Canadian Historical Review* 85, 1: 63–84.

Carson, Rachel. 1964. *Silent Spring.* Greenwich, CT: Fawcett.

Carter, Sarah. 1990. *Lost Harvests: Prairie Indian Reserve Farmers and Government Policies.* Montréal and Kingston, ON: McGill–Queen's University Press.

Carver, Humphrey. 1975. *Compassionate Landscape.* Toronto: University of Toronto Press.

Carver, Humphrey. 1978. "Building the Suburbs: A Planner's Reflections." *City Magazine* 3, 7: 40–45.

Castonguay, Stephane. 2004. "Naturalizing Federalism: Insect Outbreaks and the Centralization of Entomological Research in Canada, 1884–1914." *Canadian Historical Review* 85, 1: 1–34.

Catchpole, A. J. W., D. W. Moodie, and B. Kaye. 1970. "Content Analysis: A Method for Identification of Dates of First Freezing and First Breaking from Descriptive Accounts." *Professional Geographer* 22, 5: 252–257.

Catton, Theodore. 1997. *Inhabited Wilderness, Indians, Eskimos and National Parks in Alaska.* Albuquerque: University of New Mexico Press.

Chapman, Lyman J., and Donald F. Putnam. 1984. *The Phsyiography of Southern Ontario.* Toronto: Ontario Ministry of Natural Resources.

Chernov, I. U. U. 1985. *The Living Tundra.* Cambridge and New York: Cambridge University Press.

Chevrier, Lionel. 1959. *The St. Lawrence Seaway.* Toronto: Macmillan.

Churchill Falls, see http://ieee.ca/millennium/churchill/cf_home.html, and http://ieee.ca/millennium/churchill/cf_history.html (accessed 5 July 2005).

Clancy, Peter. 1992. "The Politics of Pulpwood Marketing in Nova Scotia, 1960–1985." Pp. 142–167 in L. A. Sandberg, ed., *Trouble in the Woods: Forest Policy and Social Conflict in Nova Scotia and New Brunswick.* Fredericton, NB: Acadiensis.

Clark, Andrew Hill. 1968. *Acadia: The Geography of Early Nova Scotia to 1760.* Madison, WI: University of Wisconsin Press.

Clark, Peter U., Shawn J. Marshall, Garry K. C. Clarke, Steven W. Hostetler, Joseph M. Licciardi, and James T. Teller. 2001. "Freshwater Forcing of Abrupt Climate Change during the Last Glaciation." *Science* 293, 5528 (13 July): 283–287.

Clarke, John. 2001. *Land, Power and Economics on the Frontier of Upper Canada.* Montréal and Kingston, ON: McGill–Queen's University Press.

Claypole, E. W. 1878. "On the Migration of Plants from Europe to America, with an Attempt to Explain Certain Phenomena Connected Therewith." Pp. 70–91 in *Third Report of the Montréal Horticultural Society and Fruit Growers Association of the Province of Québec . . . for the year 1877.* Montréal: Witness Printing House.

Claypole, E. W. 1879. "On the Migration of European Animals to America and of American Animals to Europe." Pp. 65–93 in *Fourth Report of the Montréal Horticultural Society and Fruit Growers Association of the Province of Québec . . . for the year 1878.* Montréal: Witness Printing House.

Clayton, Daniel. 2000. *Islands of Truth: The Imperial Fashioning of Vancouver Island.* Vancouver: University of British Columbia Press.

Coates, Colin M. 2000. *The Metamorphoses of Landscape and Community in Early Québec.* Montréal and Kingston, ON: McGill–Queen's University Press.

Coates, Kenneth S. 1985. *Canada's Colonies: A History of the Yukon and Northwest Territories.* Toronto: Lorimer.

Coates, Kenneth S., and William R. Morrison. 1992a. *The Alaska Highway in World War II: The U.S. Army of Occupation in Canada's Northwest.* Norman: University of Oklahoma Press.

Coates, Kenneth S., and William R. Morrison. 1992b. *The Forgotten North.* Toronto: Lorimer.

Cohen, Marjorie G. 1988. *Women's Work, Markets, and Economic Development in Nineteenth-Century Ontario.* Toronto and Buffalo, NY: University of Toronto Press.

Colpitts, George. 2002. *Game in the Garden: A Human History of Wildlife in Western Canada to 1940.* Vancouver: University of British Columbia Press.

Conrad, Margaret, and James Hiller. 2001. *Atlantic Canada.* Don Mills, ON: Oxford University Press.

Cook, Ramsay. 1987. "The Triumph and Trials of Materialism, 1900–1945." Pp. 375–466 in R. Craig Brown, *Illustrated History of Canada.* Toronto: Lester and Orpen Dennys.

Cooley, Richard A. 1967. *Alaska: A Challenge in Conservation.* Madison: University of Wisconsin Press.

Cooney, Robert. 1832. *Compendious History of the Northern Part of the Province of New Brunswick, and of the District of Gaspé, in Lower Canada.* Halifax, NS: Howe.

Copp, Terry. 1974. *The Anatomy of Poverty.* Toronto: McClelland and Stewart.

Courville, Serge. 1990a. *Entre ville et campagne: l'essor du village dans les seigneuries du Bas-Canada.* Sainte-Foy, QC: Presses de l'Université Laval.

Courville, Serge. 1990b. "Space, Territory and Culture in New France: A Geographical Perspective." Pp. 165–176 in Graeme Wynn, *People Places Patterns Processes: Geographical Perspectives on the Canadian Past.* Toronto: Copp Clark Pitman.

Courville, Serge, et al. 1996. *Atlas historique du Québec: Population et territoire.* Sainte-Foy, QC: Presses de l'Université Laval.

Courville, Serge. 2000. *Le Québec: Genèses et mutations du territoire. Synthèse de géographie historique.* Sainte-Foy, QC: Presses de l'Université Laval.

Courville, Serge, Jean-Claude Robert, and Normand Séguin. 1995. *Le pays laurentien au XIXe siècle: les morphologies de base.* Sainte-Foy, QC: Presses de l'Université Laval.

Courville, Serge, and Normand Séguin. 1989. *Rural Life in Nineteenth-Century Québec.* Trans. Sinclair Robinson. Ottawa: Canadian Historical Association.

Cowan, Helen I. 1928. *British Emigration to British North America: The First Hundred Years.* Toronto: University of Toronto Press.

Cranstone, Donald A. 2002. *A History of Mining and Mineral Exploration in Canada and Outlook for the Future.* Ottawa: Natural Resources Canada, Minerals and Metals Sector.

Creighton, Donald. 1957. *Harold Adams Innis: Portrait of a Scholar.* Toronto: University of Toronto Press.

Cronon, William. 1983. *Changes in the Land: Indians, Colonists and the Ecology of New England.* New York: Hill and Wang.

Cronon, William. 1992. "Kennecott Journey: The Paths Out of Town," in William Cronon, George Miles, and Jay Gitlin, eds. *Under an Open Sky: Rethinking America's Western Past.* New York: W.W. Norton.

Cronon, William. 1995. "The Trouble with Wilderness, or, Getting Back to the Wrong Nature." Pp. 69–90 in Cronon, ed., *Uncommon Ground: Toward Reinventing Nature.* New York: Norton.

Cronon, William. 2001. "Neither Barren nor Remote." *New York Times,* 28 February: A 19.

Crosby, Alfred W. 1986. *Ecological Imperialism: The Biological Expansion of Europe, 900–1900.* New York: Cambridge University Press.

Cross, Michael S. 1970. *The Frontier Thesis and the Canadas: The Debate on the Impact of the Canadian Environment.* Toronto: Copp Clark.

Cruikshank, Julie. 2005. *Do Glaciers Listen? Local Knowledge, Colonial Encounters, and Social Imagination.* Vancouver: University of British Columbia Press.

D

Darnell, Adrian C. ed. 1990. *The Collected Economic Articles of Harold Hotelling.* New York: Springer Verlag.

Davis, M. B. 1981. "Quaternary History and the Stability of Plant Communities." Pp. 132–153 in D. C. West, H. H. Shugart, and D. B. Botkin, eds., *Forest Succession: Concepts and Application.* New York: Springer-Verlag.

Dearden, Philip, and Bruce Mitchell, eds. 1997. *Environmental Change and Challenge: A Canadian Perspective.* Toronto: Oxford University Press.

DeLottinville, Peter. 1979. "The St. Croix Cotton Manufacturing Company and Its Influence on the St. Croix Community, 1880–1892." M.A. thesis. Dalhousie University, Halifax.

DeLottinville, Peter. 1980. "Trouble in the Hives of Industry: The Cotton Industry Comes to Milltown, New Brunswick, 1879–1892." Canadian Historical Association, *Historical Papers, Montréal* 15, 1: 100–115.

Demeritt, David. 1997. "Representing the 'True' St. Croix: Knowledge and Power in the Partition of the Northeast." *William and Mary Quarterly* 54: 515–548.

Demeritt, David. 2005. "Perspectives on Diamond's *Collapse: How Societies Choose to Fail or Succeed.*" *Current Anthropology* 46: S92–S94.

Dempsey, Hugh A., ed. 1984. *The CPR West, The Iron Road and the Making of a Nation.* Vancouver: Douglas and McIntyre.

Denys, Nicolas. 1672. *The Description and Natural History of the Coasts of North America [Acadia]*. Trans. and ed. W. F. Ganong (1908). Toronto: Champlain Society.

Derry, Margaret. 1998. "Gender Conflicts in Dairying: Ontario's Butter Industry, 1880–1920." *Ontario History* 90: 31–47.

Desbarats, Catherine. 1992. "Agriculture Within the Seigneurial Regime of Eighteenth-Century Canada: Some Thoughts on the Recent Literature." *Canadian Historical Review* 73, 1: 1–29.

Diamond, Jared M. 1997. *Guns, Germs, and Steel: The Fates of Human Societies.* New York: Norton.

Diamond, Jared M. 2005. *Collapse: How Societies Choose to Fail or Succeed.* New York: Viking.

Dibblee, Rev. F. 1818. "Diary of the Reverend Frederick Dibblee of Woodstock." Unpublished manuscript. New Brunswick Museum, St. John.

Dick, Lyle. 2001. *Muskox Land: Ellesmere Island in the Age of Contact.* Calgary, AB: University of Calgary Press.

Dickason, Olive Patricia. 1992. *Canada's First Nations. A History of Founding Peoples from Earliest Times.* Norman: University of Oklahoma Press.

Diereville, Sieur de. 1933. *Relation of the Voyage to Port Royal in Acadia.* New York: Greenwood; reprint 1968.

Dillehay, Thomas D. 1991. "The Great Debate on the First Americans." *Anthropology Today* 7, 4: 12–13.

Dillehay, Thomas D. 2001. *The Settlement of the Americas: A New Prehistory.* New York: Basic Books.

Dillehay, Thomas D. 2003. "Tracking the First Americans." *Nature*, 425 (4 September): 23–24.

Dobyns, Henry F. 1983. *Their Number Become Thinned: Native American Population Dynamics in Eastern North America.* Knoxville: University of Tennessee Press in cooperation with the Newberry Library Center for the History of the American Indian.

Donahue, Brian. 2004. The *Great Meadow: Farmers and the Land in Colonial Concord.* New Haven, CT: Yale University Press.

Dorn, Ronald I., et al. 1992. "New Approach to the Radiocarbon Dating of Rock Varnish, with Examples from Drylands." *Annals of the Association of American Geographers* 82: 136–151.

Dorsey, Kurkpatrick. 1998. *The Dawn of Conservation Diplomacy: Canadian-American Wildlife Protection Treaties during the Progressive Era.* Seattle: University of Washington Press.

Downie, David Leonard, and Terry Fenge. 2003. *Northern Lights against POPs: Combatting Toxic Threats in the Arctic.* Montréal and Kingston, ON: McGill–Queens University Press.

Drushka, Ken, and Bob Burt. 2001. "The Canadian Forest Service: Catalyst for the Forest Sector." *Forest History Today* (Spring/Fall): 19–28.

Dunlap, Thomas. 1999. *Nature and the English Diaspora: Environment and History in the United States, Canada, Australia, and New Zealand.* Cambridge and New York: Cambridge University Press.

Dunn, Charles W. 1953. *Highland Settler. A Portrait of the Scottish Gael in Nova Scotia* Toronto. University of Toronto Press.

Dyke, A. S., and V. K. Prest. 1987. "Paleogeography of Northern North America, 18,000–5,000 Years Ago." *Geological Survey of Canada, Map 1703A. Sheet 1: 18 000–12 000 years BP. Sheet 2: 11 000–8 400 years BP.* Ottawa: Geological Survey of Canada.

E

Eagle, John A. 1989. *The Canadian Pacific Railway and the Development of Western Canada 1896–1914.* Montréal and Kingston, ON: McGill–Queen's University Press.

Easterbrook, Gregg. 2003. *The Progress Paradox: How Life Gets Better While People Feel Worse.* New York: Random House.

Edwards, Gordon. 1992. "Uranium: The Deadliest Metal." *Perception Magazine* 10, 2: n.p. Available at www.ccnr.org/uranium_deadliest.html#home (accessed 5 July 2005).

Edwards, Gordon., et al. N.d. "Uranium: A Discussion Guide." National Film Board of Canada. Available (adapted) at www.ccnr.org/nfb_uranium_0.html (accessed 5 July 2006).

Ehrlich, Paul R., and Anne H. Ehrlich. 2004. *One with Nineveh: Politics, Consumption, and the Human Future.* Washington, DC: Island/Shearwater Books.

Ennals, Peter Morley. 1978. "Land and Society in Hamilton Township, Upper Canada, 1797–1861." Ph.D. diss., University of Toronto.

Environment Canada. 1984. *The Acid Rain Story.* Ottawa: Environment Canada.

Evans, Clint. 2002. *The War on Weeds in the Prairie West: An Environmental History.* Calgary, AB: University of Calgary Press.

Evans, Simon. 2000. *Cowboys, Ranchers and the Cattle Business: Cross-border Perspectives on Ranching History.* Calgary, AB: University of Calgary Press.

Evans, Simon. 2004. *Bar U and Canadian Ranching History.* Calgary, AB: University of Calgary Press.

Evenden, Matthew D. 2000. "Remaking Hells Gate: Salmon, Science and the Fraser River." *BC Studies* 127: 47–82.

Evenden, Matthew D. 2004a. "Social and Environmental Change at Hells Gate, British Columbia." *Journal of Historical Geography* 30, 1: 130–153.

Evenden, Matthew D. 2004b. *Fish vs. Power: An Environmental History of the Fraser River.* Cambridge and New York: Cambridge University Press.

Evenden, Matthew D. 2006. "Precarious Foundations: Irrigation, Environment and Social Change in the Canadian Pacific Railway's Eastern Section, 1900–1930." *Journal of Historical Geography* 32, 1: 74–95.

Ewers, John C. 1972. "Influence of the Fur Trade upon Indians of the Northern Plains." Pp. 1–26 in Malvina Bolus, ed., *People and Pelts: Selected Papers of the Second North American Fur Trade Conference.* Winnipeg, MB: Peguis.

F

Fallis, Laurence S., Jr. 1960. "The Idea of Progress in the Province of Canada: A Study in the History of Ideas." Pp. 169–183 in William Lewis Morton, ed., *The Shield of Achilles: Aspects of Canada in the Victorian Age.* Toronto: McClelland and Stewart.

Faris, James C. 1972. *Cat Harbour: A Newfoundland Fishing Settlement.* St. John's: Institute of Social and Economic Research, Memorial University of Newfoundland.

Fetherling, Doug. 2004. "One of the Least Likely Men of the Klondike." *Vancouver Sun,* February 21: F18.

Fiege, Mark. 1999. *Irrigated Eden: The Making of an Agricultural Landscape in the American West.* Seattle: University of Washington Press.

Fienup-Riordan, Ann, 1990. *Eskimo Essays: Yup'ik Lives and How We See Them.* New Brunswick, NJ: Rutgers University Press.

Fienup-Riordan, Ann. 2000. *Hunting Tradition in a Changing World: Yup'ik Lives in Alaska Today.* New Brunswick, NJ: Rutgers University Press.

Fingard, Judith. 1974. "A Winter's Tale: Contours of Pre-Industrial Poverty in British America, 1815–1860." *Canadian Historical Association, Historical Papers Toronto 1974* 9, 1: 65–94.

Fitzhugh, William W. 1985. "Early Contacts North of Newfoundland Before A.D. 1600: A Review." Pp. 23–43 in William W. Fitzhugh, ed., *Cultures in Contact: The Impact of European Contacts on Native American Cultural Institutions A.D. 1000–1800.* Washington and London: Smithsonian Institution Press.

Fladmark, Knut R. 1979. "Routes: Alternate Migration Corridors for Early Man in North America." *American Antiquity* 44, 1: 55–69.

Fording. 2002. BC Mountain Mines. 2002 Investor Day and Mine Tours. Available at: www.fording.ca/data/1/rec_docs/509_Mountain%20Mines%20020919.pdf (accessed 5 July 2006).

Foreign Affairs and International Trade Canada. Nd. "Maurice Strong." Available at: www.dfait-maeci.gc.ca/department/skelton/Strong_bio-en.asp (accessed 10 July 2006).

Forkey, Neil S. 2003. *Shaping the Upper Canadian Frontier: Environment, Society and Culture in the Trent Valley.* Calgary, AB: University of Calgary Press.

Fortuine, Robert. 1989. *Chills and Fever: Health and Disease in the Early History of Alaska.* Fairbanks: University of Alaska Press.

Foster, Janet. 1978. *Working for Wildlife: The Beginning of Preservation in Canada.* Toronto: University of Toronto Press.

Francis, Daniel, and Toby Morantz. 1983. *Partners in Furs: A History of the Fur Trade in Eastern James Bay, 1600–1870.* Montréal and Kingston, ON: McGill–Queen's University Press.

Francis, R. Douglas. 1989. *Images of the West: Changing Perceptions of the Prairies, 1690–1960.* Saskatoon, SK: Western Producer.

Frank, David. 1985. "Tradition and Culture in the Cape Breton Mining Community in the Early Twentieth Century." Pp. 203–218 in Kenneth Donovan, ed., *Cape Breton at 200: Historical Essays in Honour of the Island's Bicentennial, 1785–1985.* Sydney, NS: University College of Cape Breton Press.

Fraser, R. L. 1979. "Like Eden in Her Summer Dress. Gentry, Economy and Society: Upper Canada, 1812–1840." Ph.D. thesis, University of Toronto.

Friedman, Susan. 1996. *Marc Bloch, Sociology and Geography: Encountering Changing Disciplines.* Cambridge and New York: Cambridge University Press.

Friesen, Gerald. 1984. *The Canadian Prairies: A History.* Toronto: University of Toronto Press.

Froschauer, Karl. 1999. *White Gold: Hydroelectric Power in Canada.* Vancouver: University of British Columbia Press.

Frost, Robert. 1923. "Stopping by Woods on a Snowy Evening." In *New Hampshire: A Poem with Notes and Grace Notes.* New York: Holt.

G

Gaffield, C., and P. Gaffield, eds. 1995. *Consuming Canada: Readings in Environmental History.* Toronto: Copp Clark.

Galois, Robert. 1970. "The Cariboo Gold Rush." M.A. thesis, University of Calgary.

Galois, Robert. 1996. "Measles, 1847–50: The First Modern Epidemic in British Columbia." *BC Studies* 109 (Spring): 31–46.

Ganong, William. Francis. 1906. "On the Limits of the Great Fire of Miramichi, 1825." Natural History Society of New Brunswick, *Bulletin* 24: 410–418.

Gayler, Hugh, J. 1994. *Niagara's Changing Landscapes.* Ottawa: Carleton University Press.

Gentilcore, R. Louis. 1969. "Lines on the Land: Crown Surveys and Settlement in Upper Canada." *Ontario History* 61, 2: 57–73.

Gentilcore, R. Louis, C. "Grant Head," with a cartobibliographical essay by Joan Winearls. 1984. *Ontario's History in Maps.* Ontario Historical Studies Series. Toronto: University of Toronto Press.

Gesner, Abraham. 1849. *The Industrial Resources of Nova Scotia.* Halifax, NS: A and W MacKinlay.

Gibson, James R. 1976. *Imperial Russia in Frontier America.* New York: Oxford University Press.

Gibson, James R. 1992. *Otter Skins, Boston Ships, and China Goods. The Maritime Fur Trade of the Northwest Coast, 1785–1841.* Montréal and Kingston, ON: McGill–Queen's University Press.

Gilbert, Robert. 1994. "A Field Guide to the Glacial and Postglacial Landscapes of Southeastern Ontario and Part of Québec." Geological Survey of Canada, Bulletin 453.

Gillespie, Greg. 2002. "'I Was Well Pleased with Our Sport among the Buffalo': Big-Game Hunters, Travel Writing, and Cultural Imperialism in the British North American West, 1847–72." *Canadian Historical Review* 83, 4: 555–584.

Gillis, R. Peter. 1974. "The Ottawa Lumber Barons and the Conservation Movement, 1880–1914." *Journal of Canadian Studies* 9, 1: 14–30.

Gillis, R. Peter. 1986. "Rivers of Sawdust: The Battle over Industrial Pollution in Canada, 1865–1903." *Journal of Canadian Studies* 21, 1: 84–103.

Gillis, R. Peter., and Thomas R. Roach. 1986. *Lost Initiatives: Canada's Forest Industries, Forest Policy and Forest Conservation.* Westport, CT: Greenwood.

Girard, Michel F. 1994. *L'écologisme retrouvé: essor et déclin de la Commission de la Conservation du Canada.* Ottawa: Presses de l'Université d'Ottawa.

Godwin, C. 1938. "A Regeneration Study of Logged-Off Lands on Vancouver Island." *Forestry Chronicle* 14.

Goldring, Philip. 1986. "Inuit Economic Responses to Euro-American Contacts: Southeast Baffin Island, 1824–1940." *Canadian Historical Association. Historical Papers Winnipeg 1986* 21, 1: 146–172.

Gosse, Phillip. 1840. *The Canadian Naturalist: A Series of Conversations on the Natural History of Lower Canada.* London: Voorst.

Grand Council of the Crees. 1993. "Testimony by Grand Chief Matthew Coon Come of the Grand Council of the Crees (of Québec), before the Joint Energy Committee, State House, Boston, MA." 22 March. Available at: www.gcc.ca/archive/article.php?id=201 (accessed 12 July 2006).

Grant, John Webster. 1984. *The Moon of Wintertime: Missionaries and the Indians of Canada in Encounter since 1534.* Toronto: University of Toronto Press.

Grayson, D. K. 1989. "The Chronology of North American Late Pleistocene Extinctions." *Journal of Archeological Science* 16: 153–165.

Greenberg, Joseph H., C. G. Turner II, and S. Zegura. 1986. "The Settlement of the Americas: A Comparison of the Linguistic, Dental and Genetic Evidence." *Current Anthropology* 37, 5: 477–497.

Greenwood Museum. N.d. Greenwood Heritage Society Museum/ Archives/ Tourism Visitor Centre. Available at www.greenwoodmuseum.com/ (accessed 14 July 2006).

Greer, Allan. 1985. *Peasant, Lord and Merchant. Rural Society in Three Québec Parishes, 1740–1840.* Toronto: University of Toronto Press.

Griffiths, Naomi. 2005. *From Migrant to Acadian: A North American Border People 1604–1755.* Montréal and Kingston: McGill–Queen's University Press.

Grove, Richard H. 1995. *Green Imperialism: Colonial Expansion, Tropical Island Edens and the Origins of Environmentalism, 1600–1860.* Cambridge and New York: Cambridge University Press.

H

Hackett, F. J. Paul. 2002. *"A Very Remarkable Sickness": Epidemics in the Petit Nord, 1670 to 1846.* Winnipeg: University of Manitoba Press.

Haliburton, Robert Grant. 1862. *The Past and Future of Nova Scotia.* Halifax, NS: J.B. Strong.

Hall, D. J. 1981 and 1985. *Clifford Sifton.* 2 vols. Vancouver: University of British Columbia Press.

Handbook of North American Indians. 1984 and 1981. Vol. 5: *Arctic* (ed. David Damas); vol. 6: *Subarctic* (ed. June Helm). Washington, DC: Smithsonian Institution.

Handcock, Gordon. 1985. "The Poole Mercantile Community and the Growth of Trinity 1700–1839." *The Newfoundland Quarterly* 80, 3: 19–30.

Hardin, Garrett James. 1968. "The Tragedy of the Commons." *Science* 162: 1243–1248.

Hare, Kenneth. 1964. "A Policy for Geographical Research in Canada." *Canadian Geographer* 8, 3: 113–116.

Harley, J. Brian. 1988. "Maps, Knowledge and Power." Pp. 277–312 in Denis Cosgrove and Stephen Daniels, eds., *The Iconography of Landscape: Essays on the Symbolic Representation, Design and Use of Past Environments.* Cambridge and New York: Cambridge University Press.

Harrington, Richard. 1952. *The Face of the Arctic.* New York: Schumann.

Harris, Douglas. 1995–1996. "The 'Nlha₇'kapmx Meeting at Lytton, 1879, and the Rule of Law." *BC Studies* 108: 5–25.

Harris, Douglas. 2001. *Fish, Law and Colonialism: The Legal Capture of Salmon in British Columbia.* Toronto: University of Toronto Press.

Harris, Richard Colebrook [R. Cole Harris, pseud.]. 1966. *The Seigneurial System in Early Canada: A Geographical Study.* Madison: University of Wisconsin Press.

Harris, Richard Colebrook [R. Cole Harris, pseud.]. 1971. "Of Poverty and Helplessness in Petite Nation." *Canadian Historical Review* 52: 23–50.

Harris, Richard Colebrook [R. Cole Harris, pseud.]. 1985. "Industry and the Good Life around Idaho Peak." *Canadian Historical Review* 66, 3: 315–343; reprinted in Harris, *The Re-Settlement of British Columbia.* Vancouver: University of British Columbia Press, 1997.

Harris, Richard Colebrook [R. Cole Harris, pseud.]. 1994. "Voices of Disaster: Smallpox around the Strait of Georgia in 1782." *Ethnohistory* 41, 4: 591–626.

Harris, Richard Colebrook [R. Cole Harris, pseud.]. 1997. "The Fraser Canyon Encountered." Pp. 103–113 in Harris, *The Resettlement of British Columbia.* Vancouver: University of British Columbia Press.

Harris, Richard Colebrook [R. Cole Harris, pseud.]. 2002. *Making Native Space.* Vancouver: University of British Columbia Press.

Harris, Richard Colebrook [R. Cole Harris, pseud.], and John Warkentin. 1974. *Canada before Confederation: A Study in Historical Geography.* New York: Oxford University Press.

Harris Richard, Colebrook [R. Cole Harris, pseud.], with Chris DeFreitas and Pauline Roulston. 1975. "The Settlement of Mono Township." *Canadian Geographer* 19, 1: 1–17.

Harris, Richard, 1996. *Unplanned Suburbs: Toronto's American Tragedy, 1900 to 1950.* Baltimore, MD: John Hopkins University Press.

Harris, Richard. 2004. *Creeping Conformity: How Canada Became Suburban.* Toronto: University of Toronto Press.

Hassan, Fekri A. 1981. *Demographic Archeology.* New York: Academic.

Hatvany, Matthew. 2003. *Marshlands: Four Centuries of Environmental Change on the Shores of the St. Lawrence.* Sainte-Foy, QC: Presses de l'Université de Laval.

Haycox, Stephen. 2002a. *Alaska, An American Colony.* Seattle: University of Washington Press.

Haycox, Stephen. 2002b. *Frigid Embrace: Politics, Economics and Environment in Alaska.* Corvallis: Oregon State University Press.

Hayes, Derek. 2002. *Historical Atlas of Canada: Canada's History Illustrated with Original Maps.* Vancouver and Toronto: Douglas and McIntyre.

Hayeur, Gaetan. 2001. *Summary of Knowledge Acquired in Northern Environments from 1970 to 2000.* Montréal: Hydro-Québec.

Hayles, N. Katherine. 1995. "Searching for Common Ground." Pp. 47–63 in Michael E. Soule and Gary Lease, eds., *Reinventing Nature? Responses to Postmodern Deconstruction.* Washington, DC: Island.

Haynes, C. V., Jr. 1966. "Elephant Hunting in North America." *Scientific American* 214, 6: 104–112.

Haynes, C. V., Jr. 1969. "The Earliest Americans." *Science* 166: 709–715.

Haynes C. V., Jr. 1980. "The Clovis Culture." *Canadian Journal of Anthropology* 1: 115–121.

Head, C. Grant. 1967. "Settlement Migration in Central Bonavista Bay, Newfoundland." Pp. 92–110 in R. L. Gentilcore, ed., *Canada's Changing Geography.* Scarborough, ON: Prentice Hall.

Head, C. Grant. 1975. "An Introduction to Forest Exploitation in Nineteenth-Century Ontario." Pp. 78–112 in J. David Wood, ed., *Perspectives on Landscape and Settlement in Nineteenth Century Ontario.* Toronto: McClelland and Stewart.

Head, C. Grant. 1976. *Eighteenth Century Newfoundland: A Geographer's Perspective.* Toronto: McClelland and Stewart.

Heidenreich, Conrad E. 1971. *Huronia: A History and Geography of the Huron Indians 1600–1650.* Toronto: McClelland and Stewart.

Heidenreich, Conrad E. 2001. "The Beginning of French Exploration out of the St. Lawrence Valley: Motives, Methods and Changing Attitudes towards Native Peoples." Pp. 236– 251 in Germaine Warkentin and Carolyn Podruchny, eds., *Decentering the Renaissance: Canada and Europe in Multidisciplinary Perspective, 1500–1700.* Toronto: University of Toronto Press.

Heidenreich, Conrad E., and Arthur J. Ray. 1976. *The Early Fur Trades: A Study in Cultural Interaction.* Toronto: McClelland and Stewart.

Henige, David. 1986. "Primary Source by Primary Source? On the Role of Epidemics in New World Depopulation." *Ethnohistory* 33: 293–312.

Hetherington, Renée, J. Vaughn Barrie, Roger MacLeod, and Michael Wilson. 2004. "Quest for the Lost Land." *Geotimes.* Available at: www.agiweb.org/geotimes/feb04/feature_Quest.html#authors (accessed 30 June 2006).

Hickey, C. G. 1984. "An Examination of Processes of Cultural Change among Nineteenth-Century Copper Inuit." *Etudes/Inuit/Studies* 8, 1: 13–35.

High, Steven. 2003. *Industrial Sunset: The Making of North America's Rust Belt, 1969–1984.* Toronto: University of Toronto Press.

Hildebrand, John. 2003. "A Northern Front." *Harper's Magazine* 307, 1842: 67–76.

Hilson, Gavin M. 2000. An Examination of Environmental Performance and Eco-Efficiency in the North American Gold Mining Industry. M.A. thesis, University of Toronto.

Hinckley, T. C. 1972. *The Americanization of Alaska, 1867–1897.* Palo Alto, CA: Pacific Books.

Hirschkind, Lynn. 1983. "The Native American as Noble Savage." *Humanist* 43, 1: 16–18.

Historical Atlas of Canada. 1987–1993. Geoffrey J. Matthews, cartographer/designer. 3 vols. Vol. 1, *From the Beginning to 1800.* Richard Colebrook Harris [R. Cole Harris, pseud.], ed., 1987. Vol. 2, *The Land Transformed, 1800–1891,* R. Louis Gentilcore, ed., 1993. Vol. 3, *Addressing the Twentieth Century, 1891–1961,* Donald G. Kerr and Deryck Holdsworth, eds., 1990. Toronto: University of Toronto Press.

Hodgins, Bruce W., and Jamie Benidickson. 1989. *The Temagami Experience: Recreation, Resources, and Aboriginal Rights in the Northern Ontario Wilderness.* Toronto: University of Toronto Press.

Hoffecker, John F., W. Roger Powers, and Ted Goebel. 1993. "The Colonization of Beringia and the Peopling of the New World." *Science* 259, 5091 (1 January): 46–53.

Hoffecker, John F., and S. A. Elias. 2003. "Environment and Archeology in Beringia." *Evolutionary Anthropology* 12, 1: 34–49.

Hoffman, Stephen M. 2002. "Powering Injustice: Hydroelectric Development in Northern Manitoba." Pp. 147–170 in J. Byrne, Cecilia Martinez, and Leigh Glover, eds., *Environmental Justice: Discourses in International Political Economy.* New Brunswick, NJ, and London: Transaction.

Hogan, J. Sheridan. 1855. *Canada. An essay:* To which Was Awarded the First Prize by the Paris Exhibition Committee of Canada. Montréal: Dawson.

Holly, Donald H., Jr. 2000. "The Beothuk on the Eve of Their Extinction." *Arctic Anthropology* 27, 1: 79–85.

Holman, Andrew C. 2000. *A Sense of Their Duty: Middle-class Formation in Victorian Ontario Towns.* Toronto and Kingston, ON: McGill–Queen's University Press, Montréal and Kingston.

Hopkins, David. M., et al., eds. 1982. *Paleoecology of Beringia.* New York: Academic.

Hornaday, W. T. 1899. *The Extermination of the American Bison.* Washington, DC: Smithsonian Institution.

Hornig, James F. 1999. *Social and Environmental Impacts of the James Bay Hydroelectric Project.* Montréal and Kingston, ON: McGill–Queen's University Press.

Hornsby, Stephen J. 1992. *Nineteenth Century Cape Breton: A Historical Geography.* Montréal and Kingston, ON: McGill–Queen's University Press.

Hornsby, Stephen J. 2004. *British Atlantic, American Frontier: Spaces of Power in Early Modern British America.* Lebanon, NH: University Press of New England.

Horwood, Harold. 1959. "The People Who Were Murdered for Fun." *MacLean's Magazine,* 10 October: 27, 36, 38–43.

Howe, C. D., and J. H. White. 1913. *Trent Watershed Survey. A Reconnaissance.* Toronto: Canada Commission of Conservation.

Howison, John. 1821. *Sketches of Upper Canada, Domestic, Local and Characteristic.* 2nd ed. 1822. Edinburgh: Oliver and Boyd.

Humphreys, John. 1970. *Plaisance: Problems of Settlement at this Newfoundland Outpost of New France 1660–1690.* Ottawa: National Museums of Canada.

Hutchings, Jeffrey A. 1997. "Spatial and Temporal Variation in the Exploitation of Northern Cod, *Gadus morhua:* A Historical Perspective from 1500 to the Present." Pp. 41–68 in Daniel Vickers, ed., *Marine Resources and Human Societies in the North Atlantic Since 1500: Papers Presented at the conference entitled "Marine Resources and Human Societies in the North Atlantic Since*

1500," October 20–22, 1995. St. John's, NF: Institute of Social and Economic Research, Memorial University of Newfoundland.

Hutchings, Jeffrey A. 1999. "The Biological Collapse of Newfoundland's Northern Cod." Pp. 260–275 in Dianne Newell and Rosemary E. Ommer, eds., *Fishing Places, Fishing People: Traditions and Issues in Canadian Small-Scale Fisheries.* Toronto: University of Toronto Press.

Hutchings, Jeffrey A., and Ransom A. Myers. 1995. "The Biological Collapse of Atlantic Cod off Newfoundland and Labrador: An Exploration of Historical Changes in Exploitation, Harvesting Technology and Management." Pp. 37–93 in R. Arnason and L. Felt, eds., *The North Atlantic Fisheries: Successes, Failures and Challenges.* Charlottetown: University of Prince Edward Island Press.

I

Ingstad, Anne Stine, and Helge M. Ingstad. 1986. *The Norse Discovery of America.* 2 vols. Oslo: Universitetsforlaget.

Innis, Harold Adams. 1930. *The Fur Trade in Canada: An Introduction to Canadian Economic History.* New Haven, CT: Yale University Press.

Innis, Harold Adams. 1940. *The Cod Fisheries: The History of an International Economy.* Toronto: University of Toronto Press; reprints 1954 and 1978.

International Energy Agency. Data at www.iea.org (accessed 5 July 2006).

Irmscher, Christoph. 2001. "Nature Laughs at Our Systems: Philip Henry Gosse's *The Canadian Naturalist." Canadian Literature, A Quarterly of Criticism and Review,* 170/171 (Autumn/Winter): 58–86.

Irving, W. N. 1985. "Context and Chronology of Early Man in the Americas." *Annual Review of Anthropology* 14: 529–555.

Isenberg, Andrew. 1994. *The Destruction of the Buffalo: An Environmental History, 1720–1920.* New York: Cambridge University Press.

Islands Protection Society (B.C.). 1984. *Islands at the Edge: Preserving the Queen Charlotte Islands Wilderness.* Vancouver: Douglas and McIntyre; Seattle: University of Washington Press.

ISO [International Organization for Standardization]. 2006. Available at: www.iso .org/iso/en/aboutiso/introduction/index.html (accessed 10 July 2006).

J

Jackson, Lionel E., Jr., and Michael C. Wilson. 2004. "The Ice-Free Corridor Revisited." *Geotimes.* Available at: www.agiweb.org/geotimes/feb04/feature_Revisited.html (accessed 30 June 2006).

Jacobs, Jane. 1961. *The Death and Life of Great American Cities.* New York: Random House.

James Bay Coalition. 1991. "At James Bay [Destroying A Wilderness The Size of France]." Available at: listserv.tamu.edu/cgi/wa?A2=ind9111a&L=native-l&T=0&P=1241 (accessed 12 July 2006).

Jameson, A. 1839. *Winter Studies and Summer Rambles.* Toronto: McClelland and Stewart; reprint 1990.

Janzen, Olaf U. 1999. "The European Presence in Newfoundland, 1500–1604." Pp. 129–138 in Iona Bulgin, ed., *Cabot and His World Symposium June 1997: Papers and Presentations.* St. John's: Newfoundland Historical Society.

Janzen, Olaf U. 2002. "The French Presence in Southwestern and Western Newfoundland Before 1815." Pp. 29–49 in André Magord (directeur), *Les Franco-*

Terreneuviens de la péninsule de Port-au-Port: Évolution d'une identité franco-canadienne. Moncton, NB: Chaire d'études acadiennes, Université de Moncton.

Jasen, Patricia. 1995. *Wild Things: Nature, Culture and Tourism in Ontario 1790–1914.* Toronto: University of Toronto Press.

Jenness, Diamond. 1922. *The Life of the Copper Eskimos: Canadian Arctic Expedition, 1913–1916.* Ottawa: King's Printer.

Jenness, Diamond. 1928. *People of the Twilight.* New York: Macmillan.

Jenness, Diamond. 1964. *Eskimo Administration II. Canada.* Montréal: Arctic Institute of North America.

Jenness, Diamond. 1966. "The Administration of Northern Peoples: America's Eskimos—Pawns of History." Pp. 120–129 in R. MacDonald, ed., *The Arctic Frontier.* Toronto: University of Toronto Press.

Jennings, John, Bruce W. Hodgins, and Doreen Small, eds. 1999. *The Canoe in Canadian Cultures.* Toronto: Natural Heritage/Natural History.

Johnson, George. 1900. "Pulp Wood and Wood Pulp in Canada." Pp. 3–26 of *Paris International Exhibition, 1900. The Wood Pulp of Canada.* Ottawa: Canadian Institute for Historical Microreproductions, [CIHM/ICMH] Microfiche ser.; no. 07786.

Johnston, J. F. W. 1850. *Report on the Agricultural Capabilities of the Province of New Brunswick.* Fredericton, NB: Simpson.

Jones, Karen. 2003. "Never Cry Wolf: Science, Sentiment, and the Literary Rehabilitation of Canis Lupus." *Canadian Historical Review* 84, 1: 64–93.

Josephson, Paul. 2002. *Industrialized Nature: Brute Force Technology and the Transformation of the Natural World.* Washington, DC: Island.

Judd, Richard W. 1997. *Common Lands, Common People: The Origins of Conservation in Northern New England.* Cambridge, MA: Harvard University Press.

K

Kaika, Maria, and Eric Swyngedouw. 2001. "Fetishizing the Modern City: The Phantasmagoria of Urban Technological Networks." *International Journal of Urban and Regional Research* 24, 1: 120–138.

Kalm, Per. 1770. *Peter Kalm's Travels in North America. The English Version of 1770.* Trans. and ed. A. B. Benson. New York: Dover; reprint 1987.

Kammen, M. 2004. *A Time to Every Purpose. The Four Seasons in American Culture.* Chapel Hill: University of North Carolina Press.

Keefer, Thomas C. 1850. *A Philosophy of Railroads.* Toronto: University of Toronto Press; reprint 1972.

Keeling, Arn. 2004a. "The Effluent Society: Water Pollution and Environmental Politics in British Columbia, 1889–1980." Ph.D. diss., University of British Columbia.

Keeling, Arn. 2004b. "Sink or Swim: Water Pollution and Environmental Politics in Vancouver, 1889–1975." *BC Studies* 142–143: 69–101.

Keeling, Arn, and Robert McDonald. 2001. "The Profligate Province: Roderick Haig-Brown and the Modernizing of British Columbia." *Journal of Canadian Studies* 36, 6 (Fall): 7–23.

Keith, W. J. 1969. *Charles G. D. Roberts.* Toronto: Copp Clark.

Kelly, Kenneth. 1970. "The Evaluation of Land for Wheat Farming in Early Nineteenth Century Ontario." *Ontario History* 62, 1: 57–64.

Kelly, Kenneth. 1974a. "The Changing Attitude of Farmers to Forest in Nineteenth Century Ontario." *Ontario Geography* 8: 64–77.

Kelly, Kenneth. 1974b. "Damaged and Efficient Landscapes in Rural Southern Ontario, 1880–1900." *Ontario History* 66 (March): 1–14; reprinted on pp. 213–227 in Graeme Wynn, ed., *People Places Patterns Processes: Geographical Perspectives on the Canadian Past.* Toronto: Copp Clark Pittman, 1990.

Kelly, Kenneth. 1974c. "Practical Knowledge of Physical Geography in Southern Ontario during the Nineteenth Century." Pp. 10–17 in A. Falconer et al., eds., *Physical Geography: The Canadian Context.* Toronto: McGraw-Hill, Ryerson.

Kelly, Kenneth. 1975. "The Artificial Drainage of Land in Nineteenth Century Southern Ontario." *Canadian Geographer* 19, 3: 279–298.

Killan, Gerald. 1993. *Protected Places: A History of Ontario's Provincial Parks System.* Toronto: Dundurn.

Knight, Thomas F. 1862. *Nova Scotia and Her Resources.* Prize Essay. Halifax, NS: A. & W. MacKinlay; London: Sampson, Low, Son. Available in Canadian Institute for Historical Micro-reproductions (CIHM/ICMH) Microfiche ser.; no. 37526.

Kollin, Susan. 2001. *Nature's State: Imagining Alaska as the Last Frontier.* Chapel Hill: University of North Carolina Press.

Koppel, Tom. 2003. *Lost World: Rewriting Prehistory—How New Science Is Tracing America's Ice Age Mariners.* New York: Atria.

Kraenzel, Carl F. 1955. *The Great Plains in Transition.* Norman: University of Oklahoma Press.

Kraus, Scott D., et al. 2005. "North Atlantic Right Whales in Crisis." *Science* 309 (22 July): 561–562.

Krech, Shepard, III. 1999. *The Ecological Indian: Myth and History.* New York: Norton.

Kulchyski, Peter. 2004. "Manitoba Hydro: How to Build a Legacy of Hatred." *Canadian Dimension* 38, 3 (May/June): 24–27; also available at www.turtleisland.org/discussion/viewtopic.php?p=3783 and canadiandimension.com/articles/2004/05/01/142/ (accessed 5 July 2006).

L

Ladurie, E. L. 1971. *Times of Feast, Times of Famine: A History of Climate since the Year 1000.* Transl. B. Bray. New York: Doubleday.

Lahusen, Christian. 1996. *The Rhetoric of Moral Protest: Public Campaigns, Celebrity Endorsement and Political Mobilization.* Berlin, New York: Walter de Gruyter.

Lambert, Richard S., and Paul Pross. 1967. *Renewing Nature's Wealth: A Centennial History of the Public Management of Lands, Forests and Wildlife in Ontario, 1763–1967.* Toronto: Ontario Department of Lands and Forests.

Langton, John ("Jack"). 1988. "The Two Traditions of Geography: Historical Geography and the Study of Landscapes." *Geografiska Annaler* Ser. B, *Human Geography* 70B, 1: 17–25.

Lank, David M. 1998. *Audubon's Wilderness Palette, The Birds of Canada.* Toronto: Key Porter.

LaTour, Bruno. 1987. *Science in Action: How to Follow Scientists and Engineers through Society.* Cambridge, MA: Harvard University Press.

Le famille Plouffe. N.d. Available at: www.museum.tv/archives/etv/F/htmlF/familyplouff/familyplouff.htm (accessed 5 July 2005).

Lears, Jackson T. 1981. *No Place of Grace.* Chicago: University of Chicago Press.

Le Clerq, Father Chrestien. 1691. *New Relation of Gaspesia, with the colonies and regions of the Gaspesian Indians.* Trans. and ed. W. F. Ganong. Toronto: Champlain Society; reprint 1910.

Legget, R. F. 1976. *Canals of Canada.* Vancouver: Douglas, David and Charles.

Lehr, John. 1990. "Kinship and Society in the Ukrainian Pioneer Settlement of the Canadian West." Pp. 139–160 in Graeme Wynn, ed., *People Places Patterns Processes: Geographical Perspectives on the Canadian Past.* Toronto: Copp Clark Pittman.

Lescarbot, Marc. 1609. *The History of New France.* Trans. W. L. Grant. 3 vols. Toronto: Champlain Society; reprint 1907–1914.

Linteau, Paul-André . 1981. *Maisonneuve ou Comment des promoteurs fabriquent une ville: 1883–1918.* Montréal: Boreal Express. Trans. Robert Chodos as *The Promoters' City: Building the Industrial Town of Maisonneuve, 1883–1918* (1985). Toronto: J. Lorimer.

Linton, James M. 1984. *The Story of Wild Goose Jack: The Life and Work of Jack Miner.* Toronto: CBC Enterprises.

Little, Jack I. 1989. *Nationalism, Capitalism and Colonization in Nineteenth Century Québec: The Upper St. Francis District.* Montréal and Kingston, ON: McGill–Queen's University Press.

Little, Jack I. 1991. *Crofters and Habitants: Settler Society, Economy and Culture in a Québec township, 1848–1881.* Montréal and Kingston, ON: McGill–Queen's University Press.

Little, Jack I. 1999. "Contested Land: Squatters and Agents in the Eastern Townships of Lower Canada." *Canadian Historical Review* 80, 3: 381–412.

Liu, K. B. 1980 and 1981. "Pollen Evidence of Late-Quaternary Climatic Changes in Canada: A Review." Part I "Western Canada" and Part II "Eastern Arctic and Sub-Arctic Canada." *Ontario Geography* 15: 83–101, 17: 61–82.

Lloyd, Trevor. 1959a. "The Geographer as Citizen." *The Canadian Geographer* 13: 1–13.

Lloyd, Trevor. 1959b. "Map of the Distribution of Eskimos and Native Greenlanders in North America." *The Canadian Geographer* 13: 41–43.

London, Jack. 1910. *The Call of the Wild.* Toronto: Macmillan.

Loo, Tina. 2001. "Of Moose and Men: Hunting for Masculinities in British Columbia, 1880–1939." *Western Historical Quarterly* 32, 3 (Autumn): 296–319.

Loo, Tina. 2004. "People in the Way: Modernity, Environment and Society on the Arrow Lakes." *BC Studies* 142/143: 161–196.

Loo, Tina. 2006. *States of Nature: Conserving Canada's Wildlife in the Twentieth Century.* Vancouver: University of British Columbia Press.

Lothian, W. F. 1987. *A Brief History of Canada's National Parks.* Ottawa: Environment Canada.

Lott, Dale F. 2002. *American Bison: A Natural History.* Berkeley: University of California Press.

Lowenthal, David G. 2000. *George Perkins Marsh, Prophet of Conservation.* Seattle: University of Washington Press.

Lower, Arthur R. M. 1938. *The North American Assault on the Canadian Forest.* Toronto: Ryerson.

Lower, Arthur R. M. 1973. *"Great Britain's Woodyard": British America and the Timber Trade, 1763–1867.* Montréal and Kingston, ON: McGill–Queen's University Press.

M

Mabee, Carleton. 1961. *The Seaway Story.* New York: Macmillan.

MacEachern, Alan. 2001. *Natural Selections: National Parks in Atlantic Canada, 1935–1970.* Montréal and Kingston, ON : McGill–Queen's University Press.

MacKinnon, Robert. 1991. "Farming the Rock: The Evolution of Commercial Agriculture around St. John's, Newfoundland, to 1945." *Acadiensis* 20, 2: 32–61.

MacKinnon, Robert. 2003. "Roads, Cart Tracks, and Bridle Paths: Land Transportation and the Domestic Economy of Mid-Nineteenth-Century Eastern British North America," *Canadian Historical Review* 84 (2): 177–216.

MacLachlan, Ian. 2002. *Kill and Chill: Restructuring Canada's Beef Commodity Chain.* Toronto: University of Toronto Press.

MacMillan, Ken. 2001. "Discourse on History, Geography, and Law: John Dee and the Limits of the British Empire, 1576–80." *Canadian Journal of History* 36, 1: 1–25.

MacNeil, Alan R. 1985. *A Reconsideration of the State of Agriculture in Eastern Nova Scotia, 1791–1861.* M.A. Thesis, Queen's University, Kingston, ON.

MacNeil, Alan R. 1986. "Cultural Stereotypes and Highland Farming in Eastern Nova Scotia, 1827–1861." *Histoire Sociale–Social History* 19: 39–56.

Macpherson Alan G. 1972. "People in Transition: The Broken Mosaic" Pp. 46–72 in Macpherson, ed., *The Atlantic Provinces.* Toronto: University of Toronto Press.

Macpherson, Joyce Brown. 2005. "The Vegetational History of St. John's." Pp. 19–36 in Alan G. Macpherson, ed., *Four Centuries and the City: Perspectives on the Historical Geography of St. John's.* St. John's: Department of Geography, Memorial University of Newfoundland.

Magnusson, Warren, and Karena Shaw, eds. 2003. *A Political Space: Reading the Global through Clayoquot Sound.* Minneapolis: University of Minnesota Press.

Mancke, Elizabeth. 2005. "Spaces of Power in the Early Modern Northeast." Pp. 32–49 in Stephen J. Hornsby and John G. Reid, eds., *New England and the Maritime Provinces: Connections and Comparisons.* Montréal and Kingston, ON: McGill–Queen's University Press.

Maniates, M. 2002. "Individualization: Plant a Tree, Buy a Bike, Save the World." Pp. 43–66 in Thomas Princen, Michael Maniates, and Ken Conca, eds., *Confronting Consumption.* Cambridge, MA: MIT Press.

Mannion, John J. 1976. *Point Lance in Transition: The Transformation of a Newfoundland Outport.* Toronto: McClelland and Stewart.

Mannion John J. 1982. "The Waterford Merchants and the Irish-Newfoundland Provisions Trade 1770–1820." Pp. 178–203 in Donald Akenson, ed., *Canadian Papers in Rural History,* vol. 3. Gananoque, ON: Langdale.

Mannion, John J. 2000. "Victualling a Fishery: Newfoundland Diet and the Origins of the Irish Provisions Trade, 1675–1700." *International Journal of Maritime History* 12, 1: 1–60.

Manore, Jean. 1999. *Cross-currents: Hydroelectricity and the Engineering of Northern Ontario.* Waterloo, ON: Wilfrid Laurier University Press.

Marchak, M. P. 1983. *Green Gold: The Forest Industry in British Columbia.* Vancouver: University of British Columbia Press.

Marcus, Alan. R. 1995. *Relocating Eden: The Image and Politics of Inuit Exile in the Canadian Arctic.* Hanover, NH, and London: University Press of New England.

Marsh, G. P. 1864. *Man and Nature; or Physical Geography as Modified by Human Action.* New York: Scribners.

Marshall, Ingeborge. 1996. *A History and Ethnography of the Beothuk.* Montréal and Kingston, ON: McGill–Queen's University Press.

Martell, J. S. 1940. "The Achievements of Agricola and the Agricultural Societies, 1818–1825," *Provincial Archives of Nova Scotia, Bulletin* 2, 2. Halifax, NS: PANS.

Martin, Calvin. 1974. "The European Impact on the Culture of a Northeastern Algonquian Tribe: An Ecological Interpretation." *William and Mary Quarterly* 31, 1: 3–26.

Martin, Calvin. 1975. "The Four Lives of a Micmac Copper Pot." *Ethnohistory* 22, 2 (Spring): 111–133.

Martin, Calvin. 1978. *Keepers of the Game.* Berkeley: University of California Press.

Martin, P. S. 1973. "The Discovery of America." *Science* 179: 969–974.

Matthews, David Ralph. 1995. "Commons Versus Open Access: The Collapse of Canada's East Coast Cod Fishery." *Ecologist* 25, 2–3 (March/April, May/June): 86–96.

McCalla, Douglas. 1993. *Planting the Province: The Economic History of Upper Canada.* Toronto: University of Toronto Press.

McCann, L. D. 1981. "The Mercantile-Industrial Transition in the Metals Towns of Pictou County." *Acadiensis* 10, 2: 29–64.

Mc Elreath, R. 2005. "Perspectives on Diamond's *Collapse: How Societies Choose to Fail or Succeed.*" *Current Anthropology* 46: S96–S97.

McGhee, Robert. 1972. *Copper Eskimo Prehistory.* Ottawa: National Museums of Canada, National Museum of Man, Publications in Archeology, No. 2.

McGhee, Robert. 1984. "Contact between Native North Americans and the Medieval Norse: A Review of the Evidence." *American Antiquity* 49, 1: 4–26.

McGhee, Robert. 1987. "The Vinland Map: Hoax or History?" *The Beaver* 67, 2: 37–44.

McGhee, Robert. 2005. *The Last Imaginary Place: A Human History of the Arctic World.* Oxford: Oxford University Press.

McGovern, Thomas H. 1980–1981. "The Vinland Adventure: A North Atlantic Perspective." *North American Archaeologist* 2, 4: 285–308.

McGovern, Thomas H., Gerald Bigelow, Thomas Amorosi, and David Russell. 1988. "Northern Islands, Human Error, and Environmental Degradation: A View of Social and Ecological Change in the Medieval North Atlantic." *Human Ecology* 16, 3: 225–270.

McIlwraith, Thomas. 1886. *The Birds of Ontario, Being a Concise Account of Every Species of Bird known to have been Found in Ontario . . .* 2nd ed. 1894. Toronto: W. Briggs.

McIlwraith, Thomas F. 1990. "The Adequacy of Rural Roads in the Era before Railways: An Illustration from Upper Canada." Pp. 196–212 in Graeme Wynn, ed., 1990. *People Places Patterns Processes: Geographical Perspectives on the Canadian Past.* Toronto: Copp Clark Pittman.

McIlwraith, Thomas F. 1997. *Looking for Old Ontario: Two Centuries of Landscape Change.* Toronto: University of Toronto Press.

McIlwraith, Thomas F. 2001. "British North America, 1768–1867." Pp. 207–234 in McIlwraith and Edward K. Muller, eds., *North America: the Historical Geography of a Changing Continent.* Lanham, MD: Rowman and Littlefield.

McKenzie, Judith I. 2002. *Environmental Politics in Canada, Managing the Commons into the Twenty-First Century.* Don Mills, ON: Oxford University Press.

McLaren, John P. S. 1984. "The Tribulations of Antoine Ratté: A Case Study of the Environmental Regulation of the Canadian Lumbering Industry in the Nineteenth Century." *University of New Brunswick Law Journal* 33: 203–259.

McLaren, John P. S., A. R. Buck, and N. E. Wright. 2005. *Despotic Dominion: Property Rights in British Settler Society.* Vancouver: University of British Columbia Press.

McNabb, Debra. 1986. *Land and Families in Horton Township, N.S., 1760–1830.* M.A. thesis. University of British Columbia, also available in 1988 microform, Ottawa: National Library of Canada.

McNeill, John R. 2000. *Something New under the Sun: An Environmental History of the Twentieth Century World.* New York: Norton.

McNeill, John R. 2003. "Observations on the Nature and Culture of Environmental History." *History and Theory* 42: 5–43.

Meggs, Geoff. 1995. *Salmon: The Decline of the British Columbia Fisheries.* Vancouver: Douglas and McIntyre.

Meinig, Donald W. 2004. *The Shaping of America: A Geographical Perspective on 500 Years of History: Volume 4: Global America, 1915–2000.* New Haven, CT: Yale University Press.

Meltzer, D. J. 1989. "Why Don't We Know When the First People Came to North America?" *American Antiquity* 54, 3: 471–490.

Meltzer, D. J. 1996. *Search for the First Americans.* Washington, DC: Smithsonian Institution.

Meltzer, D. J., and J. I. Mead. 1985. "Dating Late Pleistocene Extinctions: Theoretical Issues, Analytical Biases, and Substantive Results." Pp. 145–174 in J. I. Mead and D. J. Melzer, eds., *Environments and Extinctions: Man in Late Glacial North America.* Orono, ME: Center for the Study of Early Man.

Metcalfe, E. Bennett. 1985. *A Man of Some Importance, The Life of Roderick Langmere Haig-Brown.* Vancouver: Wood.

M'Gonigle, Michael, and Wendy Wickwire. 1988. *Stein, the Way of the River.* Vancouver: Talonbooks.

Miller, Virginia P. 1976. "Aboriginal Micmac Population. A Review of the Evidence." *Ethnohistory* 23, 2: 117–128.

Millward, Hugh. 1985. "Mine Locations and the Sequence of Coal Exploitation on the Sydney Coalfield, 1720–1980." Pp. 183–202 in Kenneth Donovan, ed., *Cape Breton at 200: Historical Essays in Honour of the Island's Bicentennial, 1785–1985.* Sydney, NS: University College of Cape Breton Press.

Mineral Resources Education Program of British Columbia, "Coal Mountain," at: www.bcminerals.ca/files/bc_mine_information/000128.php (accessed 5 July 2005).

MiningWatch Canada. 2001. *Mining in Remote Areas: Issues and Impacts* at www.miningwatch.ca/updir/mine_impacts_kit.pdf (accessed 4 July 2006).

MiningWatch Canada. 2005. Elliot Lake Uranium Mines, Tuesday September 13, at www.miningwatch.ca/index.php?/Uranium/Elliot_Lake_Uranium_ (accessed 5 July 2006).

Moodie, S. 1852. *Roughing It in the Bush.* London: R. Bentley.

Moore, Christopher. 1987. "Colonization and Conflict: New France and Its Rivals." Pp. 95–180 in R. C. Brown, ed., *The Illustrated History of Canada.* Toronto: Lester and Orpen Dennys.

Moorsom, W. S. 1830. *Letters from Nova Scotia: Comprising Sketches of a Young Country.* London: H. Colburn and R. Bentley.

Morantz, Toby. 2001. "Plunder or Harmony? On Merging European and Native Views of Early Contact." Pp. 48–67 in Germaine Warkentin and Carolyn Podruchny, eds., *Decentering the Renaissance: Canada and Europe in Multidisciplinary Perspective, 1500–1700.* Toronto: University of Toronto Press.

Morse, Kathryn. 2003. *The Nature of Gold: An Environmental History of the Klondike Gold Rush.* Seattle: University of Washington Press.

Morton, Desmond. 1987. "Strains of Affluence 1945–1996." Pp. 467–562 in R. C. Brown, ed., *The Illustrated History of Canada.* Toronto: Lester and Orpen Dennys.

Morton, W. L. 1946. "Marginal." *Manitoba Arts Review* 5: 26–31.

Mouat, Jeremy. 2000. *Metal Mining in Canada, 1840–1950.* Ottawa: National Museum of Science and Technology.

Mouat, Jeremy, and Logan W. Hovis. 1984. *The History of Mining in British Columbia: A Bibliography.* Vancouver: University of British Columbia.

Mowat, Farley. 1959. *The Desperate People.* Boston: Little, Brown.

Mowat, Farley. 1968. *People of the Deer.* New York: Pyramid.

Muir, John. 1915. *Travels in Alaska.* Boston: Houghton Mifflin.

Mulvihill, Peter R., Douglas C. Barker, and William R. Morrison. 2001. "A Conceptual Framework for Environmental History in Canada's North." *Environmental History* 6, 4: 611–626.

Mumford, Lewis. 1934. *Technics and Civilization.* New York and Burlingame, FL: Harcourt, Brace and World.

Munton, Don. 2002. "Fumes, forests and further studies: Environmental science and policy inaction in Ontario." *Journal of Canadian Studies* 37, 2: 130–163.

Murphy, T. 1987. "Potato Capitalism: McCain and Industrial Farming in New Brunswick." Pp. 21–22 in Gary Burrill and I. McKay, eds., *People, Resources and Power.* Fredericton, NB: Acadiensis.

Myers, Ransom A. 2001. "Testing Ecological Models: The Influence of Catch Rates on Settlement of Fishermen in Newfoundland, 1710–1833." Pp. 13–30 in Paul Holm, Tim D. Smith, and David J. Starkey, eds., *The Exploited Seas: New Directions for Maritime Environmental History.* St. Johns, NF: International Maritime Economic History Association.

N

National Gallery of Canada. 1995. *The Group of Seven: Art for a Nation.* Ottawa: National Gallery of Canada.

Natural Resources Canada. N.d. "Trailblazers: Elihu Stewart." Available at: www.nrcan.gc.ca/inter/trailblazers/elihustewart_e.html (accessed 10 July 2005).

Neill, Robin. 1972. *A New Theory of Value; the Canadian Economics of H. A. Innis.* Toronto; Buffalo: University of Toronto Press.

Nelles, H. Viv. 1974. *The Politics of Development.* Toronto: Macmillan.

Nelson, J. G. 1976. *Man's Impact on the Western Canadian Landscape.* Toronto: McClelland and Stewart.

Nelson, J. Gordon. 2000. "Tourism and National Parks in North America: An Overview." Pp. 303–322 in Richard W. Butler and Stephen W. Boyd, eds., *Tourism and National Parks; Issues and Implications.* Chichester, UK, and Toronto: Wiley.

New Brunswick Royal Commission on Agriculture. 1909. *Report of the Agricultural Commission, 1909.* Fredericton. Printed by Order of the Legislature. Available at: www.lib.unb.ca/Texts/NBHistory/Commissions/ES47/pdf/es47r0.pdf (accessed 10 July 2006).

Newell, Dianne. 1993. *Tangled Webs of History: Indians and the Law in Canada's Pacific Coast Fisheries.* Toronto: University of Toronto Press.

Newell, Dianne, ed. 1989. *The Development of the Pacific Salmon-Canning Industry: A Grown Man's Game.* Montréal and Kingston, ON: McGill–Queen's University Press.

Newell, Dianne, and Rosemary E. Ommer, eds. 1999. *Fishing Places, Fishing People: Traditions and Issues in Canadian Small-Scale Fisheries.* Toronto: University of Toronto Press.

Norrie, Kenneth, and Doug Owram. 1990. *A History of the Canadian Economy.* Toronto: Harcourt Brace Jovanovich Canada.

Norrie, Kenneth, and Rick Szostak. 2005. "Allocating Property Rights over Shoreline: Institutional Change in the Newfoundland Inshore Fishery." *Newfoundland and Labrador Studies* 20, 2: 233–263.

Notzke, Claudia. 1994. *Aboriginal Peoples and Natural Resources in Canada.* North York, ON: Captus.

Nova Scotia. House of Assembly. *Journals,* 1847, 1885.

O

Ommer, Rosemary E. 1991. *From Outpost to Outport: A Structural Analysis of the Jersey-Gaspé Cod Fishery, 1767–1886.* Montréal and Kingston, ON: McGill–Queen's University Press.

Ommer, Rosemary E. 1999. "Rosie's Cove: Settlement Morphology, History, Economy and Culture in a Newfoundland Outport." Pp. 17–31 in Dianne Newell and Rosemary E. Ommer, eds., *Fishing Places, Fishing People: Traditions and Issues in Canadian Small-Scale Fisheries.* Toronto: University of Toronto Press.

Ondaatje, Michael. 1987. *In the Skin of a Lion.* Toronto: McClelland and Stewart.

Ontario Department of Agriculture. 1952. *Ontario Soils: Their Use, Management and Improvement.* Bulletin 492. Toronto.

Organization for Economic Co-operation and Development. 1994. *The Environmental Effects of Trade.* Paris: Organization for Economic Co-operation and Development (OECD).

Ower, John. 1971. "Portraits of the Landscape as Poet: Canadian Nature as Aesthetic Symbol in Three Confederation Writers." *Journal of Canadian Studies* 6: 27–32.

Owram, Douglas. 1980. *Promise of Eden, The Canadian Expansionist Movement and the Idea of the West 1856–1900.* Toronto: University of Toronto Press.

P

Page, Robert. 1986. *Northern Development: The Canadian Dilemma.* Toronto: McClelland and Stewart.

Parr, Joy. 1990. *The Gender of Breadwinners.* Toronto: University of Toronto Press.

Parr, Joy. 2001. "Notes for a More Sensuous History of Twentieth Century Canada: The Timely, the Tacit and the Material Body." *Canadian Historical Review* 82, 4: 720–745.

Parr, Joy. 2004. "Lostscapes: Found Sources in Search of a Fitting Representation." *Journal of the Association for History and Computing* 7, 1. Available at http://mcel.pacificu.edu/jahc/JAHCVII1/ARTICLES/parr1.html (accessed 5 July 2006).

Pastore, Ralph. 1987. "Fishermen, Furriers, and Beothuks: The Economy of Extinction." *Man in the Northeast* 33: 47–62.

Pastore, Ralph. 1989. "The Collapse of the Beothuk World." *Acadiensis* 19, 1: 52–71.

Pastore, Ralph. 1994. "The Sixteenth Century. Aboriginal Peoples and European Contact." Pp. 22–39 in P. A. Buckner and J. G. Reid, eds., *The Atlantic Region to Confederation: A History.* Toronto: University of Toronto Press.

Paterson, Donald G., and William L. Marr. 1980. *Canada: An Economic History.* Toronto: Macmillan.

Pearse, Peter H. 1974. *The MacKenzie Pipeline: Arctic Gas and Canadian Energy Policy.* Toronto: McClelland and Stewart.

Pearson, N. 1957. "Hell Is a Suburb." *Community Planning Review* 7, 3: 124–128.

Pepin, P.-Y. 1968. *Life and Poverty in the Maritimes.* ARDA Research Report No. RE-3. Ottawa: Ministry of Forestry and Rural Development.

Perley, Moses H. 1852. *Reports on the Sea and River Fisheries of New Brunswick.* Fredericton, NB: Simpson.

Peters, B. Guy, and John Pierre. 1998. "Governance without Government? Rethinking Public Administration." *Journal of Public Administration Research and Theory* 8, 2: 223–243.

Phillips, R. A. J. 1959. "Slum Dwellers of the Wide-Open Spaces." *Weekend Magazine* 9, 15: 20.

Pielou, E. Chris. 1991. *After the Ice Age: The Return of Life to Glaciated North America.* Chicago: University of Chicago Press.

Polanyi , Karl. 1944. *The Great Transformation.* New York: Octagon; reprint 1975.

Polanyi, Karl. 1968. *Primitive, Archaic, and Modern Economies: Essays of Karl Polanyi.* Ed. George Dalton. Garden City, NY: Anchor.

Pollan, Michael. 2006. *The Omnivore's Dilemma: A Natural History of Four Meals.* New York: Penguin.

Pope, Charles. 2005. "Senate OKs oil drilling in Alaska's ANWR. Democrats vow that the fight is not over for wilderness area." *Seattle Post-Intelligencer.* March 17. Available at: http://seattlepi.nwsource.com/local/216352_anwr17.html (accessed 8 July 2006).

Pope, Peter. 1997. *The Many Landfalls of John Cabot.* Toronto: University of Toronto Press.

Pope, Peter. "Early Estimates: Assessments of Catches in the Newfoundland Cod Fishery, 1660–1690." Pp. 7–40 in Daniel Vickers, ed., *Marine Resources and Human Societies in the North Atlantic since 1500: Papers presented at the conference entitled "Marine Resources and Human Societies in the North Atlantic since 1500," October 20–22, 1995.* St. John's, NF: The Institute of Social and Economic Research, Memorial University of Newfoundland.

Pope, Peter. 2003. "Outport Economics: Culture and Agriculture in Later Seventeenth-Century Newfoundland." *Newfoundland Studies* 19, 1: 153–186.

Pope, Peter. 2004. *Fish into Wine: The Newfoundland Plantation in the Seventeenth Century.* Chapel Hill: University of North Carolina Press.

Potyondi, Barry. 1995. *In Palliser's Triangle: Living in the Grasslands, 1850–1930.* Saskatoon, SK: Purich.

Pratt, Larry, and Ian T. Urquhart. 1994. *The Last Great Forest: Japanese Multinationals and Alberta's Northern Forests.* Edmonton, AB: NE West.

Proulx, Jean-Pierre. 1986. *Whaling in the North Atlantic from the Earliest Times to the Mid-19th Century.* Ottawa: Parks Canada.

Proulx, Jean-Pierre. 1993. *Basque Whaling in Labrador in the 16th Century.* Ottawa: National Historic Sites.

Prowse, D. W. 1895. *History of Newfoundland.* London: MacMillan.

Pynn, Larry. 2004. "Spear Points Tell Story of B.C.'s First Residents," *Vancouver Sun,* 19 January.

Q

Quammen, David. 1998. "Planet of Weeds: Tallying the Losses of Earth's Animals and Plants," *Harper's Magazine* (October): 57–69. Also published as "The Weeds Shall Inherit the Earth," *The Independent* (London), November 22, 30–39.

Quinn, David. 1961. "The Argument for the English Discovery of America between 1480 and 1494." *Geographical Journal* 127: 277–285

Quinn. David. 1988. "Review Essay—Norse America: Reports and Reassessments." *Journal of American Studies* 22, 2: 269–273.

R

Radforth, Ian. 1987. *Bushworkers and Bosses: Logging in Northern Ontario, 1900–1980.* Toronto: University of Toronto Press.

Rajala, Richard. 1998. *Logging the Pacific Rainforest.* Vancouver: University of British Columbia Press.

Ralston, Keith. 1968–1969. "Patterns of Trade and Investment on the Pacific Coast, 1867–92: The Case of the British Columbia Salmon Canning Industry." *BC Studies* 1: 37–45.

Ray, Arthur J. 1966. "Diffusion of Diseases in the Western Interior of Canada, 1830–1850." *Geographical Review* 66: 139–157.

Ray, Arthur J. 1974. *Indians in the Fur Trade: Their Role as Trappers, Hunters, and Middlemen in the Lands Southwest of Hudson Bay, 1660–1870.* Toronto: University of Toronto Press.

Ray, Arthur J. 1975. "Some Conservation Schemes of the Hudson's Bay Company, 1821–50: An Examination of the Problems of Resource Management in the Fur Trade." *Journal of Historical Geography* 1, 1: 49–68.

Ray, Arthur J. 1980. "Indians as Consumers in the Eighteenth Century." Pp. 255–268 in Carol M. Judd and Arthur J. Ray, eds., *Old Trails and New Directions: Papers of the Third North American Fur Trade Conference.* Toronto: University of Toronto Press.

Ray, Arthur J. 1985. "Buying and Selling Hudson's Bay Company Furs in the Eighteenth Century." Pp. 95–115 in Duncan Cameron, ed., *Explorations in Canadian Economic History in Honour of Irene M. Spry.* Ottawa: University of Ottawa Press.

Ray, Arthur. J. 1987. "When Two Worlds Met." Pp. 17–104 in R. C. Brown, ed., *The Illustrated History of Canada.* Toronto: Lester and Orpen Dennys.

Ray, A. J. 1996. *I Have Lived Here since the World Began: An Illustrated History of Canada's Native Peoples.* Toronto: Lester.

Ray, Arthur J., and Donald B. Freeman. 1978. *Give Us Good Measure: An Economic Analysis of Relations between the Indians and the Hudson's Bay Company before 1763.* Toronto: University of Toronto Press.

Reichwein, Pearl Ann. 1996. "Beyond the Visionary Mountains: The Alpine Club of Canada and the Canadian National Park Idea, 1906–1969." Ph.D. diss., Carleton University, Ottawa.

Renouf, Priscilla. 1999. "Prehistory of Newfoundland Hunter-gatherers: Extinctions or Adaptations?" *World Archaeology* 30, 3: 403–420.

Richardson, Mary, Joan Sherman, and Michael Gismondi. 1993. *Winning Back the Words: Confronting Experts in an Environmental Public Hearing.* Toronto: Garamond.

Riddell, William A. 1981. *Regina: From Pile O' Bones to Queen City of the Plains.* Burlington, ON: Windsor.

Ritchie, J. C., and G. M. MacDonald. 1986. "The Patterns of Postglacial Spread of White Spruce," *Journal of Biogeography* 13: 527–540.

Roach, Thomas R. 1987. "The Pulpwood Trade and the Settlers of New Ontario, 1919–1938." *Journal of Canadian Studies* 22, 3: 78–88.

Roach, Thomas R. 1996. *Newsprint. Canadian Supply and American Demand.* Durham, NC: Forest History Society.

Rome, Adam. 2001. *The Bulldozer in the Countryside.* Cambridge and New York: Cambridge University Press.

Ross, Eric. 1970. *Beyond the River and the Bay: Some Observations on the State of the Canadian Northwest in 1811 with a View to Providing the Intending Settler with an Intimate Knowledge of that Country.* Toronto: University of Toronto Press.

Ross, Eric. 1991. *Full of Hope and Promise: the Canadas in 1841.* Montréal and Kingston, ON: McGill–Queen's University Press.

Ross, Ken. 2000. *Environmental Conflict in Alaska.* Boulder: University Press of Colorado.

Ross, W. Gillies. 1975. *Whalers and Eskimos: Hudson Bay 1860–1915.* Ottawa: National Museum of Man.

Ross, W. Gillies. 1979. "The Annual Catch of Greenland Bowhead Whales in Waters North of Canada 1719–1915: A Preliminary Compilation." *Arctic* 32, 2: 91–121.

Ross, W. Gillies. 1985. *Arctic Whalers, Icy Seas: Narratives of the Davis Strait Whale Fishery.* Toronto: Irwin.

Ross, W. Gillies. 1997. *This Distant and Unsurveyed Country: A Woman's Winter at Baffin Island, 1857–58.* Montréal and Kingston, ON: McGill–Queen's University Press.

Rotstein, Abraham. 1970. "Karl Polanyi's Concept of Non-Market Trade." *Journal of Economic History* 30: 117–130.

Roy, Gabrielle. 1947. *The Tin Flute.* New York: Reynal and Hitchcock.

Roy, Louis, and Michel Verdon. 2003. "East-Farnham's Agriculture in 1871. Ethnicity, Circumstances and Economic Rationale in the Eastern Townships of Lower Canada." *Canadian Historical Review* 84, 3: 355–393.

Ruddick, J. A. 1909. "Some of the Essentials to Success in Co-operative Dairying, with Special Reference to Nova Scotia." Pp. 103–112 in *Annual Report of the Secretary for Agriculture, Nova Scotia . . . 1908.* Halifax, NS: King's Printer.

Russell, Peter A. 1983. "Forest into Farmland: Upper Canadian Clearing Rates 1822–39." *Agricultural History* 57: 326–339.

Russell, Peter A. 1985. "Rates of Clearing Land in Upper Canada: An Index of Individual Economic Success and Problems with Its Measurement." *Bulletin of Canadian Studies* 9, 1 (Spring): 34–47.

Rutherford, R. Paul. 1990. *When Television Was Young: Primetime Canada 1952–1967.* Toronto: University of Toronto Press.

Ryan, Shannon. 1983. "Fishery to Colony: A Newfoundland Watershed, 1793–1815." *Acadiensis* 12, 2: 34–52.

Ryan, Shannon. 1986. *Fish out of Water: The Newfoundland Saltfish Trade 1814–1914.* St. John's, NL: Breakwater.

Ryan, Shannon. 1994. *The Ice Hunters: A History of Newfoundland Sealing to 1914.* St. John's: Breakwater.

S

Salisbury, Richard F. 1986. *A Homeland for the Cree: Regional Development in James Bay, 1971–1981.* Montréal and Kingston, ON: McGill–Queen's University Press.

Sandberg, Anders. 1992. "Forest Policy in Nova Scotia: The Big Lease, Cape Breton Island, 1899–1960." Pp. 65–89 in L. A. Sandberg, ed., *Trouble in the Woods: Forest*

Policy and Social Conflict in Nova Scotia and New Brunswick. Fredericton, NB: Acadiensis.

Sandberg, Anders, and Peter Clancy. 2002. "Politics, Science and the Spruce Budworm in New Brunswick and Nova Scotia." *Journal of Canadian Studies* 37, 2: 164–191.

Sandlos, John. 2001. "From the Outside Looking In, Aesthetics, Politics and Wildlife Conservation in the Canadian North." *Environmental History* 6, 1: 6–31.

Sandlos, John. 2002. "Where the Scientists Roam: Ecology, Management and Bison in Northern Canada." *Journal of Canadian Studies* 37, 2: 93–129.

Sanger, Chesley. 1977. "The Evolution of Sealing and the Spread of Permanent Settlement in Northeastern Newfoundland." Pp. 136–151 in John Mannion, ed., *The Peopling of Newfoundland: Essays in Historical Geography.* St. John's: Institute of Social and Economic Research.

Sansom, Joseph. 1817. *Sketches of Lower Canada: Historical and Descriptive.* New York: Kirk and Mercein. Available at: www.canadiana.org/ECO/mtq?doc=40352 (accessed 5 July 2006).

Saunders, S. A. 1932. *The Economic Welfare of the Maritime Provinces.* Wolfville, NS : Acadia University.

Schmitt, Daniel M., David D. Grimble, and Janet L. Searcy. 1984. *Managing the Spruce Budworm in Eastern North America.* Washington, DC: U.S. Department of Agriculture.

Schwartz, Joan M. 1977–1978. "The Photographic Record of Pre-Confederation British Columbia." *Archivaria* 5: 17–44.

Science Council of Canada. 1983. *Canada's Threatened Forests. A Statement by the Science Council.* Ottawa: Science Council of Canada.

Scott, Geoffrey J. 1995. *Canada's Vegetation. A World Perspective.* Montréal and Kingston, ON: McGill–Queen's University Press.

Scott, James C. 1998. *Seeing Like a State.* New Haven, CT, and London: Yale University Press.

Seaver, Kirsten A. 1996. *The Frozen Echo: Greenland and the Exploration of North America ca. A.D. 1000–1500.* Stanford, CA: Stanford University Press.

Seed, Patricia. 1992. "Taking Possession and Reading Texts: Establishing the Authority of Overseas Empires." *William and Mary Quarterly,* 3rd ser., 49, No. 2: 183–209.

Seed, Patricia. 1995. *Ceremonies of Possession in Europe's Conquest of the New World,* 1492–1640. Cambridge, New York: Cambridge University Press.

Sennett, Richard. 1970. *The Uses of Disorder; Personal Identity and City Life.* New York: Knopf.

Service, Robert. 1987. *Dan McGrew, Sam McGee and Other Great Service.* Toronto: McGraw-Hill Ryerson.

Sherman, Joan, and Mike Gismondi. 1997. "Jock Talk, Goldfish, Horselogging, and Star Wars." *Alternatives* 23, 1: 14–20. Also available at: www.athabascau.ca/html/staff/academic/gismondi/goldfish.htm#15 (accessed 10 July 2006).

Sherwood, Morgan. 1981. *Big Game in Alaska, A History of Wildlife and People.* New Haven, CT: Yale University Press.

Sierra Club. N.d. Hudson Bay/James Bay Watershed Ecoregion. Available at: www.sierraclub.org/ecoregions/hudsonbay.asp (accessed 7 August 2006).

Smith, Donald B. 1990. *From the Land of the Shadows: The Making of Grey Owl.* Saskatoon, SK: Western Producer Prairie.

Smith, Philip E. L. 1987. "In Winter Quarters." *Newfoundland Studies* 3, 1: 1–36.

Smith, Philip E. L. 1995. "Transhumance among European Settlers in Atlantic Canada." *Geographical Journal* 141, 1: 79–86.

Smith, Ray. 1969. *Cape Breton Is the Thought Control Centre of Canada.* Toronto: Anansi.

Smith, Titus. 1835. "Conclusions on the Results on the Vegetation of Nova Scotia, and on Vegetation in General, and on Man in General, of Certain Natural and Artificial Causes Deemed to Actuate and Affect Them." *Magazine of Natural History and Journal of Zoology, Botany, Mineralogy, Geology and Meteorology* 8: 641–662.

Smith, William H. 1846. *Smith's Canadian Gazetteer; Comprising Statistical and General Information Respecting All Parts of the Upper Province, or Canada West* . . . Toronto: H. and W. Rowsell; reprint 1970: Toronto: Coles.

Snow, Dean R. 1989. *The Archeology of North America.* New York: Chelsea House.

Snow, Dean R., and Kim M. Lanphear. 1988. "European Contact and Indian Depopulation in the Northeast: The Timing of the First Epidemics." *Ethnohistory* 35, 1: 15–33.

Snow, Dean R., and Kim M. Lanphear. 1989. "'More Methodological Perspectives': A Rejoinder to Dobyns." *Ethnohistory* 36, 3: 285–299.

Spry, Irene M., and John Palliser, with an Introduction and Notes by Irene M. Spry. 1968. *The Papers of the Palliser Expedition, 1857–1860.* Toronto: Champlain Society.

Stalker, Archibald MacSween. 1969. "Geology and Age of Early Man Site at Taber, Alberta." *American Antiquity* 34, 4: 425–428.

Stalker, Archibald MacSween. 1977. "Indications of Wisconsin and Earlier Man from the Southwest Canadian Prairies." *Annals of the New York Academy of Science* 288: 119–136.

Stalker, Archibald MacSween. 1983. "Detailed Stratigraphy of the Woodpecker Island Section and Commentary on the Taber Child Bones." *Canadian Journal of Archeology* 7: 209–222.

Steele, J., J. Adams, and T. Sluckin. 1998. "Modelling Paleoindian Dispersals." *World Archeology* 30, 2: 286–305.

Stefansson, Vilhjalmur. 1921. *The Friendly Arctic. The Story of Five Years in Polar Regions,* New York: Macmillan.

Stefansson, Vilhjalmur. 1962. *My Life with the Eskimo.* New York: Macmillan.

Stegner, Wallace. 1962. *Wolf Willow; A History, a Story, and a Memory of the Last Plains Frontier.* New York: Viking.

Steinberg, Theodore. 2002. *Down to Earth: Nature's Role in American History.* New York: Oxford University Press.

Stoll, Steven. 2002. *Larding the Earth: Soil and Society in Nineteenth Century America.* New York: Hill and Wang.

Stone, K., and B. Boyd. 2001. "Coal," *Canadian Minerals Yearbook 2001.* Ottawa: Minerals and Metals Sector, Natural Resources Canada.

Storck, Peter L. 1982. "Paleo-Indian Settlement Patterns." *Canadian Journal of Archeology* 6, 1: 1–32.

Storck, Peter L. 1984. "Research into Paleo-Indian Occupations of Ontario: A Review." *Ontario Archeology* 41: 3–28.

Storck, Peter L. 2004. *Journey to the Ice Age. Discovering an Ancient World.* Vancouver: University of British Columbia Press.

Strickland, Samuel. 1853. *Twenty-seven Years in Canada West; or, The Experience of an Early Settler.* London: R. Bentley.

Sundick, Robert I. 1980. "The Skeletal Remains from the Taber Child Site, Taber, Alberta." *Canadian Review of Physical Anthropology/ Revue Canadienne d'anthropologie physique* 2: 1–6.

Suzuki, David. 2000. "Walkerton Should Be a Wake-Up Call," in *Sciene Matters*, 07 June. Available at www.davidsuzuki.org/About_us/Dr_David_Suzuki/Article_Archives/weekly06070001.asp (accessed 14 September 2006).

Swift, J. 1983. *Cut and Run: The Assault on Canada's Forests.* Toronto: Between the Lines.

T

Tainter, Joseph. 1988. *The Collapse of Complex Societies.* Cambridge and New York: Cambridge University Press.

Taylor, Joseph E., III. 2002. "The Historical Roots of the Canadian-American Salmon Wars." Pp. 155–182 in John Findlay and Ken Coates, eds., *Parallel Destinies: Canadian-American Relations West of the Rockies.* Seattle: University of Washington Press.

Teacherserve. N.d. National Humanities Centre, Nature Transformed Series: "Paleoindians and the Great Pleistocene Die-Off." Available at www.nhc.rtp.nc.us/tserve/nattrans/ntecoindian/essays/pleistocene.htm (accessed 5 July 2006).

Tester, Frank J., and Peter Kulchyski. 1994. *Tammarniit (Mistakes): Inuit Relocation in the Eastern Arctic, 1939–63.* Vancouver: University of British Columbia Press.

Thistle, John. 2004. "As Free of Fish as a Billiard Ball Is of Hair: Dealing with the Depletion in the Pacific Halibut Fishery, 1899–1924." *BC Studies* 142/143: 105–125.

Thomas, Lowell J. 1957. *The Story of the St. Lawrence Seaway.* Buffalo, NY: Stewart.

Thompson, E. P. 1967. "Time, Work Discipline and Industrial Capitalism." *Past and Present* 38: 56–97.

Thompson, John Herd. 1998. *Forging the Prairie West.* Don Mills, ON: Oxford University Press.

Thoreau, Henry David. 1866. *A Yankee in Canada.* Montréal: Harvest House; reprint 1961.

Thornton, R. 1997. "Aboriginal North American Population and Rates of Decline, ca A.D. 1500–1900." *Current Anthropology* 38, 2: 310–315.

Thwaite, Ann. 2002. *Glimpses of the Wonderful: The Life of Philip Henry Gosse.* London: Faber and Faber.

Thwaites, Rueben G. 1896–1901. *The Jesuit Relations and Allied Documents: Travels and Explorations of the Jesuit Missionaries of New France, 1610–1791.* 73 vols. Cleveland, OH: Burrows.

Tobias, John. 1983. "Canada's Subjugation of the Plains Cree, 1879–1885." *Canadian Historical Review* 64, 3: 519–548.

Torroni, A., et al. 1992. "Native American Mitochondrial DNA Analysis Indicates That the Amerind and Nadene Populations Were Founded by Two Independent Migrations." *Genetics* 130: 153–162.

Traill, Catharine Parr. 1836. *The Backwoods of Canada: Being Letters from the Wife of an Emigrant Officer, Illustrative of the Domestic Economy of British America.* London: Charles Knight.

Traill, Catharine Parr. 1852. *Canadian Crusoes: A Tale of the Rice Lake Plains.* London: Arthur Hall, Virtue.

Trigger, Bruce G. 1976. *The Children of Aataentsic: A History of the Huron People to 1660.* Montréal and Kingston, ON: McGill–Queen's University Press.

Trigger, Bruce G. 1985. *Natives and Newcomers: Canada's 'Heroic Age' Reconsidered.* Montréal and Kingston, ON: McGill–Queen's University Press.

Trimble, Stanley W. 1974. *Man-induced Soil Erosion on the Southern Piedmont, 1700–1970*. Ankeny, IA: Soil Conservation Society of America.

Trimble, Stanley W. 1983. "A Sediment Budget for Coon Creek, Wisconsin, 1853–1975." *American Journal of Science* 283: 454–474.

Trimble, Stanley W. 1999. "Rates of Soil Erosion." *Science* 286: 1477–1478.

Trimble, Stanley W, and P. Crosson. 2000. "U.S. Soil Erosion Rates—Myth or Reality." *Science* 289: 248–250.

Tuck, James A. 1976. *Newfoundland and Labrador Prehistory*. Toronto: Van Nostrand Reinhold.

Tuck, James A., and Robert Grenier. 1981. "A 16th Century Basque Whaling Station in Labrador." *Scientific American* 245 (5 November): 180–190.

Tuck, James A., and Robert Grenier. 1989. *Red Bay, Labrador: World Whaling Capital, A.D. 1550–1600*. St. John's, NL: Atlantic Archeology.

Tuck James A., and Ralph Pastore. 1985. "A Nice Place to Visit, but . . . : Prehistoric Human Extinctions on the Island of Newfoundland." *Canadian Journal of Archaeology* 9, 1: 69–80.

Tucker, Richard P. 2000. *Insatiable Appetite: the United States and the Ecological Degradation of the Tropical World*. Berkeley: University of California Press.

Turchin, Peter. 2003. *Historical Dynamics: Why States Rise and Fall*. Princeton, NJ: Princeton University Press.

Turgeon, Laurier. 1995. "Fluctuations in Cod and Whale Stocks in the North Atlantic During the Eighteenth Century." Pp. 41–68 in Daniel Vickers, ed., *Marine Resources and Human Societies in the North Atlantic Since 1500: Papers presented at the conference entitled "Marine Resources and Human Societies in the North Atlantic Since 1500," October 20–22, 1995*. St. John's, NF: The Institute of Social and Economic Research, Memorial University of Newfoundland.

Turgeon, Laurier. 1997. "The Tale of the Kettle: Odyssey of an Intercultural Object." *Ethnohistory* 44, 1 (Winter): 1–29.

Turgeon, Laurier. 1998. "French Fishers, Fur Traders, and Amerindians during the Sixteenth Century: History and Archeology." *The William and Mary Quarterly* 3, 55 (4 October): 585–610.

Turner, C. G., II. 1986. "The First Americans: The Dental Evidence." *National Geographic Research* 2: 37–46.

U

Upton, Leslie F. S. 1977. "The Extermination of the Beothuks of Newfoundland." *The Canadian Historical Review* 58: 133–153.

Urquhart, Jane. 1997. *The Underpainter*. Toronto: McClelland and Stewart.

U.S. Department of Justice. 1997. "Alaska Mining Company Agrees to $4.7 Million Environmental Settlement," at www.usdoj.gov/opa/pr/1997/July97/294enr.htm (accessed 5 July 2006).

V

Valverde, Marianne. 1991. *The Age of Light, Soap and Water*. Toronto: McClelland and Stewart.

van Kirk, Sylvia. 1980. *Many Tender Ties: Women in Fur-Trade Society in Western Canada, 1670–1870*. Winnipeg, MB: Watson and Dwyer.

van Nus, W. 1984. "The Fate of City Beautiful Thought in Canada, 1893–1930." Pp. 167–186 in Gilbert A. Stelter and Alan F. Artibise, eds., *The Canadian City: Essays in Urban and Social History*. Ottawa, ON: Carleton University Press.

Vickers, Daniel. 1990. "Competency and Competition: Economic Culture in Early America." *William and Mary Quarterly 3*, 47: 3–29.

Vickers, Daniel. 1996. "The Price of Fish: A Price Index for Cod, 1505–1892." *Acadiensis* 25, 2: 62–81.

Villagarcia, M. G., R. L. Haedrich, and J. Fischer. 1999. "Groundfish Assemblages of Eastern Canada Examined over Two Decades." Pp. 239–259 in Dianne Newell and Rosemary E. Ommer, *Fishing Places, Fishing People: Traditions and Issues in Canadian Small-Scale Fisheries.* Toronto: University of Toronto Press.

Vogel, R. 1969. "One Hundred and Thirty years of Plant Succession in a Southeastern Wisconsin Lowland." *Ecology* 50, 2: 248–255.

W

Wadland, John Henry. 1978. *Ernest Thompson Seton: Man in Nature in the Progressive Era.* New York: Arno.

Waiser, W. A. 1989. *The Field Naturalist: John Macoun, The Geological Survey, and Natural Science.* Toronto: University of Toronto Press.

Waite, Peter B. 1987. "Between Three Oceans: Challenges of a Continental Destiny." Pp. 277–376 in R. C. Brown, ed., *The Illustrated History of Canada.* Toronto: Lester and Orpen Dennys.

Waldram, James B. 1988. *As Long as the Rivers Run: Hydroelectric Development and Native Communities in Western Canada.* Winnipeg: University of Manitoba Press.

Wallace, Iain. 1998. "Canadian Shield: Development of a Resource Frontier." Pp. 227–267 in Larry McCann and Angus Gunn, eds., *Heartland and Hinterland: A Regional Geography of Canada*, 3d ed. Scarborough, ON: Prentice Hall Canada.

Warecki, George. 2000. *Protecting Ontario's Wilderness: A History of Changing Ideas and Preservation Politics, 1927–1973.* New York: Peter Lang.

Warkentin, John. 1964. *The Western Interior of Canada.* Toronto: McClelland and Stewart.

Warkentin, John H. 1967. "Western Canada in 1886." Pp. 56–82 in R. L. Gentilcore, ed., *Canada's Changing Geography.* Scarborough, ON: Prentice Hall.

Warner, Sam Bass. 1962. *Streetcar Suburbs.* Boston: Harvard University Press.

Warren, Louis S. 2003. *American Environmental History.* Oxford: Blackwell.

Watson, Alexander John. 2006. *Marginal Man: The Dark Vision of Harold Innis.* Toronto: University of Toronto Press.

Weaver, John C. 2003. *The Great Land Rush and the Making of the Modern World, 1650–1900.* Montréal and Kingston, ON: McGill–Queen's University Press.

Webb, Walter P. 1931. *The Great Plains.* Boston: Ginn.

Webb, Walter P. 1952. *The Great Frontier.* Boston: Houghton Mifflin.

Weiger, Axel. 1990. *Agrarkolonisation, Landnutzung und Kulturlandschftsverfall in der Provinz New Brunswick (Kanada).* Aachen, Germany: Geographisches Institut.

Weir, Thomas R. 1972. "The Population." Pp. 85–98 in Peter J. Smith, *The Prairie Provinces.* Toronto: University of Toronto Press.

Weisman, Jonathan. 2005. "GOP to Strike Arctic Drilling from House Bill." *Washington Post*, November 10: A04. Available at: www.washingtonpost.com/wp-dyn/content/article/2005/11/09/AR2005110901930.html (accessed 8 July 2006).

Western Canada Wilderness Committee. N.d. WCWC website available at: www.wildernesscommittee.org/ (accessed 10 July 2006).

Weyler, Rex. 2004. *Greenpeace. How a Group of Ecologists, Journalists and Visionaries Changed the World.* Vancouver: Raincoast.

White, Richard. 1991. *The Middle Ground: Indians, Empires and Republics in the Great Lakes Region, 1650–1815.* Cambridge: Cambridge University Press.

White, Richard. 1995. *Organic Machine: The Remaking of the Columbia River.* New York: Hill and Wang.

Whitley, David S., and Ronald I. Dorn. 1993. "New Perspectives on the Clovis vs. Pre-Clovis Controversy." *American Antiquity* 58, 4: 626–647.

Williams, R. C., A. G. Steinberg, H. Gershowitz, P. H. Bennett, W. C. Knowler, D. J. Pettitt, W. Butler, R. Baird, L. Dowda-Rea, T. A. Burch, et al. 1985. "GM Allotypes in Native Americans: Evidence for Three Distinct Migrations across the Bering Land Bridge." *American Journal of Physical Anthropology* 66: 1–19.

Wilson, Michael C., David W. Harvey, and Richard G. Forbis. 1983. "Geoarchaeological Investigations of the Age and Context of the Stalker (Taber Child) Site, DlPa 4, Alberta." *Canadian Journal of Archeology/ Journal Canadien d'archeologie* 7, 2: 179–183.

Wilson, Michael C., and J. A. Burns. 1999. "Searching for the Earliest Canadians: Wide Corridors, Narrow Doorways, Small Windows." Pp. 213–248 in R. Bonnichsen and K. L. Turnmire, eds., *Ice Age People of North America: Environments, Origins and Adaptations.* Corvallis, OR: Oregon State University Press and Center for the Study of the First Americans.

Wilson, J. 2004. "For the Birds? Neoliberalism and the Protection of Biodiversity in British Columbia." *BC Studies* 142/143: 241–277.

Withrow, William Henry. 1889. *Our Own Country: Canada, Scenic and Descriptive.* Toronto: W. Briggs.

Wonders, William C. 1962. "Our Northward Course." *The Canadian Geographer* 6, 3–4: 96–105.

Wonders, William C. 1971. *Canada's Changing North.* Toronto: McClelland and Stewart.

Wonders, William C. 1972. "The Future of Northern Canada." Pp. 137–46 in *The North–Le Nord.* Published for the 22nd International Geographical Congress, Montréal and Toronto: University of Toronto Press.

Wood, J. David. 2000. *Making Ontario: Agricultural Colonization and Landscape Re-creation before the Railway.* Montréal and Kingston, ON: McGill–Queen's University Press.

Wood, J. David., ed. 1975. *Perspectives on Landscape and Settlement in Nineteenth Century Ontario.* Toronto: McClelland and Stewart.

Woodsworth, J. S. 1911. *My Neighbour.* Toronto: University of Toronto Press.

Worster, Donald. 1990. "Transformations of the Earth: Toward an Agroecological Perspective in History." *Journal of American History* 76, 4 (March): 1087–1106, 1090–1091.

Worster, Donald. 1992. "Alaska: The Underworld Erupts." Pp. 154–224 in Worster, *Under Western Skies: Nature and History in the American West.* New York: Oxford University Press.

Wright, Ronald. 2004. *A Short History of Progress.* Toronto: Anansi.

Wynn, Graeme. 1969. "The Utilisation of the Chignecto Marshlands of Nova Scotia and New Brunswick, 1750–1800." M.A. research paper, University of Toronto.

Wynn, Graeme. 1977. "Administration in Adversity: The Deputy Surveyors and Control of the New Brunswick Crown Forest before 1844." *Acadiensis* 6, 1: 49–65.

Wynn, Graeme. 1979a. "Late Eighteenth-Century Agriculture on the Bay of Fundy Marshlands." *Acadiensis* 8, 2: 80–89.

Wynn, Graeme. 1979b. "Notes on Society and Environment in Old Ontario." *Journal of Social History* 13, 1: 49–65.

Wynn, Graeme. 1980. "'Deplorably Dark and Demoralized Lumberers'? Rhetoric and Reality in Early Nineteenth-Century New Brunswick." *Journal of Forest History* (October): 168–187.

Wynn, Graeme. 1981. *Timber Colony: A Historical Geography of Early Nineteenth Century New Brunswick.* Toronto: University of Toronto Press.

Wynn, Graeme. 1982. "The Maritimes: The Geography of Fragmentation and Underdevelopment." Pp. 156–213 in L. D. McCann, ed., *Heartland and Hinterland: A Geography of Canada.* Toronto: Prentice Hall.

Wynn, Graeme. 1985. "Hail the Pine." *Horizon Canada* 37: 872–877.

Wynn, Graeme. 1987a. "Forging a Canadian Nation." Pp. 373–409 in R. D. Mitchell and P. A. Groves, eds., *North America: The Historical Geography of a Changing Continent.* Lanham, MD: Rowman and Littlefield.

Wynn, Graeme. 1987b. "On the Margins of Empire, 1760–1840." Pp. 189–278 in R. C. Brown, ed., *The Illustrated History of Canada.* Toronto: Lester and Orpen Dennys.

Wynn, Graeme. 1988. "A Share of the Necessaries of Life." Pp. 17–52 in B. Fleming, ed., *Beyond Anger and Longing.* Fredericton, NB: Mount Allison University/Acadiensis Press.

Wynn, Graeme. 1990a. "Exciting a Spirit of Emulation among the 'Plodholes': Agricultural Reform in Pre-Confederation Nova Scotia." *Acadiensis* (Autumn): 5–51.

Wynn, Graeme. 2002. "Ontario: A Fine Country, Not Half Like England." *Canadian Geographer* 46, 4: 371–376.

Wynn, Graeme. 2003. "Hacia una historia ambiental de los bosques de pino de la Norteamérica nororiental (1700–1900)" [Toward an Environmental History of the Pine Forests of Northeastern America, 1700–1900]. Pp. 125–140 in Bernardo Garcia Martinez y Maria del Rosario Prieto, eds., *Estudios sobre historia y ambiente en America II.* Mexico: El Colegio de México, Instituto Panaméricano de Geografía e Historia.

Wynn, Graeme. 2004. "On Heroes, Hero-Worship, and the Heroic in Environmental History." *Environment and History* 10, 2 (May 2004): 133–151.

Wynn, Graeme. 2004. "'Shall We Linger Along Ambitionless?' Environmental Perspectives on British Columbia." *BC Studies* 142/143: 5–67.

Wynn, Graeme, and Robert MacKinnon. N.d. "The Countryside of Atlantic Canada." Unpublished manuscript.

Wynn Graeme, ed. 1990b. *People Places Patterns Processes: Geographical Perspectives on the Canadian Past.* Toronto: Copp Clark Pittman.

Wynn, Graeme, and T. Oke, eds. 1992. *Vancouver and Its Region.* Vancouver: University of British Columbia Press.

Y

Young, John. 1822. *Letters of Agricola on the Principles of Vegetation and Tillage.* Halifax, NS: Holland.

Z

Zaslow, Morris. 1971. *The Opening of the Canadian North, 1870–1914.* Toronto: McClelland and Stewart.

Zaslow, Morris. 1975. *Reading the Rocks: The Story of the Geological Survey of Canada.* Toronto: Macmillan.

Zaslow, Morris. 1988. *The Northward Expansion of Canada 1914–1967*. Toronto: McClelland and Stewart.

Zebra Mussels. 2006. Available at: www.octopus.gma.org/surfing/human/zebra.html (accessed 10 July 2006).

Zelko, Frank. 2004. "Making Greenpeace: The Development of Direct Action Environmentalism in British Columbia." BC Studies 142/143 (Summer/Fall): 197–239.

Zeller, Suzanne. 1987. Inventing Canada. Toronto: University of Toronto Press.

Zeller, Suzanne. 1997. "Nature's Gullivers and Crusoes: The Scientific Exploration of British North America, 1800–1870." Pp. 190–243 in John L. Allen, ed., *North American Exploration*, vol. 3: *A Continent Comprehended*. Lincoln, Nebraska: University of Nebraska Press.

Zeller, Suzanne. 1998. "Classical Codes: Biogeographical Assessments of Environment in Victorian Canada," *Journal of Historical Geography* 24, 1: 20–35.

INDEX

ABOUT THE AUTHOR

MUCH OF GRAEME WYNN'S scholarly career has been directed to understanding human transformations of the earth. When he began his work, as a graduate student, research on this topic was seen as part of a venerable, albeit fading, geographical tradition; in recent years it has been given new impetus as "environmental history." The core of his work has always been interdisciplinary, rooted in geography and history and engaged with the environmental sciences. Over time his early interests in eastern Canada broadened to encompass New Zealand and the rest of Canada, as well as other areas of European settlement overseas. In each of these realms a fair part of his work turns, in one way or another, on the histories and geographies of forest exploitation, conservation, preservation, and management. His academic writing has been directed over the years to both specialist scholars and the educated lay public (through such contributions as the extended chapter in *The Illustrated History of Canada*) in the conviction that it is important to communicate the fruits of academic research to an audience beyond the academy.

Born and raised to the age of fifteen in South Africa, Graeme Wynn completed his schooling in England and took his first degree from the University of Sheffield (where he also played Varsity cricket and rugby) in 1968. He submitted his doctorate to the University of Toronto in 1973 and joined the Department of Geography of the University of Canterbury, New Zealand. Since 1976, he has been a faculty member in the University of British Columbia, where he also served six-year terms as Associate Dean of Arts and Head of Geography between 1990 and 2002. He is currently, and again, Head of his department. He is the author of a hundred academic publications (books, edited works, articles, and review essays) and almost a score of items published in atlases, dictionaries, and encyclopedias, as well as some fifty book reviews. He has supervised sixteen MA students and eight PhD students to completion and currently supervises one MA and seven PhD students.

He came to Canada as a Commonwealth Scholar and among other awards and fellowships he has held a Canada Council Killam Research Fellowship (1988–1990); the Senior Visiting Research Fellowship in St. John's College, Oxford (1998); and the Quatercentenary Visiting Research Fellowship in Emmanuel College, Cambridge (2003). He received the UBC Killam Research Prize in 2000. His publications include *Timber Colony* (1981), *People Places Patterns Processes: Geographical Perspectives on the Canadian Past* (ed., 1990), *Vancouver and Its Region* (ed. With T. Oke, 1992), and *A Scholar's Guide to Geographical Writing on the American and Canadian Past* (1993, with M. Conzen and T. Rumney). Most recently he was Guest Editor of a special double issue of *BC Studies*, "On the Environment," published in 2004. He is currently General Editor of the Nature/History/Society series, published by UBC Press, and coeditor of the *Journal of Historical Geography*.

CPSIA information can be obtained at www.ICGtesting.com
Printed in the USA
LVOW03*1941051213

364043LV00013B/530/P